COLLECTING OLD MAPS

PLANISPHERE
REPRESENTANT TOUT L'ETENDUE DU MONDE.
DANS L'ORDRE QU'ON A SUIVI DANS CE LIVRE.

N. LES CHIFFRES SE RAPORTENT AUX CARTES ET AUX PAGES DES DESCRIPTIONS.

A. AMSTERDAM, Chez LOUIS RENARD, Avec Privilège de Nosseigneurs les Etats de Hollande et de Westfrise.

Collecting Old Maps

Francis J. Manasek

TERRA NOVA PRESS

ISBN 0-9649000-6-8

First edition 1998

Frontispiece: *Planisphere Representant Toute L'etendue Du Monde.* Original frontispiece to Louis Renard's *Atlas de la navigation et du commerce...* Amsterdam: 1715. Copperplate engraving, overall approximately 17 x 10 inches (44 x 26 cm). Uncolored.

TERRA NOVA PRESS

G.B. Manasek, Inc.
Norwich, Vermont, 05055-1204
USA

To Anne and Jared. My two best friends.

CONTENTS

ACKNOWLEDGMENTS

In any work such as the present, it should appear obvious to the reader that many people, in addition to the author, have contributed ideas, suggestions, criticism and assistance. As indeed is the case.

My colleagues in the map trade offered support and assurance. When my enthusiasm for this project waned episodically, their "When is it coming out?" and similar questions restored my flagging energies. I know of no other business where such a high level of collegiality exists. The assistance given me, without thought of return, speaks toward the high standards held by members of this trade and is but one more reason why I hold my colleagues in such high regard.

Tom Suárez read an early draft and, despite its flaws, continued to be supportive. He loaned me many maps for the illustrations and answered innumerable questions and gave of his opinions readily. Rick Casten blue-pencilled a draft with his insightful comments, many of which were incorporated and which certainly improved the text. George Robinson's comments materially helped with my thoughts about the direction of the book, and his generous loan of material is appreciated. George and Mary Ritzlin's thorough reading removed many residual ambiguities, and they kindly loaned me maps from their inventory to use as illustrations. Bob Augustyn (Martayan Lan, New York) also kindly loaned some of the images used in this work. Chris Watters' help, especially with those parts of the text dealing with the nineteenth century, gave more balance to the work by convincing me to flesh out the later periods. Jonathan Potter took time out from a busy fair to go over the manuscript and he also answered, patiently, my questions about some of the maps and shared with me his extensive knowledge. Lee Jackson kindly provided me with a map to be used as an illustration. Geof Dreher and Sidney Knafel graciously shared their collections and each provided me with a map illustration.

Alan Berolzheimer helped with the compilations of the dictionary and bibliography. His thorough copy-editing and proofreading was an essential aid.

My wife, Anne Pearson, put up with a lot while this beast was gestating. I owe her big time.

F.J. Manasek
Norwich, 1997

ix

INTRODUCTION

Collecting old maps is a splendid hobby.

One doesn't need great means to build a great collection. Most dealers will agree that maps are undervalued, even after the recent price rises, in comparison to other antiques or art. Whether you collect 16th- and 17th-century world maps, 18th-century maps of France, 19th-century maps of Africa, or maps of Ohio, the hobby is fascinating and rewarding. However, the vastness and technical complexity of some of the aspects of old map collecting have proved daunting to many beginning collectors and the difficulty of "finding out" has dissuaded many from starting to collect maps.

However, how to start?

Go to any shop selling antiques or old books and you are likely to see an old map or two. Usually the dealer knows little about them but has put some kind of price on them that may seem high or may seem low. Who knows? Also, what other maps might there be? Where does one find them and how can one begin to learn to recognize them, understand them and value them?

There are already many books about maps. Some of them are lavish coffee-table books filled with folding pages of color reproductions of famous trophy maps, manuscript maps, and wall maps that can cost many thousands of dollars. Others are detailed descriptions of maps and the history of the mapping of different continents and seemingly endless lists of arcane detail about individual mapmakers.

But there are few serious books about map *collecting;* books that can lead the beginner into this hobby and guide him or her through the early pitfalls that many others have experienced. In writing this book, I have kept this in mind and have deviated from previous attempts in several major ways. Most other books about old maps were written from the standpoint of the mapping of different parts of the world, the maps produced by this effort, and why they are important to the history of cartography. These books are technically very good, and contain much detailed information about what the maps show, mapping history and the mapmakers who made these maps. However, they tend to ignore the nuts and bolts of *collecting* and the problems associated with it. In this book, I approach the hobby from the standpoint of the *collector* and have made the collector and the collection the common thread that ties the work together. I have included a basic background of knowledge necessary to provide the collector, and certainly also the dealer, with an adequate vocabulary and an adequate overview of the field. Hopefully, where I discuss some of the "nuts and bolts" issues, I will help map a path through the *Terra Incognita* of the early stages of map collecting.

I have also deviated from previous books in another important way. It's fun to read about the world's great rare maps, or, as I call them, *trophy maps*. However, many of us cannot afford to build collections of maps costing $5,000 and up. Also, what good is it to the beginner to look at a book filled with examples of maps that exist largely in museums? Accordingly, in this book I commit the heresy of writing for *collectors who want to collect maps that are collectable.* I illustrate the book with many maps that can be bought from most dealers rather easily and affordably.

Although map collecting has traditionally involved antiquarian material, generally pre-1800, the last few decades have seen a burgeoning interest in 19th-century and, dare we say it, early-20th-century maps. While collecting such things as gasoline company road maps is, *sensu strictu*, map collecting, these activities overlap to a large degree the areas of collecting ephemera. Nonetheless, I shall attempt to integrate them into the field of antiquarian map collecting as it is more generally perceived.

I have included a healthy amount of technical detail for those who might be interested. Much of this material is difficult to obtain elsewhere; some of it is entirely new and based upon my own work in my personal laboratory. It is not necessary to absorb this material in order to enjoy maps or map collecting, but it is there if you want it. I do believe that it provides a firm foundation for long-term enjoyment of this fascinating hobby.

Finally, the important thing is to learn about map collecting and to enjoy the hobby.

F.J. Manasek
Norwich, Vermont

Part I

Collecting old maps

Chapter 1

Before You Begin...

Why collect old maps?

There is one major answer. Enjoyment.

Collections of old maps provide their owners with a great deal of enjoyment. Enjoyment in the chase after a great specimen, enjoyment in the beauty and joy of ownership, and enjoyment in the knowledge derived. (Note that I have, even this early in the book, omitted financial gain from this discussion, since as I will point out later, I do not recommend maps as financial vehicles.) As in all such pastimes, most collectors get more enjoyment if they know more about their subject, but the field of old maps is so vast and diverse that it is impossible to summarize it in a single book. This is precisely why it is a richly rewarding hobby: one never grows out of it. The more one learns, the more there is to know and the more meaning and enjoyment one derives from the maps collected.

Although many collectors were attracted initially to old maps after seeing a particularly beautiful colored specimen, beauty is certainly not the only vector that transmits the map-collecting virus. Some maps are, indeed, breathtaking in their beauty, but there are many other reasons for collecting them, reasons as varied as the individual collectors themselves. They include, among others, personal interest in a region or period.

Collectors often begin by buying one or two old maps as impulse purchases and then, as they learn more about the technical aspects of the field, they tend to focus their collections. Map collecting is much like stamp collecting. The collector can be as sophisticated or unsophisticated as he pleases. Some stamp collectors put their stamps in an album and occasionally look at them; others pore over them, learning everything about their designers, subjects, reasons for being issued, perforations and watermarks. Some delight in acquiring nice examples of current commemoratives; others fight it out in the auction rooms for the upside-down airmail or the Graf Zeppelin or penny-black. It's the same with map collectors. Some map collectors buy a few maps to hang on the wall; others become obsessed and add to their house to hold the collection. Several I know have built separate buildings for their collections.

THE COST

"Ah," you say. "But it must cost a fortune to collect old maps."

Yes and no. It is much more difficult to spend a fortune on a good map collection than on a comparably important collection of Impressionist paintings or old master prints or, for that matter, 18th-century American furniture. But it can be done. My firm has dealt with heads of state, captains of industry and individuals of great private means who have built, or are building, collections of staggering monetary value. We also deal with individuals of modest means who are building collections of staggering interest around themes or areas that are meaningful to them personally, but do not necessarily cost a great deal of money. Although such collections may not require vast amounts of money they are every bit as difficult to assemble, and just as rewarding and enjoyable. They become collections of great *personal* value.

If the collector does not chase fashion, eminently collectable maps in the low hundred dollar price range (or even lower) suddenly appear. Surprising as it may seem, one hundred dollars is still the price range of many 16th-century maps, and one certainly finds many 19th-century maps in this price range. Many people find it hard to believe that 16th-century maps can be so inexpensive. My advice? Talk to a good map dealer about this. You'll be pleasantly surprised.

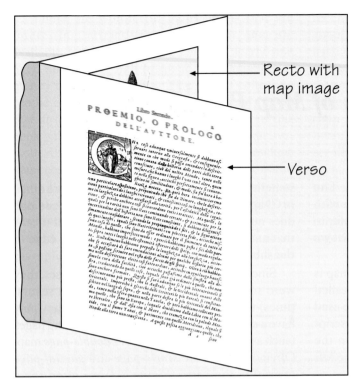

Recto with map image

Verso

Figure 2.2 The relationship between *recto* and *verso* is shown in this diagram. If the text on the verso is particularly bold it may be visible from the recto and result in *show-through*. The *binder's stub* is still in place, glued to one side of the centerfold on the map's verso.

consequently, these maps do not have **stitch holes** in the centerfold. Unfortunately, the binder's guard was often glued on with a glue that age-darkened and discolored the paper, hence we often find *centerfold darkening.*

We measure map size by measuring the *image,* generally height first. Technically, this is the *neatline size* (neatline is defined below and illustrated in Figure 2.3). Many dealers and certainly most technical works on maps use the metric system and express dimensions in centimeters, but the English system has a habit of hanging on in American catalogues. Map sizes are sometimes given as *folio* or *quarto* or *octavo.* Strictly speaking, these terms refer to the number of times a printed sheet of paper was folded to produce a book page, but they have come into general use to mean dimensions. This book is roughly *quarto* size, *folio* is about twice the size and *octavo* about half.

Printed maps are printed from a *plate* that might be copper *(copperplate),* as is the case for most engraved maps, or steel in the case of some 19th-century engraved *(steelplate)* maps, or stone in the case of lithographed maps. *Woodblock maps* are, of course, printed from a carved wooden block.

Printed maps do not generally bleed out to the edge of the page, but usually have a line that delimits their image. This line is called the **neatline.** The area encompassed by the neatline is smaller than the plate from which the image was printed. If the map was engraved, then it was printed by forcing the paper under high pressure onto the inked, engraved copperplate. This enabled it to pick up the inked image from the plate, but in the process it compressed the fibers of the paper that were in contact with the copperplate. Along the edges of the plate this left an indentation, called the *platemark,* visible best under an oblique, or raking, light. Platemarks are useful in identifying maps as being engraved (rarely does a lithographed map have one) and can be used to actually identify the original plate from which the map was printed. Indeed, in at least one instance, the width of the distance between the neatline and the platemark identified a map as a forgery. Clever makers of map reproductions often will use paper that has fake platemarks. The writer remembers, among his many very pleasant times in Paris, wandering the left bank of the Seine and marveling at how so many of the prints displayed along the quay seemed to have deep platemarks, even though they were all recent lithographed copies.

Figure 2.3 This cartouche is from a copperplate engraved map by Porcacchi, 1590, shown actual size. Note the platemark along the bottom (arrows) and the neatline (arrowheads).

18

Figure 2.4 A full-size example of a cartouch on an Ortelius map of 1590. Copperplate engraving, this exemplar has original color. Note the engraved *privilege* in the lower right.

Between the platemark or the neatline and the edge of the paper is the *margin.* Collectors consider wide margins better than narrow margins and good wide margins often command a price premium. Unfortunately, some maps were too big for the paper on which they were originally printed and started with very narrow margins. Every time a book or atlas was rebound, the binder was likely to plane the edges to achieve a uniform appearance. (This dreadful practice is still common and I would advise anyone having a book rebound to give explicit written instructions to the binder *not* to plane the edges.) Each planing removed a bit from the margin. If the margin was narrow to start with, then it became even more narrow, and sometimes the plane bit into the image. In that case the map

Figure 2.5 The cartouche in this exemplar (a mid-17th-century copperplate engraved map by Blaeu) has original color. Shown actual size. Note the platemark along the right side (arrows). This map has the maker's name engraved below the cartouche.

is *trimmed* into the platemark, neatline, or image, creating a serious defect. The damaged map can be *remargined*. (See Chapter 6.)

Many maps have a bit of text in the lower margin identifying the engraver, publisher, or other person involved in the map's production. See Appendix D for more details and an explanation.

CARTOUCHES

Within the neatline is the map itself. Most old maps have a *cartouche* (Figure 2.1), the (sometimes) decorative device that encloses the title and other information (Figure 2.4). This is the first place to look for the title and mapmaker (see Figure C.1.) The publisher is often listed within the cartouche as well. Cartouches may be very elaborate as in figures 2.4, 2.5, 2.6 and 2.7, or they may be a very simple oval or rectangle (Figures 2.8, 2.9) as in many 19th century maps (as well as some 16th century ones.) Cartouches are a study in themselves and they are often rich in iconography. Some mapmakers used very specific images that were put on maps of the New World, South America, Africa, Asia or Europe. Cartouches can be found *fully colored* or *uncolored* or *partially*

colored. Some mapmakers, such as Homann and Seutter, rarely colored their cartouches. This style, where only the geographical areas are colored, is called *body color*. Often, in order to sell maps into the decorative market, their cartouches have been colored recently. To the practiced eye, they do not look good.

Cartouches seem to attract the initial attention of many budding collectors. Often it is possible to identify the maker of a map simply on the basis of the design of the cartouche. Recognizing their importance, I have photographed some representative different styles of cartouches and include them here in an effort to show their remarkable diversity.

20

Figure 2.6 A Coronelli copperplate engraving from 1690. Actual size. The cartouche in this exemplar is uncolored, as is usual for most Coronelli maps. In my opinion, later color does not improve the appearance of these maps and deletes the power and dignity of the black/white cartouche. Often large and powerfully engraved, the cartouche became an integral part of the overall design of the map.

Fig. 2.7 Homann, c. 1720. 65% actual size. Copperplate engraving. The cartouches in Homann and Seutter maps were almost always uncolored, even if the rest of the map was colored originally. If one sees these cartouches colored, one must suspect later color.

DISTANCE SCALES

Maps are made to different scales and generally the scale is shown graphically by means of graduated distance bars. These can be enclosed in ornate decoration, as we see in a Jansson map (Figure 2.10), or displayed rather simply as in the 19th-century map by John Cary (Figure 2.9).

Frequently, especially on older maps, the distance scale shows bars for many of the different units of measurement then in use (see Figure 2.9).

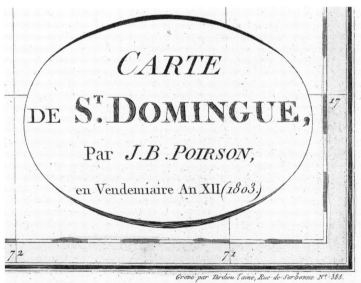

Figure 2.8 Many 19th-century cartouches were simple ovals or rectangles containing the necessary text, as in this 1803 copperplate engraved map by Poirson, shown actual size. The trend was away from embellishment and decoration. The date in this cartouche is given in the year of the Republic (XII) following the French Revolution as well as in the more conventional *anno domini*.

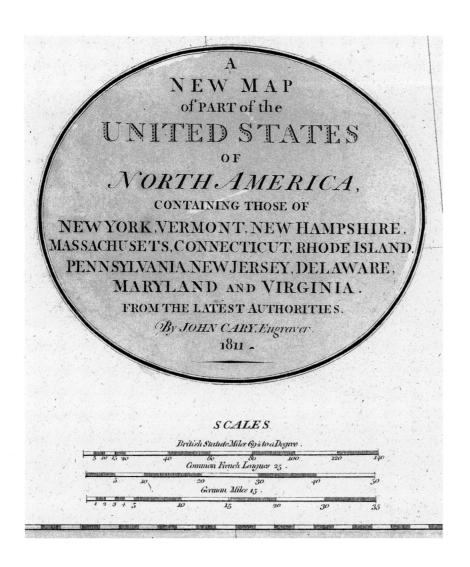

Figure 2.9 John Cary used simple, oval cartouches in his copperplate engraved maps. This is typical of the less ornate designs found on 19th-century maps. This map had full original hand color and the oval cartouche had a pale wash. Note the scale beneath the cartouche. This map was published in an atlas in London in 1812 (the plate bears an 1811 completion date). The size of the map image is 18 x 20.25 inches (46 x 52 cm). Although not shown here, this map has an unusually wide platemark (3 cm; about 1.25 inches).

Figure 2.10 This full-size distance scale from a Jansson map, *Nova Anglia, Novum Belgium et Virginia,* of 1636 (first state of three) is enclosed in an ornate design. The bold arrows point to the *platemark* which typically for maps of this period is quite narrow. The dark printed line just inside the platemark is the *neatline*. The platemark is indicative of an intaglio impression (see Chapter 4) and its width (distance between it and the neatline) can be used as an important quantitative characteristic of a map.

ANGLIAE, SCOTIAE, ET HIBERNIAE, SIVE
BRITANNICAR:INSVLARVM DESCRIPTIO.

Figure 2.11 (left) This copperplate engraved map by Ortelius shows all of Great Britain. North is at the right, and the map seems to have been done "on its side" in order to accommodate the shape of the paper. The notion that north "belongs" at the top was not yet fixed. *Angliae, Scotiae, Et Hiberniae, Sive Britannicar: Insularum Descriptio.* 13.5 x 19.5 inches (34 x 49.5 cm). This exemplar had full original hand color.

Figure 2.12 (above) This full-size compass rose is from the Mercator/Hondius *Virginiae Item et Floridae...* of 1638. It shows a 16-point compass card with a prominent north indicator. Note the moire pattern to the engraved sea. This pattern is typical for the Mercator-derived maps.

Figure 2.13 (below) This illustration shows *Chorus,* the Northwest Wind, full-size as depicted on a woodblock map of the world by Münster, c. 1550. Note the prominent neatline at the left. Windheads can be found as decorations on many early world maps, often appearing in the areas between the map's outer perimeter and the neatline, and, as is the case here, pillowed by clouds.

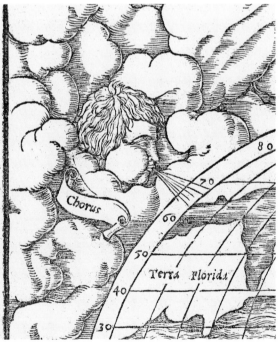

24

DIRECTIONS

It is simply an evolved Western convention that we now generally show north at the top of a map. Many maps, especially earlier ones, did not conform to this convention and these now appear a bit strange to us. Ortelius' map of Great Britain, (Figure 2.11) showing all of Great Britain "on its side" with north at the right, is a splendid example of such a map.

Japanese woodblock maps, which were intended to be viewed on the floor, often have *radial symmetry* and, in these maps, there is no true "top" or "bottom."

Cardinal directions were often indicated by means of a device called a *compass rose.* In Figure 2.12, the compass rose (sometimes called a *windrose)* represents a compass card and shows the directions as well as the points. Compass roses are often elaborate, decorative, nicely colored embellishments. They may be associated with *wind-heads,* the mythical figures seen puffing and blowing onto the map. The wind-heads are named (see our Latin dictionary) and I show a photograph of *Chorus,* the Northwest wind (Figure 2.13).

FIGURES ON MAPS

Some maps of the 17th century have small figures comprising the border (see Figure 2.14). These figures, often etched rather than engraved (see Chapter 4), generally show native peoples in their native costumes, nobility, famous individuals and scenes considered typical of the regions shown on the map. Some of the scenes may be allegorical, and may represent seasons, planets or some other subject.

Maps with such figured borders are much in demand, especially those by Blaeu, Speed and Jansson. These maps are known as *cartes à figures.* Speed, Blaeu, Jansson and Visscher, among others, produced such figured maps and they are much sought after by today's collectors.

FOLDING MAPS

Folding maps are among the more interesting map subtypes. These maps were folded to increase their portability and also as a way of dealing with otherwise excessively large formats. Some folding

Figure 2.14
Several figured panels from W. Blaeu's *Africa* are shown here. The top row contains views of ports and the side panels have figures of natives in their typical dress. Maps with these figures are known as *cartes à figures* and can command substantial prices.

Courtesy: T. Súarez

maps, such as those made in Japan and to a lesser extent, in China, are simple sheets of paper that are folded into rectangles. In the West, this type of map is a common style for city plans designed for use by travelers and for the much-collected American state and county maps, the so-called *pocket maps.*

Folding maps are often glued into stiff boards or covers. These are called *self-covers* and they may have *printed* or *manuscript labels.* The covers were designed to protect the map while it was carried about (see Figure 2.15).

Another form of protection is the *slipcase* (Figure 2.16), a separate pouch into which the folded map slides. In order to facilitate sliding the map into or out of the tight-fitting slipcase, maps often had yet another fitted piece, called a *chemise.* The chemise was often lost, but if the map fits too loosely into its slipcase, we can surmise that it had originally a chemise.

Chemises are most often associated with those folding maps that were *linen-backed* (Figures 2.16, 2.17, 2.18). One problem with folded paper maps

25

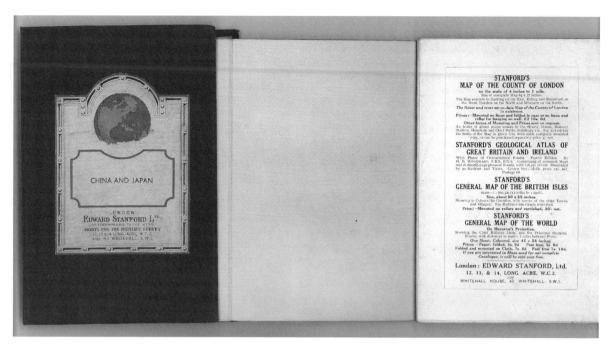

Figure 2.15 (above) This photograph shows a partly opened linen-backed folding map that was issued with a self-cover. *Stanford's Map Of China And Japan with adjacent parts of Soviet Russia, India, Burma, &c.* has a printed paper label on the front of the self-cover with the short-title *China And Japan*. There is an additional printed paper label advertisement on the inside, pasted directly on the linen backing. Further unfolding reveals the lithographed map, published in London c. 1925. Overall dimensions are about 26.5 x 40 inches (67.3 x 101.6 cm); each panel measures 6.5 x 4.5 inches (16.5 x 11.4 cm).

Figure 2.16 (above) The linen-backed folding map (far right) fits into a chemise (in the middle) which in turn slips into the slipcase (left). Both case and chemise are covered in marbled paper. Knowledge of different marbled papers is useful in determining if such cases and chemises are original. The map is an engraved city plan of Vienna, with original hand color, published by Artaria und Comp. in 1830.

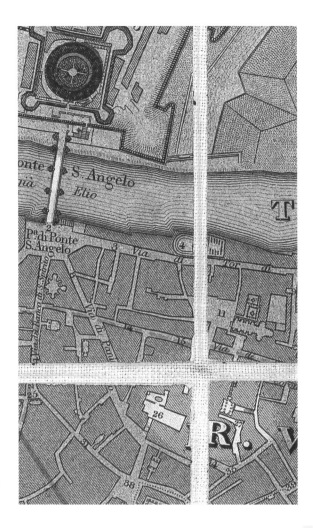

Figure 2.17 (left) This photograph (enlarged 2X) of a portion of four panels from William Harwood's *A Topographical Plan of Modern Rome with the New Additions* (published in London in 1865) illustrates the principle of the dissected linen-backed map. Note that the individual paper panels are separated by a gap and all the folding takes place along the linen; the paper is never bent and consequently the paper fibers do not undergo strain each time the map is folded or unfolded. Overall dimensions are 22 x 26.5 inches (55.9 x 67.3 cm); each panel measures 5.5 x 8.75 inches (14 x 24.8 cm). This finely engraved map is a nice example of a 19th-century folding city plan made for travelers.

Figure 2.18 (below) The English mapmaker Aaron Arrowsmith produced linen-backed folding maps, often issued folded to a uniform 10 x 8 inch (25.5 x 20.5 cm) size with a blue silk binding sewn around the edges. In this photograph we see the silk binding (arrows) as well as a printed paper label showing the outline of the map. A previous owner has pasted a filing tab to the map. This is one of a set of six (plus a key sheet) copperplate engraved maps titled *Map of France, Belgium And Part Of Switzerland. From Cassini, National Atlas, Ferrari, Weiss &c.* Published in London in 1817.

was that the paper in the folds would eventually fail and the maps would come apart. Failure usually occurred first at the intersections of folds. In order to avoid this problem, mapmakers would often *dissect* the printed map, usually into rectangles, and then mount the rectangles on linen, leaving a small gap between adjacent paper edges, as seen in Figure 2.17. Thus, when the map was closed, only the linen was folded and the paper part remained flat. Since linen is more resilient than paper and can be folded countless times without damage, such dissected, linen-backed folding maps usually survived use and abuse that their unbacked counterparts could not. I do note that the Japanese woodblock printed maps were, by and large, folding maps, but they were printed on paper with very long fibers that was relatively resistant to breaking at the folds.

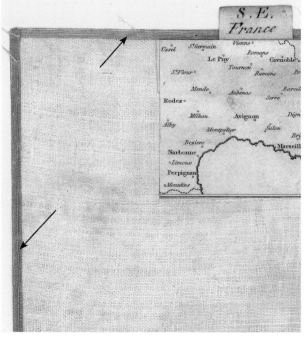

Linen-backed folding maps were particularly useful for travellers and such maps were produced well into the 20th century, when they gradually were replaced by more ephemeral, unbacked maps.

Many durable American pocket maps were printed on banknote paper which resisted breaking at folds, but many of these maps today do show cracking and breaking, especially at the fold intersections.

Curious miscellany – 2

An unusual image relating to maps or mapmaking

Atlases often contained images that are difficult to classify except in the broad category of ***cartographical curiosa.*** I have scattered a few of these throughout this book. This one is by Seutter, and represents Daniel's interpretation of King Nebuchadnezzar's dream (Dan. 2:39-40). History's great empires are on the breastplate; Eastern empires on the left leg, Western ones listed on the right. There are three others in this series. Published in Augsburg, c. 1730. Approximately 22.5 x 19.25 inches (57 x 49 cm).

Chapter 3

Kinds of Maps

We should probably not be overly smug about maps with errors. It is far too easy, from the vantage point of our satellite images, high-altitude aerial photographs and satellite positioning, for us to sneer about the now-quaint concepts embodied on many old maps. What we need to do, I think, is marvel at the overall reliability and accuracy of most early maps, rather than focus on their errors. It took less than a hundred years from the time the New World was discovered, to produce quite accurate maps of the new continents and most of the Caribbean. We can quibble about the time frame and what constitutes a "good map" but certainly Americas maps of the early 1500s were surprisingly good, and by the 1580s the maps of Ortelius are actually quite accurate.

Consider that accurate longitude measurements were nearly impossible in the absence of good timepieces; consider that

tiny ships traveling at a few knots took months to complete even a brief round trip, let alone do any serious exploring in, for example, the Pacific. For all recorded history, up to our own lifetimes, all accurate terrestrial mapping was done by triangulation. This is basically the sort of thing surveyors still do on a local level, using concepts derived from Euclid and the ancient Egyptians. The vastness of India, Siberia and our own continent were measured by men who walked across them, using transit, compass, level and chain. It is mind-boggling to think that it was on foot, through uncharted regions, that such great cartographical triumphs as laying out the Pacific

Figure 3.1 Printed in 1472, this woodblock map is called a "T-O" map. The known realms of the Earth, Asia, Africa and Europe were placed in a circle and surrounded by the ocean sea. The "T" consists of the Mediterranean Sea, the Nile and the River Don (here collectively called the Mediterranean). Noah's three sons are located in their respective domains and the cardinal directions are given around the outside of the circle. The map is 2.4 inches in diameter (6.5 cm) and is the work of Isidore of Seville, in his *Etymologiae sive originum libri XX.*

Courtesy: S. Knafel

29

railway or the Trans-Siberian railway route were determined. Quite literally, giant triangles were plotted and measured all across our continent, from Atlantic to Pacific! If we consider how laborious these techniques truly are, then it is all the more surprising that mapping the world with any degree of coincidence with objective reality was even remotely possible. And after obtaining all this knowledge, somehow it had to be put on pergamen or paper, silk or birchbark, so that others could see and use it. These resulting pictorial representations are, of course, maps.

In this chapter we will survey the basics of *what different types of maps look like*. Much of this chapter is benignly technical. It is not difficult, but it does contain some basic knowledge that anyone serious about map collecting ultimately needs to know. If you are only casually interested in maps, for example you want an occasional map for decorative purposes, then this chapter can be skimmed over. However, even in this case, it contains useful information that will ultimately help even nonspecialists make better decisions regarding the maps they own or have for sale. If you are not technically inclined, don't worry. Skim the chapter to become familiar with its various terms and then go on to read about the rest of map collecting. You can always come back to this chapter as you wish, to help clarify things that come up later in the book.

Maps, as we will consider them for the purposes of this book, are representations (analogues or, if you will, models) of physical structures, usually the earth or parts of it, or the heavens and its stars, planets and systems. Maps are most often flat representations, usually on paper. Sometimes they are three-dimensional, such as **relief** maps that try to show the relative heights of mountains and other terrain. Sometimes they attempt to mimic the

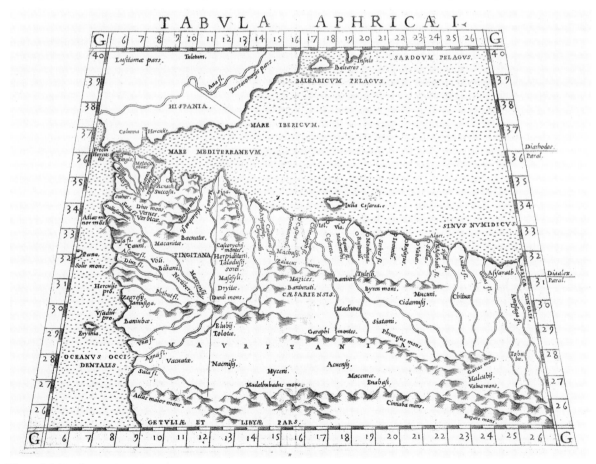

Figure 3.2 The first (1561) edition of Ruscelli's translation of Ptolemy's Geography, *La Geografia Di Claudio Tolomeo....*contained copperplate engraved maps such as the one shown here. Printed in Venice, the map shows northern Africa, part of the Mediterranean Sea and the southern part of Spain. This is a typical trapezoidal Ptolemaic projection. The base of this image is about 10 inches (25 cm) wide and the image is about 7 inches (18 cm) high.

30

actual object being mapped, such as a globe that represents the earth or another planet. Since this is a book on map collecting, we will discuss primarily the most widely collected form of maps: flat paper maps showing parts of the earth's surface.

Whatever they attempt to show, be it the coastline of Zanzibar or the distribution of smallpox in central Africa, most maps purport to be descriptors, or representatives of an independently verifiable reality. If a map that was made to be informative (as opposed to primarily decorative) is clearly not a good model, or representative, of something objectively perceived, then it is not a good map. Here, of course, is a catch. In order to make this judgment about a map, we need to know why it was made and we need to understand what the perceived objective reality was at the time it was made.

There is a type of medieval European map called the "T-O" map (pronounced "tee-oh") because it appears as a large letter T circumscribed by a large circle, or letter O (see Figure 3.1). Original "T-O" maps are quite expensive. If we just look at this kind of map as a representation of geographic reality from our present cultural perspective, then it is a poor map indeed. However, if we perceive it, correctly, as a representation of a cosmography, an intellectual system that was a combination of demonstrable reality and of belief, embodying a conceptual relationship as well as a geographic relationship, then we can understand the map in a different sense. No longer is it a woefully inadequate "map" in the modern sense, but it has greater meaning.

WHY DO MAPS HAVE DIFFERENT SHAPES AND APPEARANCES?

Since our paper maps are flat and the earth is approximately spherical, the flat map cannot represent the earth's surface exactly. The mapmaker must somehow distort the (roughly) spherical surface of the world to the planar one of our paper maps. Such distortions, or planigraphic (plani=flat; graphic=picture) representations of a spherical surface, are called *projections*. A projection is basically a technique for transforming a spherical coordinate grid to a planar one. Imagine, for example, that we draw the outlines of the continents on the surface of a transparent globe and then put a light at the center. The outlines of the continents can be seen projected onto a flat sheet of white paper held close to the globe, but will look like a distorted version

Figure 3.3 Peter Apianus' little book, *Cosmographia...* published in 1574 had this diagram illustrating the Ptolemaic projection. The original page size is about 8 x 6 inches (20.3 x 15.2 cm). Among the names we recognize on the schematic map are Vienna, Prague, Venice, Ingoldstadt. Note the *worming* in the upper left and the *show-through* from the *verso* in the lower half. (See Glossary for terms.)

31

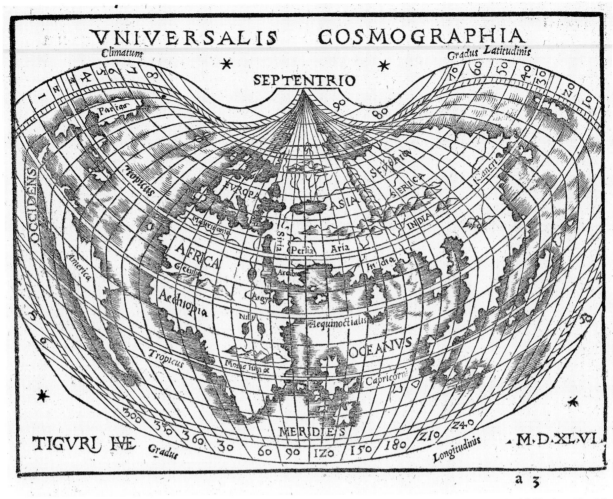

Figure 3.4 Appearing in Honter's *Rudimenta Cosmographica,* this little woodblock map (measuring about 4.75 x 6.25 inches, 12 x 16 cm) is an example of a small cordiform (heart-shaped) map. The map was published from 1546 through 1595 in various editions of this book done in Zürich, Antwerp, Basle and Prague. A new block was cut in 1561 but, even so, some of the later editions of the book contained maps printed from the original block.

of the outlines drawn on the globe itself. We have made a *projection* of the continents.

There are many ways projections can be made, usually depending upon the purpose of the map. Some of the projections most likely to be encountered by the collector are described briefly.

PTOLEMAIC PROJECTION

The maps of Claudius Ptolemy (c. 150 AD) covered the world known to the ancients. They were, of course, manuscript maps, since they predate printing. Our knowledge of them comes from successive copies of the manuscripts and, finally, printed versions. Ptolemy suggested how the world should be divided into individual maps and introduced *latitude* and *longitude*. Latitude divides the north-south line into degrees; longitude does the same for the east-west line. Latitude starts at the equator and progresses to 90° at each pole; longitude starts at some **prime meridian** and proceeds *westward* from there. Ptolemy used both Alexandria (Egypt) and the Fortunate (Canary) Isles as prime meridians. Meridians, in Ptolemaic projections, converged as they extended northward or southward from the equator. They did not converge, however at either the magnetic pole or the geographic pole. As would be the case in map segments meant to be pasted onto globes (**gores**). Nonetheless, Ptolemaic maps are easily recognized by their characteristic trapezoidal appearance,

32

60.70.80.90.where the Pole Antarctike is, and maketh the Figure as you said of halfe an hart.

Spoud. This can I practise by my selfe at an other season:wherefore I praye you procede to the finishinge of this Mappe.

Philo. Then taking the Clothe or Parchemente, in whiche you will describe the Paralleles, and Meridiane Circles:you shall reduce all the Circles with theyr diuisions, whiche you made in A.B.C.into this seconde Mappe, the Center of whiche is.K.by the healpe of your compasse ,firste drawinge a righte line.K.L. the middes of whiche shall be M.and this line muste be in lengthe equall to the Line.D.F.in the first Mappe. Then placing th'one ende of the compasse in K. extende th'other vnto.M.and prot[...] che shall represente th'Equinoctiall[...] vnto the Circle. A.B.C. after ta[...] passe the distaunce of euerye Arck[...] firste Mappe, and wyth th'one fote[...] (placinge th'other foote in.K.)dra[...] circuit, as one of these shalbe foure ti[...] one of th'other in the firste Mappe.

Moreouer you shall diuide th'E[...] to .360.equall portions ,suche as ar[...] B.C.In like sorte th'other Parallele[...] [...]all, and Southe from th'Equinoc[...] from euerye diuision of one Parallele[...] lynes as you did from. D.to. H. in

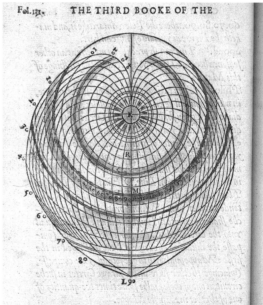

your Carde will not onlye growe to the forme of a harte, but also of a double herte one within an other , as thys demonstratiõ,& figure sheweth. Last you shal drawe the Tropickes of Cancer and Capricorny , the Circles Articke and Ant arctike, makinge them double lines for the easier knowinge them from th'other paralleles.

Figure 3.5 William Cunningham's *The Cosmographical Glasse...* published in London in 1559, had these directions for laying out a map on the cordiform projection. This book is particularly interesting because it is one of the few with such information in English. Note the *worming* in the lower left and the *show-through* of the cordiform grid which is visible, faintly, in the text. The cordiform grid (inset, reduced) appears on the *verso* of the text page and is reproduced here as an inset, reduced in size.

33

resembling orange-peel segments or partial globe gores (Figures 3.2, 3.3).

Ptolemy's *Geographia* covered the known world. Although there is scholarly dispute about just how much Ptolemy actually contributed to this work, maps of this type are universally called Ptolemaic maps. Many early atlases, or "geographies," contained two sets of maps, the Ptolemaic maps and the "modern" maps. These maps were produced in quantity through the 16th century and the last edition of Ptolemy was published in 1730. Quite a run if one considers that this represents about 1600 years of copying! With the exception of some of the great and important printed editions, such as the 1513 Strasbourg and the 1482 and 1486 Ulm editions, Ptolemaic maps are relatively inexpensive. They seem to be undercollected, largely I suspect, because they are not perceived by modern collectors to be attractive. Rarely colored when printed, one must be suspicious that colored Ptolemaics have later color. Moreover, many of these maps are small. The relatively plentiful editions of Magini (1550 through about 1620) and Ruscelli (1561 to 1599) contain maps of modest size. Also, since all Ptolemaic maps are derivative copies of earlier manuscript maps, they are perceived as having less cartographic significance. I think that these maps will become more collected in the near future. It would appear that the factors listed above combine to make them a bargain in today's market.

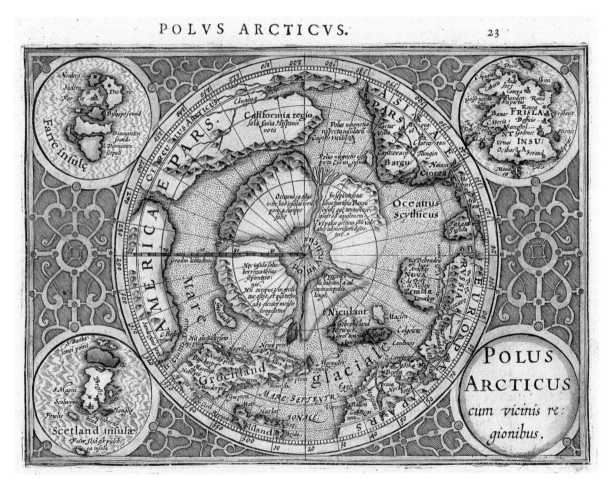

Figure 3.6 (above) A north polar projection map is shown close to full size (7.25 x 5.25 inches, 18.5 x 13 cm). It was published in an early-17th-century edition of the Mercator/Hondius *Atlas Minor,* and is a nice example of a copperplate engraved map. This map had full original color.

Figure 3.8 (right) A world map on the Mercator projection, by Bonne, and printed in France in the late 18th century is shown on the next page. A small map, it nonetheless shows clearly the ever-widening distance between parallels as one approaches the poles. Original size is about 8.25 x 12.5 inches (21 x 31.7 cm).

34

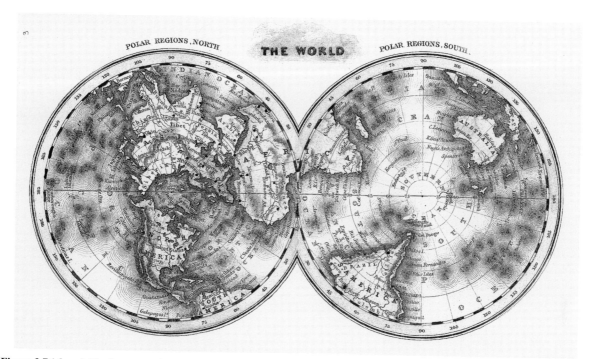

Figure 3.7 (above) Neatly engraved and published by Carey and Lea, in Philadelphia in 1832, this double hemisphere polar projection world map is shown full size. It is a close copy of the English map in the Starling *Royal Cabinet Atlas* of the same period. This copy had later hand color. Each hemisphere is 3 inches (7.6 cm) in diameter.

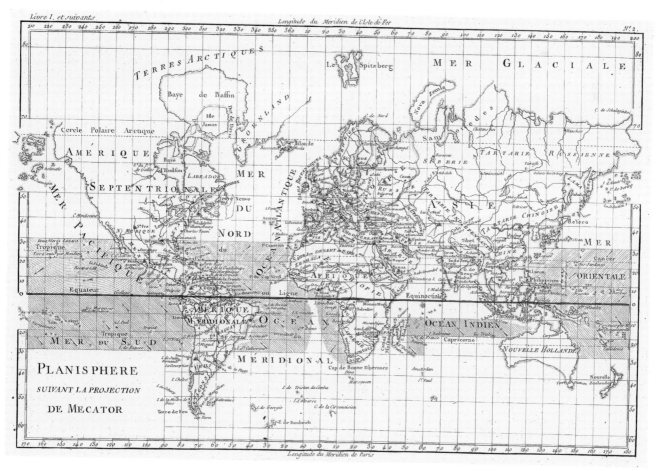

CORDIFORM PROJECTION

The *cordiform* (cordi=heart; form=shape) projection is literally a heart-shaped projection used in some early world maps. It is one way to make an equal-area representation of a globe. These maps are usually quite early and scarce. Some of the notable early cordiform maps were those of Bernard Sylvanus (1511), Peter Apianus (1530), Mercator (1538), Ortelius (1564) and Franco (c. 1586). These are all rare and expensive. We illustrate the more available Honter (1546 or later) map in Figure 3.4. This map was published in Zürich in *Rudimenta Cosmographica*, which went through many editions. The block was recut in 1561. (For a discussion of woodblocks and recutting them, see Chapter 4.) Early directions for laying out a cordiform map are shown in Figure 3.5.

POLAR PROJECTIONS

Polar projections are essentially hemispheric projections with one of the poles at or near the center. These projections show the world as though viewed from a great distance. Variations of this type of projection include those that have Jerusalem, London or some other (very important) city as the center. Polar maps have a wonderful and interesting history. In medieval times it was thought that there were whirlpools at the poles where the oceans' water flowed into the earth. As late as the early 17th century Mercator's map of the North Pole showed the four apocryphal islands and the central whirlpool. This is perhaps one of the most attractive polar maps ever made. It appeared also as a miniature in the small Mercator/Hondius atlas (Figure 3.6). The large folio north polar maps by Hondius, for example, do not refer to these putative phenomena, but are restricted to actual known, rather than hypothetical, geographical data. We note, however, that the imaginary, oft-shown Antarctic landmass turned out to be real.

The golden age of polar exploration in the 19th century is well represented in maps such as the wonderfully detailed maps by Justus Perthes in the *Stieler Hand-atlas*. These maps are relatively inexpensive, but quite important in the history of polar exploration.

MERCATOR'S PROJECTION

The Mercator "projection" is technically not a projection at all, but a mathematical transform. There is no way that a simple projection of a spherical surface can result in a Mercator-style map. This map's main virtue is the fact that straight lines represent compass courses. This is a most useful feature for navigators. It makes it easy if one can connect two points with a straight line and read the compass course directly. Most other projections cannot do this because their longitudes are not parallel but rather converge on the poles. The straight line connecting two points and

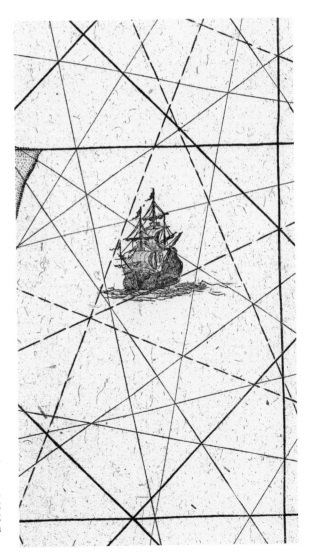

Figure 3.9 The ocean is crisscrossed with *rhumb lines* in this detail, shown full size, from a van Keulen sea chart of 1684. A sailing ship is seen, appearing a bit like a spider working its web. Sailing ships seem to enhance the value, or at least the collectability, of a map.

36

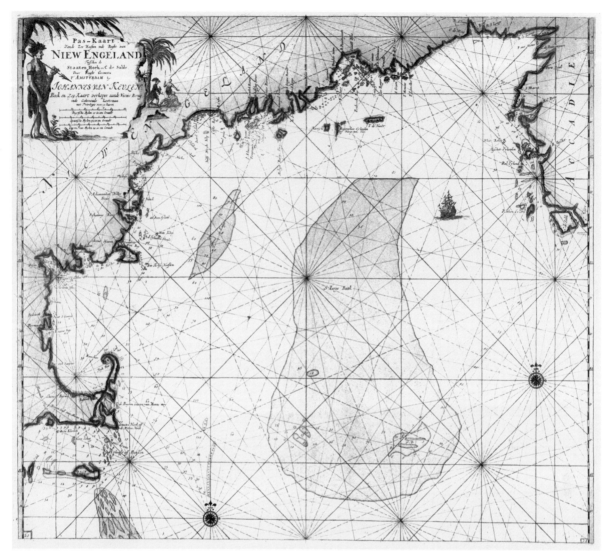

Figure 3.10 *Pas-Kaart Vande Zee Kusted inde Boght van Niew Engeland Tusschen de Staaten Hock en C. de Sable.* By Johannes van Keulen. Copperplate engraved sea chart. Fine contemporary full color. 20.25 x 23 inches (51.5 x 58.5 cm). Published in 1684. This is a fine example of a great sea chart: utilitarian yet beautiful. Note the rhumb lines (see Glossary) crisscrossing the ocean and the clearly shown fishing banks. The image is a bit too large for the paper on which it was printed and the margins were consequently very narrow. Many copies have trimmed margins. In this exemplar, the margins were complete, but close, all around. Scarce in early color and fine condition.

representing a compass course is called a **loxodrome.** Loxodromes intersect the meridians at equal angles.

On the earth, longitudes converge on the poles. On Mercator's projection they remain parallel. In order to convert the actual converging longitudes to parallel (non-converging) lines other distortions had to be introduced. This is why latitude lines become farther apart nearer the poles in a Mercator map, with the resulting visual distortion of high latitude landmasses. I remember how surprised I was, when, as a child I first learned that Greenland, appearing so large on a Mercator map, is relatively small.

CHARTS

Maps specialized for use in sailing (or flying) are called *charts*. Charts may differ from land maps not only in their type of projection, but also in the type of detail shown. In general sailors don't need to know much about the interior of landmasses, but

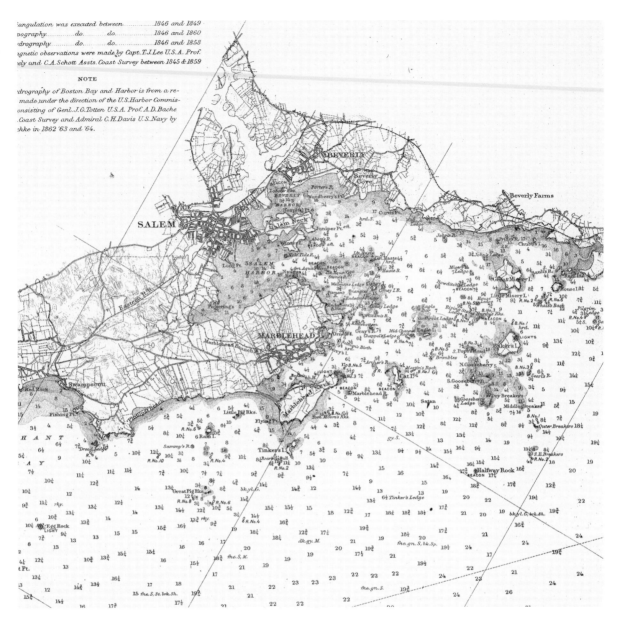

Figure 3.11 This detail, shown in full size, is from an 1866 chart by the Survey of the Coast of the United States. The entire chart is 36 x 38.5 inches (91.4 x 97.8 cm). The versions printed for the Survey's annual report to the Congress are quite common and inexpensive, but generally need extensive conservation work because they were printed on very poor paper. Color on such charts is often a modern addition.

coastal detail and land profiles are of utmost importance.

Sea charts. The earliest extant sea charts are the *portolans* which were manuscript coastal charts that showed detail graphically. Such maps were meant to replace or supplement the verbal descriptions of coast pilots. Portolans are outside the scope of this book.

Sea charts are basically coastline maps with additional information essential for traversing large areas of featureless sea. This information may include currents, magnetic deviation and winds. Where applicable, charts also give the depth (*soundings*) of the sea bottom.

Early sea charts were often embellished with nicely engraved images of sea creatures and sailing ships. Compass headings, compass (wind) roses and windheads (Chapter 2) are often shown on these charts. Many early sea charts and maps with large areas of ocean are often spiderwebbed with *rhumb lines* radiating from the compass rose, making them particularly attractive to many collectors. More modern charts generally do not have such embellishments, but are more utilitarian. Figure 3.11 shows a part of a 19th-century American coastal chart. Detail of the Massachusetts towns of Salem and Marblehead are shown and the navigational detail is excellent. This chart was made by the Coastal Survey. One edition was printed on robust paper for use aboard ship and another on very thin, fragile paper for inclusion in the annual report to Congress. The latter are more common.

Aviation charts. As soon as controllable powered flight was a reality, aerial charts were developed and they were standardized by international convention. Over the years scale and information shown have changed. Some of the more interesting charts are those that show the ground as seen by the pilot or navigator. These 20th-century charts are becoming collectable and some are quite scarce. I show part of a Sectional Chart I personally used in the late 1960s (Figure 3.12). In addition to showing landmarks, these charts include information essential for aerial navigation, including height of the land, radio navigation aids, and prominent artificial structures such as railroads, roads and outdoor theatres. They are all depicted in stylized ways that make it easy for the aviator to recognize features on the ground from elevations of several thousand feet.

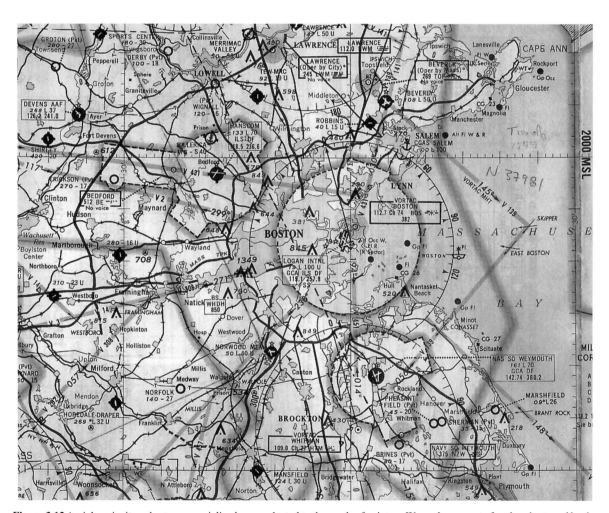

Figure 3.12 Aerial navigaiton charts are specialized maps adapted to the needs of aviators. We see here a part of such a chart used by the author in the late 1960s. Known as a *sectional* chart, it provides topographical information in a stylized way useful to aviators and also superimposes information necessary for safe flying.

TOWN PLANS AND VIEWS

The *bird's-eye view* or *three-quarter view* is an oblique view of a town showing both the layout in semi-plan fashion and the buildings and surroundings in perspective. This was an enormously popular type of view, and is still used widely. Among the earliest such printed views are those in the *Liber Cronicarum* (Nuremberg Chronicle) of 1493; an example is shown in Figure 3.13.

If we look straight down on a city or town and draw a map of its streets, parks and blocks, such a map is called a *plan* (see Figure 3.14). The folded maps we carry with us when we find our way through a strange city are common examples of such plans, and perhaps this is why most of us think of these straight-down, or *normal*, views when we think of the mapping of cities.

Major cities were well mapped and one can assemble splendid city plan series from the 15th or 16th centuries to the present. And they need not be overly expensive. Even such wonderful city plans as those done in London by the S.D.U.K. (Society for the Diffusion of Useful Knowledge)

in the early to mid-19th century, an example of which is shown in Figure 3.14, are modestly priced.

A *panorama* is a view that takes in a greater amount of azimuth. A panorama can be an extravagant production, sometimes taking in a full 360 degrees, or can be many feet long and rolled up, scroll-like (Figure 3.15). Panoramas are often three-quarter views as well.

GORES

Globes are spherical representations of the world or celestial spheres. Most globes were made by printing maps on flat sheets of paper, cutting them out, and pasting them onto the surface of a blank sphere. These spheres were either solid, in the case of very small globes, or, more often, elaborate built-up hollow spheres. Globes were technically quite sophisticated. Modern X-ray analysis has revealed a lot of the details of their construction. X-ray analysis is also used to identify recent fakes.

Globe makers had a peculiar problem; they needed to print their maps in such a way that the flat paper could be deformed to a spherical surface. They solved the problem by printing roughly

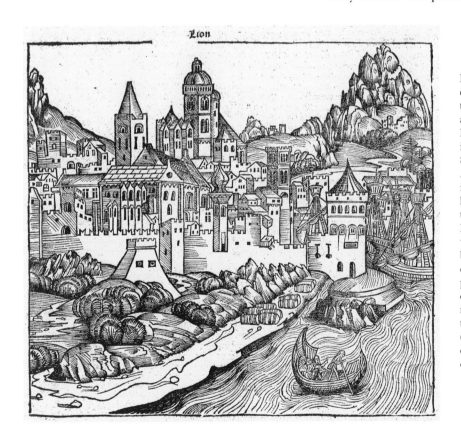

Figure 3.13 Among the earliest printed town views is this woodblock city view that appeared in Schedel's Nuremberg Chronicle, printed in 1493. Original size is about 8 x 9 inches (20 x 22.5 cm). Although some of these illustrations were originally hand colored, most (including this one) were uncolored. Many of these three-quarter views were generic and used by the publishers with different names in different parts of the book. These very early town views are still readily available. Some of them are large two-sheet views (see Maps 4, 5, page 125); others are smaller than this one.

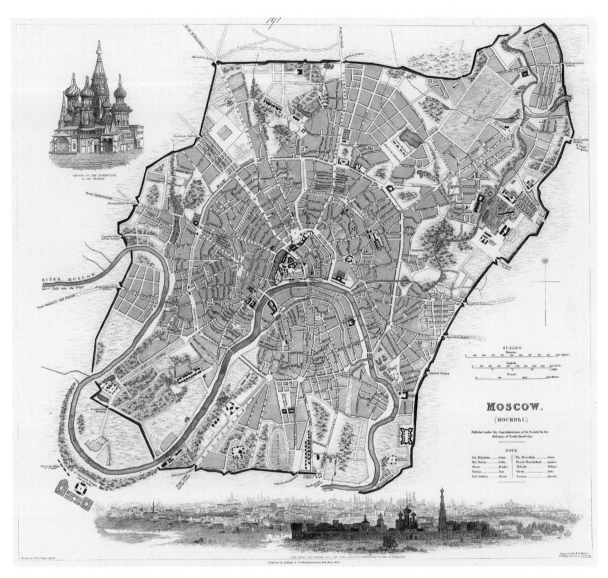

Figure 3.14. The S.D.U.K. (Society for the Diffusion of Useful Knowledge) published this copperplate engraved city plan, dated 1836. A very detailed plan measuring 12.75 x 14 inches (32.5 x 36 cm) with simple outline hand color, it is embellished with vignettes showing St. Basil and a view overlooking the city. Such plans exist for major cities and show them just before the industrial revolution changed the world's cities forever.

triangular segments called *gores* (Figure 3.16). Gores were cut out, wetted to make them pliable and glued to the surface of the ball that was to become the globe. For colored globes, the gores were generally colored *after* being placed on the globe. Thus, most of the single gores, or sets, on the market should be uncolored, since they were not used in making actual globes. If colored, we can often surmise that the color is later. Relatively scarce, unusual in appearance, and generally considered attractive, gores are collected avidly.

Those of the late 17th century are very desirable, especially the ones by the great Italian maker, Coronelli. A single gore showing a highly collected region, such as insular California or the American Great Lakes, can cost several thousand dollars.

Celestial globes were made in the same way, but would often be shown with the constellations "reversed." We, the viewers, are used to looking at the stars from inside the "celestial sphere," but when we look at a celestial globe, we are looking at them from outside the sphere.

41

Figure 3.15 This wood engraved panorama *Grand Panorama of London and the River Thames* measures some 18 feet in length. The image is 5 inches (12.5 cm) high and was printed on several sheets that were joined to produce a long scroll. The entire panorama is wound on the wooden spindle seen at the right and had a self-cover with a string tie. This mid-19th-century example must have been produced in large numbers but has become scarce relatively recently.

OTHER WORLDS

The mapping of the surfaces of the Moon and other planets is an interesting subspecialty of map collecting that merits discussion.

The invention of the telescope in the early 17th century made it possible to discern detail on other planets as well as the earth's moon (see Figure C.9). For the next 350 years, earth-bound observers, using visual techniques and later, photographic techniques, sought increasingly finer detail on these surfaces. There are superb atlases and individual maps available. The 19th century is particularly rich in moon maps and planet maps, but the earlier work is very important and also quite attractive. The small woodblock reproduced here in actual size (Figure 3.17) is taken from a work by

Galileo *(Lettera Del Sig. Galileo Galilei al Padre Christoforo Grienberger Della Compagnia di Giesu. In materia delle Montuosità della Luna)* published in Bologna in 1655. It is primitive, even for that era.

Allegorical representations of the constellations (Figure 3.18) are widely collected and, when attractive, can be quite expensive. Some of the most beautiful celestial maps are those by Cellarius (Map 54, page 173), published first in the mid-17th century, and reissued by Valk and Schenck in the early 18th century. In addition to star maps with constellation figures, the Cellarius atlas contained particularly dramatic solar system maps. The Cellarius plates and, to a lesser extent, those in the Valk and Schenck reissue, are much in demand by decorators, but their prices make them less attractive to casual collectors or astronomers. The

Figure 3.16 This fine, single half-gore is one of several that were made for a Coronelli globe. It is an uncolored copperplate engraving, 18 inches (46 cm) high, 11 inches (28 cm) wide at the top and 6.5 inches 16.5 cm) wide at the bottom. It was made for a globe approximately 43 inches (110 cm) in diameter. The meridian at the right is the old prime meridian passing through the Azores. This gore was printed in Venice in the late 17th century. Note that it was printed on a flat, rectangular sheet of paper. To use on a globe, the gore would be cut out, moistened to increase the paper's flexibility, and pasted to the surface of the globe.

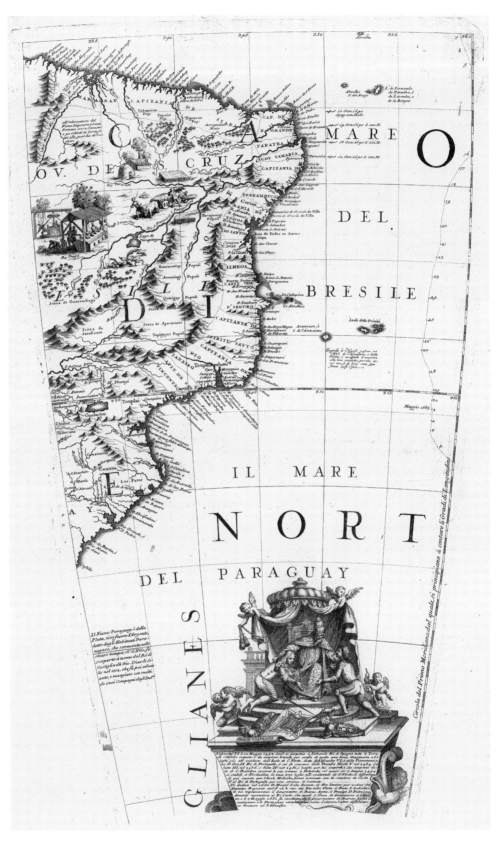

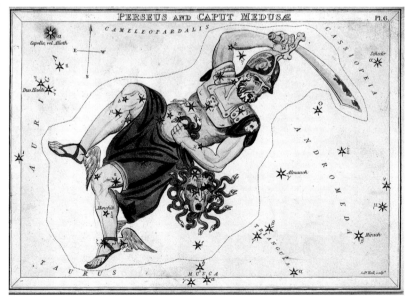

liſſima, & vn giorno auanti, & vno doppo il ple-
nari montuoſità già indubitabilmente ſi ſpargono,
l'ultima circonferenza Lunare; ma perche in tali
ure, & adom-
n ſcorcio me-
, & incurua-
ella Luna ap-
ghe mà ſtret-
lla preſente
le medeſime
, che nella
edute in fac-
o grandiſſime
, quanto per
icino all' vl-
nare doue ſi
quaſi in pro-
arghezza, & appàriſcàno lunghe ſi, mà ſtretté,
iſſimo ſe gli eleua il raggio viſuale, mà trasferen-
dole

Figure 3.17 Galileo made the first moon maps by drawing what he saw through his telescope. The woodblock map above is shown actual size.

American star maps by Burritt, published in the 1800s, are quite attractive and priced modestly.

By the mid-19th century most star atlases had stopped portraying the allegorical constellation figures and the atlases became less decorative and more "scientific."

CURIOUS MAPS

There are many whimsical maps that attract the attention of collectors; these are often

Figure 3.18 Constellation maps were popular in the 19th century. This hand-colored lithographed example was printed on card stock measuring 5.5 x 7.75 inches (13.5 x 20 cm). The major stars were punched out so that the card, when held to a light, would simulate the actual star patterns. A set of 32 such cards, published in London in 1823, was sold as *Urania's Mirror.* Complete sets in original boxes are very scarce.

classified under the heading of *cartographic curiosa*. Many of these maps are quite charming and are collectable just because they are, from our perspective, so curious. Kircher's map of Atlantis (Figure 3.19) is a good example. Another example of a map based on erroneous geographical concepts is the copperplate engraved map shown in Figure 3.20. This map embodies most of the erroneous concepts ever imagined about the American Northwest. There is a huge "Sea of the West" as well as several other curious formations purportedly discovered by the fictitious Admiral de Fonte. It is really quite interesting from the standpoint of the mapping of the western part of North America and is still priced very affordably. Maps showing California as an island are collected avidly and although insular California is a curious cartographic notion, these maps are not generally lumped in the category with fantasy maps.

Among the more famous cartographic curiosa are those maps that have been transformed, by the cartographer, to fit (more or less) into the forms of humans or other animals. A number of different maps were made showing Belgium as a lion; these are the *Leo Belgicus* maps. The first was made by Baron Michael von Eytzinger in 1583.

Bünting (who made the cloverleaf world map illustrated in Part II) also made a map of Asia in the form of Pegasus, the flying horse, and a "Europe as a virgin" similar to the map shown in Figure 3.21. This kind of portrayal of female

44

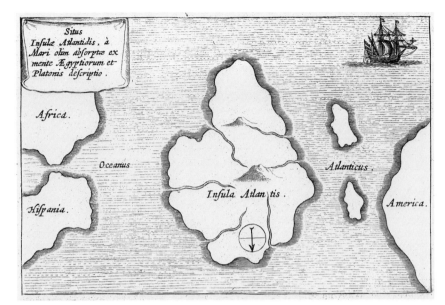

Figure 3.19 A fantasy map, this copperplate engraving is from a Dutch edition of Kircher's *Mundus Subterraneous* published in Amsterdam in the latter 17th century. South is at the top and the legendary land of Atlantis is prominent in the Atlantic between the Old and the New Worlds. Original size is 4 x 6.25 inches (10.2 x 15.5 cm).

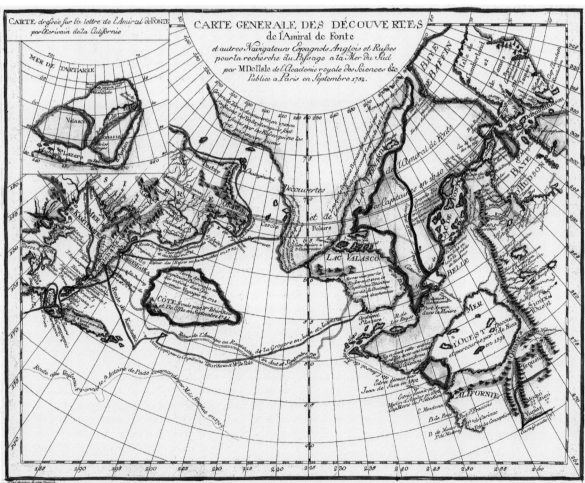

Figure 3.20 The history of cartography is replete with erroneous geography. This late-18th-century copperplate engraved map, 11.5 x 14.5 inches (29 x 37 cm), shows the discoveries of the fictitious Admiral de Fonte. It depicts a marvelously confused geography of the American west coast, including the fantasy "Sea of the West." Later hand color.

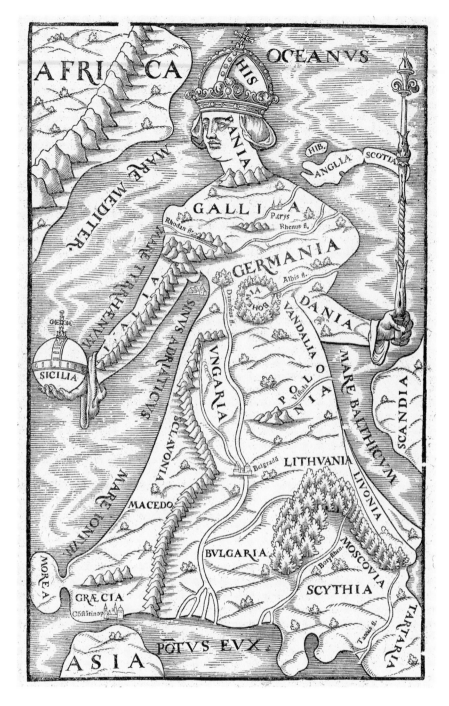

Figure 3.21 Maps sometimes show an anthropomorphized landform. Europe has been portrayed both as a woman and as a virgin. This woodblock, which appeared in editions of Münster's *Cosmographia* published in Basle in 1588 and later, is a typical example of "Europe as a woman," although it has been argued by some that the figure represents Charles of Spain. The size of the original is about 10.25 x 6.5 inches (26 x 16 cm). The globe in her right hand (at the end of Italy) may be a representation of a "T-O" map (see Figure 3.1).

In the late 1700s, Bowles & Carver of London produced maps of England and Wales and Scotland in the form of distorted humans. The practice of depicting maps as animals or people continued into the 20th century with large lithograph maps showing, for example, the Russian Bear.

The fantasy world of Schlaraffenland (Figure 3.22) has been mapped with a high degree of wit and satire. Schlaraffenland is the Utopia of German legend. It is here where, in the children's version, sausages grow on trees and tables set themselves. The perception of Schlaraffenland provided by the map in Figure 3.22 is quite a bit more adult-oriented. An enormously detailed map, we see such places as *Terra Sancta Incognita*, and from there we can travel south to *Mammonia* and continue on down to *Republica Venerea*. We note also that the land straddles the

Europe was made first by Bucius in 1537, but popularized by Münster (Figure 3.21). Münster's map appeared between 1588 and 1628.

The "Europe as a woman" maps are modestly priced and generally available without too much difficulty. The *Leo Belgicus* maps that show Belgium as a lion are, in comparison, relatively expensive.

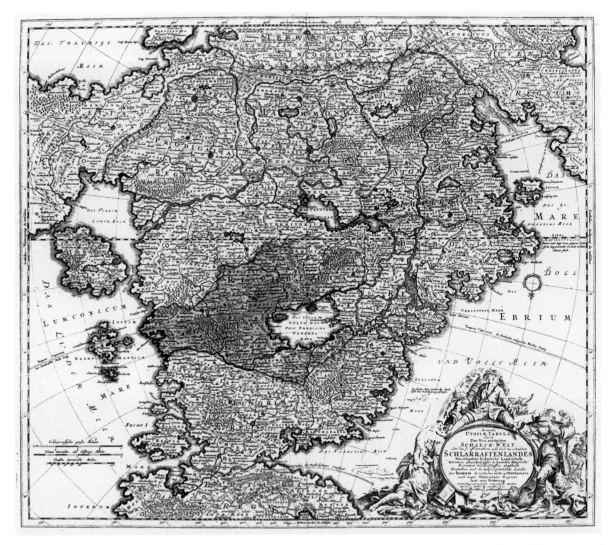

Figure 3.22 This copperplate engraved map, by Homann, *Accurata Utopiæ Tabula Das ist Schalck-Welt oder des so offtbenanten, und doch nie erkanten Schlarraffenlandes....,* shows the fictitious "Schlaraffenland." Published in Nürnberg, c. 1710. It has fine full original hand color. 19.25 x 22.5 inches (49 x 57 cm). Curiously, Homann spelled "Schlaraffenland" with a double "r" whereas other versions of this map have the correct spelling.

equator between about 360 and 540 degrees longitude. The cartouche has a splendid scene of debauchery and carnal excess. An exceptional map, indeed.

Map games

It should not be surprising that maps were utilized in games and play. Indeed, the first jigsaw puzzles were maps, and were produced to help teach geography to children. These late-18th-century puzzles are extremely scarce and very few come onto the market. Early map jigsaw puzzles used existing maps which were pasted onto wooden slabs and sawed into pieces. Puzzle makers frequently used outdated maps for many years and it can be difficult dating a puzzle solely on the basis of the map. Some of the later puzzles, especially those made for advertising purposes, used maps made expressly for the puzzle.

It is no longer generally feasible to build a large collection of 18th-century and early-19th-century jigsaw puzzle maps because of their scarcity. Map jigsaw puzzles are still made today, and collections of puzzles dating from about the mid-19th century to modern times can be assembled readily.

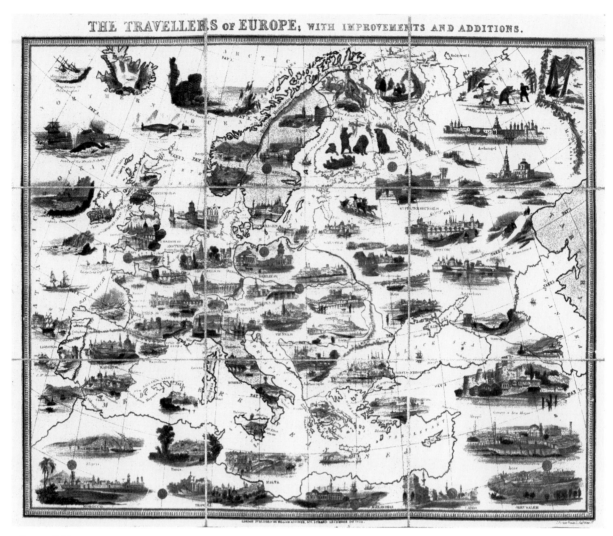

Figure 3.23 Linen-backed folding map games are collected widely. The one shown here, *The Travellers of Europe: With Improvements And Additions*. Published in London in 1852 by William Spooner, it measures about 19 x 23 inches (48.4 x 58.5 cm). The map is lithographed and has full original hand color. In stiff self-covers with original printed label. This appears to be one of the last editions of this map game. The map has wonderful vignettes of "typical" native activities such as whaling and bear-hunting.

Map games (Figure 3.23) are usually folding linen-backed maps with places for players to advance their tokens. Some games had the rules printed on the surface, others had the rules in separate booklets. A curious fact is that since dice were frowned upon in Victorian times, many of the games required the use of spinners or teetotums. Some of the games included pasteboard cut-out spinners.

Map games were, much like the map jigsaw puzzles, made for use as teaching tools. However, unlike map jigsaw puzzles, map games used maps made especially for the purposes of the game and often included ethnological (see Figure 3.23) or commercial information. Map games have been collected widely and, as a consequence, also have become rather scarce in recent years.

ROLLER MAPS

Roller maps are large-scale maps meant to be displayed on a wall (not all *wall maps* have rollers), generally used for teaching purposes. Some firms that produced folding maps also printed wall versions.

Early wall maps, such as those by Blaeu, are very scarce and very expensive. These were

48

engraved, printed on multiple sheets of paper and then often laid on a linen backing.

Many 19th-century wall maps (see Figure 3.24) are still available and still relatively inexpensive. The difficulty of collecting such large maps (they can be six feet by six feet or larger!) helps to keep the price down. Made to be used in schools or lecture rooms, they often were backed with linen to make them more robust. Many of these are beautiful, with vivid color. They became common in the 19th century, possibly because advances in lithography made them less expensive to produce. After mounting on linen, they were generally (but not always) varnished. The varnish has usually browned (see Figure C.10) and cracked, and the browning may hide the brilliant colors underneath.

American roller maps typically have a wooden molding across the top and a dowel along the bottom. The dowel often sported a turned wooden finial at each end (see also Figure C.10). The wooden components are integral parts and if they are missing or not original, the value is somewhat less.

Repair or conservation work on large wall maps is very expensive. There are relatively few conservators who have the equipment to wash and soak such large pieces. However, even simply removing the browned varnish is often a major cosmetic improvement, since the color can then be seen in its original brilliance.

Unless the map is of particular importance, I advise not buying one that needs restoration or serious repair. Missing surfaces, torn or punctured linen backing and heavy water stains are all serious problems and difficult to repair in these large maps.

Courtesy: Chris Watters

Figure 3.24 *Pictorial Map of The United States.* This roller map, by Ensign, Bridgman and Fanning, is printed on paper, linen-backed and brightly hand-colored. Although the varnish has browned with age, it has served to protect both the bright colors underneath and the surface from minor abrasion. The colors regain their brilliance if the varnish is removed, but any more extensive repair or restoration is very expensive due to the large size of these maps. The condition of this map, shown unrestored, is quite good with only minor discoloration and surface defects. Actual size of the image plus margins is 30 x 44 inches (76 x 112 cm). Image alone is 28 x 39.5 inches.

This large copperplate engraved image, approximately 22.5 x 19.25 inches (57 x 49 cm) is by Seutter, and was published in Augsburg, c. 1730. It depicts a pope, and inscribed on him are the names of all the popes from Peter into the 18th century. The full original hand color on these images is superb.

Chapter 4

How Maps Look

Maps can be printed by different processes and each process imparts a distinctive appearance to the image. We can learn a lot by examining carefully the image that is a map. Maps were either drawn by hand (manuscript maps) or printed. Manuscript maps are, obviously, unique. Printed maps were made in some definable number and are, more or less, invariate copies. It is useful to learn to distinguish the different processes used to make maps. This knowledge can help us make

judgments about a map's technical quality, whether it is an early or late impression and, even, if it is genuine.

MANUSCRIPT MAPS

A *manuscript* (abbreviated: mss.) is anything written by hand (manu = hand). The earliest maps were, of course, manuscripts. Manuscript maps can be quite beautiful, as in the case of *portolans,*

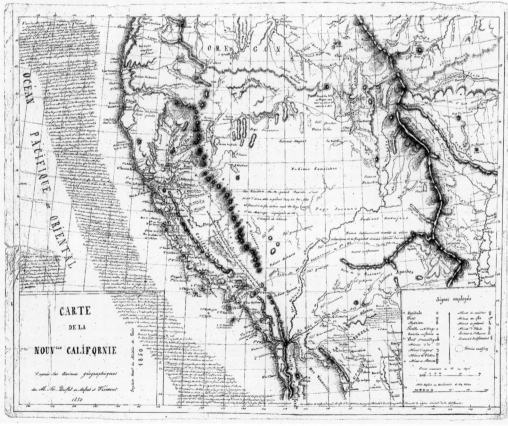

Courtesy: G. Dreher

Figure 4.1 This fine example of a mid-19th-century manuscript map on paper, by Duflot de Mofras, appears to be an unpublished revision of an earlier map. Discovered in Switzerland, it is now in an American collection. 14.75 x 18.75 inches (37.5 x 47.5 cm).

which are highly decorative early navigational charts. These charts, most often of the Mediterranean regions, are rare and very expensive today. Most of the ones we see on today's markets were made in the 15th and 16th centuries.

Manuscript maps are still made today; they are often either individual efforts to produce a decorative item, or the projects of school children. Indeed, some of the earlier 19th-century school projects are quite collectable and expensive today. Explorer's maps, especially those drawn while on expedition, are also manuscript and can be very valuable. Unfortunately, many manuscript maps are modern copies, some of which are meant to deceive. It takes a great deal of experience and knowledge to ascertain the authenticity of a manuscript map. They are best left to the expert.

PRINTED MAPS

The invention of printing in the 15th century (in the West) was a revolution almost beyond comprehension. For the first time in human history it was possible to produce invariate copies of an image, indeed any number of invariate copies, be it text or drawing, and disseminate them. Quite literally, two people on opposite sides of the earth could, for the first time, have the same visual and textual information and therefore be exposed to invariate artifacts of the same culture and intellectual tradition.

In the case of maps, this meant that the same view of the world could be shared universally. From the practical standpoint of today's map collector or dealer, it means that we can compare the images of most maps. Most printed maps have a history; we know who made them, who printed them and who sold them. We can also compare their conditions, their colors, and when we discuss them, others know of what we speak because we are dealing with maps that other people have seen and with which they are familiar.

There are three general classes of printing.

1. Relief. The image is formed by pressing paper against a *raised* inked surface such as type, the familiar rubber stamp, or a carved wooden block. These images can be printed on a normal letterpress. In relief printing, a map and any type on the same page can be printed in one operation.

Letterpress, also a relief process, refers to text printed from movable type.

2. Intaglio. This is the mechanical reverse of the above. Examples are *etching* and *engraving.* The image is formed by cutting a groove in a plate, usually copper. The plate is inked and the ink is wiped off the high, uncut parts of the plate. This leaves only the grooves filled with ink. Paper is pressed onto the surface of the plate, forcing the fibers into the grooves where they take up the ink. The high, unincised parts of the plate, wiped free of ink, do not leave an image. Because of the high pressure required to force paper into the incised grooves, intaglio images require a special press. Accordingly, if a map is to be printed on the same page as letterpress, the sheet must be passed through two separate presses.

3. Planigraphic. In this process ink is not held by either raised or depressed parts of the plate, but lies on a flat surface. The image rather than being the result of elevation differences, results from the way the plate was treated chemically. An example is *lithography.*

The three different ways of printing an image produce lines on paper that have different characteristic qualities. *By examining the image, we can determine the process by which it was produced.*

WOODCUT MAPS (RELIEF)

A woodcut map is printed from a carved wooden block. The image is carved on the block's side, into horizontal grain, by cutting away all areas that should not print, leaving behind raised lines. It is these lines that accept the ink and against which the paper is pressed during printing. A wooden block is very much like an office rubber stamp or the common linoleum block technique used to produce illustrations. In this type of printing the printed line is slightly indented in the paper, much the same way a piece of type indents the paper with which it comes into contact.

Woodblock printed maps (see Figure 4.2) were common in the Western world from the 15th century through the early 17th century, when the technique fell into general disuse. Early printed books were illustrated with woodblock images and it seems natural that the early maps would be made using this technique. Often the blocks were cut to accept movable type and we note several maps that were labeled with different fonts at different times in their printing history.

The woodblock line is characteristic of a relief impression (Figure 4.3). If well-inked, the line exhibits an uneven distribution of ink, heavier around the edge. We see the same phenomenon in letterpress printing from large type.

Woodblock maps often have heavy lines. Although those woodblock maps made in the West generally lack the fluidity that an engraved map demonstrates, they have a peculiar charm.

Woodblock maps produced in the Orient often have a somewhat different appearance. Figure 4.4 shows a portion of a Japanese map. The lines on these maps are surprisingly fluid and sometimes superficially resemble engraved lines. The author used to deal extensively in Japanese woodblock maps and encountered several that had been described in the Japanese literature as being examples of early Meiji period copperplate engraving, but were, in reality, well-carved woodblocks designed to copy the appearance of the engraved line.

Woodblock printing is different from *wood engraving,* which is a 19th-century development (invented in England c. 1800). In wood engraving, the image is cut into the end grain of a block of

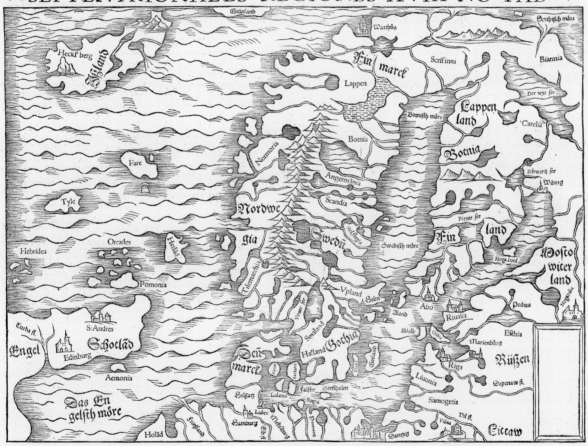

Figure 4.2 Typical of the appearance of a woodblock map, this one is by Sebastian Münster and was printed in Basle in 1540. The image measures about 9.75 x 13.3 inches (23 x 34 cm) and is printed on laid paper with letterpress text on the verso. Note the overall simplicity of lines that characterize most Western woodblock printed maps. Some of the text, such as "S: Andres;" "Lappen" and "Hetläd" appear to be printed from movable type set into the block, whereas the larger place names may be cut directly into the wooden block itself. On the next page **(Figure 4.3)** we have enlarged a portion of the map and have deliberately photographed it so that the faint show-through from the text on the verso can be seen. In this enlargement (4X original size) we can see the characteristic line of the typical woodcut. Because of the relative coarseness and wide spacing of these lines, they are relatively easy to duplicate by modern means and forgeries of some of the more valuable woodblock maps have been made.

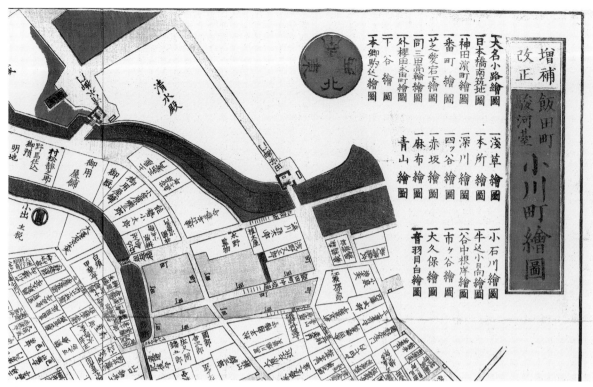

Figure 4.4 The woodblock technique was widely used in the Orient, and we see here part of a fine example of a Japanese woodblock printed map, *Surugadai Ogawamachi Ezu;* carved in 1849 (Kaei 2) this impression was made in 1860 (Ansei 7, or Manen 1). Published by Owariya Seishichi. Overall size is 13.5 x 27 inches (34.5 x 68.5 cm). Japanese maps were printed from wooden blocks, using a separate block for each color; this map required four separate impressions. This city plan is one of a set covering all of Edo (Tokyo).

wood, not the side. The cutter uses a tool called a **burin** and can produce much finer lines than is possible with classical woodblock technique. Wood engraving arrived late in the history of map making and was not much used, even in the 19th century. Its forté was in the production of illustrations for newspapers, books and magazines. A very few maps were produced by wood engraving, mostly in less expensive textbooks and in 19th-century magazines.

ENGRAVING (INTAGLIO)

As woodblock printing waned in the 16th century, maps were produced increasingly by the process of **engraving** (see Figures 4.5, 4.6). An engraving is made by incising a deep line in a metal (usually copper) plate with a sharp instrument called a burin, or **graver.** In order to print the plate, ink is rubbed over the entire surface and then wiped off the high parts. Thus, the ink is left in the engraved lines only. Dampened paper is laid over the inked plate and the sandwich of paper/plate is run through a roller press. The press looks much like the mangle of an old-style clothes washer. The rollers exert great pressure on the paper, forcing it into the incised lines of the plate where it takes up the ink. This is called **intaglio** printing. *Since the paper is forced into the engraved line, the line is slightly raised on the surface of the paper.* If we examine an engraving with a magnifier and oblique light, we can easily see the raised nature of the engraved line.

The intense pressure needed to print an engraving forces the dampened paper down around the edge of the metal plate, creating the **platemark.** The platemark, unless it is fake, is diagnostic of intaglio printing. Because the paper must be dampened to receive an impression from an intaglio plate, prints from the same plate may vary in size by as much as several percent.

Engravings produce very attractive images, due in part to the fact that the printed engraved line is slightly raised above the paper surface. This produces an overall three-dimensionality to the image that is lacking in other reproductive techniques. Once you recognize this, you will be able to spot an engraving from across the room.

Since most maps produced as reproductions are not engravings, the difference between the original and the reproduction is usually dramatic.

Steelplate engraving is a variant of copperplate engraving. In steelplate, the line is incised on an iron or soft steel plate rather than the softer copper. This makes it possible to print many more impressions before the plate wears. The lines engraved on steel plates are "harder" in appearance than those printed from copper plates and can be spaced more closely. In order to increase their wear properties, copper plates were often electroplated with iron. These techniques, devised in the nineteenth century, were used also for producing engraved views and illustrations, as well as many nineteenth-century atlases.

It is of interest that many 19th-century atlases that were identified as "steelplate" atlases on the title page were actually printed by means of lithography. Close examination of the printed image indicates that the maps were not the result of intaglio printing, but rather planigraphic techniques. The original images were engraved, but the images were then transferred to lithographic plates for printing. This is called a *transfer lithograph.*

It is important to be able to identify the type of printing used in these images for one-self, since many such maps contain statements claiming they are engraved.

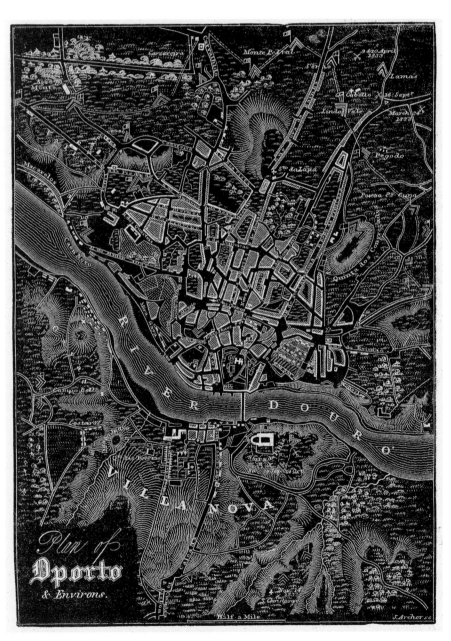

Figure 4.5 *Wood-engraving* was not often used for the production of maps, and images such as the city plan shown here are un-common but in-expensive. The plan measures 9 x 6.5 inches (23 x 16.5 cm) and is from Pinnock's *Guide to Knowledge* published in London in the 1830s. This is also an example of a *white-line* image in which the engraved line is not printed since the ink remains on the high parts. Such white-line maps are unusual; very few publishers produced them.

Figure 4.6 The engraving is superb on this copperplate map by Wytfliet, published shortly after 1597. It is the second state (see Glossary) of this map, and measures about 9 x 11.25 inches (23 x 28.5 cm). The Wytfliet maps are from the earliest atlas of the New World and are much sought after by collectors. The map was produced entirely by engraving on copperplate, including the sea represented in a moire pattern. These maps were rarely colored and the illustrations here are from an uncolored exemplar. In **Figure 4.7** shown on the next page, we have enlarged a portion of the map including the northeast part of Jamaica. In this 4X enlargement of the original size, we can see the distinct quality of the engraved line. Dark lines are formed by incising the plate and a single incision creates a dark line (the incised line catches ink that is then transferred to the paper), whereas in the case of woodcut (see Figure 4.3) it requires *two* cuts to create a dark line (material must be removed from either side of the remaining wood). If we look carefully at the 4X enlargement we can see little specks that represent pits and fissures in the surface of the copper plate. Each such "defect" catches ink that is transferred to the paper. If we examine the rare examples of copper plates that survive from this era, we indeed see all these little surface imperfections. The inset in the upper left of Figure 4.7 is a 10X enlargement of the original image. Here, we clearly see the dominant characteristic of the *engraved* line; a long taper. The line begins as a thick incision and tapers to a fine point. Contrast this to the lines seen in the woodblock (Figures 4.2, 4.3) and the etchings (Figures 4.8, 4.9). We also note that the text on engraved maps was generally incised directly into the same plate as the image. The engraver cut a "mirror image" text and the character of these letters is often diagnostic of the period and country of origin. It is important to acquire a visual familiarity with the appearance of these engraved lines in order to recognize reproductions and forgeries. Note in Figure 4.7 that the lines are clean and distinct; lines on intaglio plates made by photographic processes lose this distinct property. I emphasize that it is to the collector's advantage to examine the detail on maps with care and modest magnification.

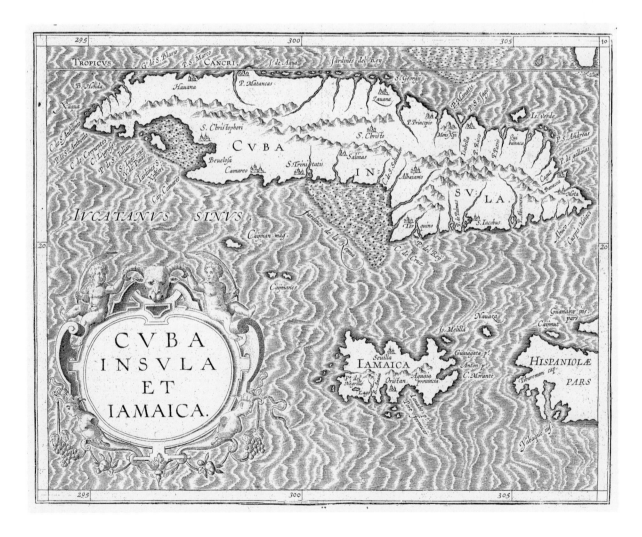

57

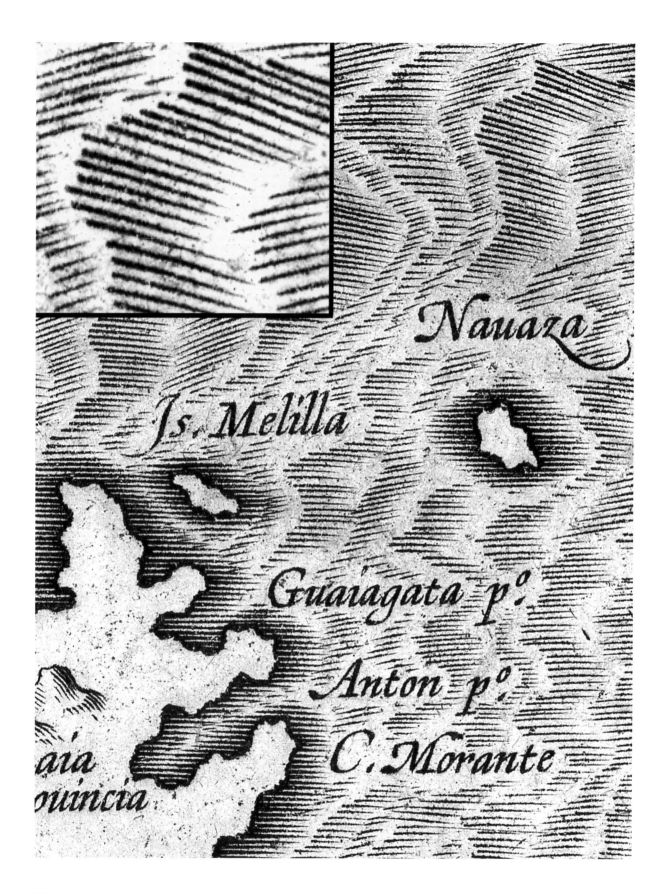

Nauaza

Is. Melilla

Guaiagata p.º

Anton p.º

C. Morante

aia
ouincia

58

ETCHING (INTAGLIO)

Etching is a technique in which a coating of wax or some other acid-resistant material (this material is called a *resist)* is put on the face of a metal (generally copper) plate. A design is scratched with a needle through the wax down to the underlying metal. The plate is then placed in acid; the acid attacks the metal where it is exposed by the needle line but not where it remains covered by the resistant wax. Thus, instead of physically digging out a line with a burin, as in engraving, the line is formed by the reaction of acid with the metal. Etched lines are more fluid in appearance than engraved lines, since the etcher has more freedom of movement with the needle than has the engraver with the burin. Consequently the overall appearance of an etching is qualitatively different than the appearance of an engraved image. We can look at individual lines and often determine if they were made by etching (Figure 4.8) or engraving. The etched line usually has a blunt end, whereas the engraved line is tapered (Figure 4.7).

Because of the relative ease and rapidity of creating an image by etching, the overall layout of a plate was often etched and then the detail was added by engraving. Thus, if we examine a large image we can often see both etched and engraved lines. I have seen this admixture of etched and engraved lines in maps, which suggests that here, too, etching was used chiefly to rough out a map which would then later be finished by the burin.

Etched lines are unmistakably present in the small peripheral figures and views on maps of the continents by Blaeu. Many cartouches, when examined carefully, reveal their predominantly etched nature (Figure 4.9).

Despite the use of the etching technique on various parts of a map, and its employment by Hollar, there are few early maps done entirely by the etching process. The Torniello world map (1609 and later, Figures 4.10 and 4.11), often cited as an example of an etched map, contains obvious engraved lines, some of which look like etched lines.

LITHOGRAPHY (PLANIGRAPHIC)

Lithography (litho = stone) is a reproduction technique invented around 1800. In its original form, lithography utilized heavy, thick slabs of stone to hold the image. The image was drawn on the stone with a wax or grease crayon. The stone was then wetted with water and an oil-based ink applied. The ink would not adhere to the wet stone, but would adhere to the greasy line drawn by the crayon. The inked image would be transferred to a sheet of paper pressed down on the stone. This kind of printing (Figure 4.12) is called *planigraphic* since there is no relief to the printing surface. Unlike either woodblock (relief, or *cameo*) or engraving (intaglio), lithography is a flat process that neither indents nor raises the paper where it is printed. Generally no platemark is produced.

Figure 4.8 This micrograph of a group of *etched* lines was made using a strong oblique light coming in from the upper left. Note that etched lines have blunt ends. The arrows point to several examples; you can find many more. Where etched or engraved lines intersect, the intersection is clean and sharp. Such is not the case in photoetched reproductions, where the intersections and space between close lines are often filled in and indistinct. Compare these lines to the engraved ones in Figure 4.7.

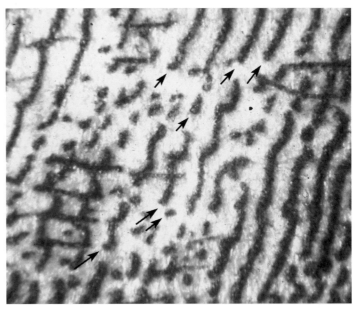

59

Figure 4.9 An enlarged view of part of the cartouche of *Hind, Hindoostan, Or India...* published in 1800 by William Faden reveals interesting information. Although the map itself is engraved, the cartouche is predominantly etched. The etched image looks very different than the engraved one: the lines are more "squiggly" and have blunt ends. It is quite possible that the etched cartouche was touched up, before printing, by judicious use of the burin.

60

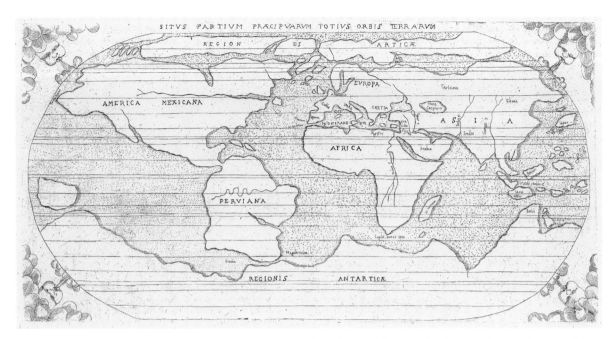

Figures 4.10 (above) 4.11 (below) This world map, by Augustino Torniello (1611), has only simple, somewhat crudely executed lines. It has been described as an example of an etched map, but it contains unmistakably engraved lines, particularly obvious in the windheads. Original size is 19 x 37 cm (7.25 x 14.5 inches). The arrows in the enlargement below point to the blunt ends of lines. This is characteristic of an etched line, but in this case it is probably due to clumsy use of the burin. The map contains lines unequivocally made by the burin.

Lithographs have a characteristic "flat" appearance, dissimilar to the fine visual texture inherent in an engraved image. The 19th century witnessed the decline of the engraved map and its virtual replacement by the lithographed map. Many of the American state maps from the Mitchell and Johnson atlases were lithographed. Unlike an engraved copper plate, a lithograph stone essentially does not wear out and changes to the image are relatively easy to make. Images can be "lithographed" from many substances other than stone. Metal, particularly zinc, has come into wide use.

Offset printing is a modern technique, invented in 1904. In offset printing, an image from a metal or sometimes even a paper plate is transferred to a second drum. It is this secondary, or "offset" image that is then transferred to a sheet of paper. This type of printing is very fast and inexpensive.

HALFTONE PROCESSES

Up to now we have discussed images made up of lines. An image can also be printed to have tone and look somewhat like a gray-scale photograph or a painting. This is accomplished by breaking up the image into tiny dots by photographing it through a screen. Now consisting of thousands of dots rather than continuous tones or unbroken lines, shades of gray are mimicked by changing the density or size (or both) of the dots: pale areas have fewer and/or smaller dots than dark areas. The image is then transferred photographically and

Figure 4.12 A highly enlarged part of a chromolithographed map of Ireland, done in Germany by Justus Perthes for the 1911 edition of *Stielers Hand-Atlas*. The atlas is advertised as having copperplate maps, but the actual images are lithographed. The very high-quality printed color is achieved by printing it in narrow stripes rather than in dots. This is most apparent in the ocean. Where the stripes intersect, the color is more intense. The hachures indicating mountains are also colored. This atlas, in print for over half a century, has some of the finest lithographed maps ever produced, yet the images are "flat" when compared to engraved and hand-colored maps.

chemically to a plate from which it is printed. If the screen is fine enough, that is, there is a great enough number of dots per inch, the image will have the appearance of a continuous tone. However, close examination will invariably reveal the dotted nature of the image.

Images that are broken into discrete dots by means of a screen can be printed by *gravure* (machine-printed) or *relief halftone*. Images produced by these different processes can be distinguished by their different dot characteristics, but this is not of great consequence in the old map world.

Halftone lithography also breaks up an image into dots, but it is then printed planigraphically. Most modern magazines and books are illustrated by halftone processes. Look at the illustrations in this book with a magnifying glass. They were printed by offset lithography using a screen of 150 lines per inch (the dots are produced by a line screen, but that is a technical aspect we do not need

to address). Finer line screens can be used, making the dots harder to detect, but also much harder to print without the ink clogging the spaces between dots and making the image appear muddy.

Another term for this appearance is *dithering,* used quite often in computerspeak, where in order to make a laser printer produce a more acceptable image of a continuous-tone scan, the computer dithers, or breaks the image up into patterns that can be printed more readily than a continuous tone.

Halftone printing is a dead giveaway that a map has been printed recently, say in the past hundred-plus years. Any map that purports to be older but whose image is made up of halftone dots is not genuine. Most of the decorative reproductions of famous and beautiful maps are made with this technique and are therefore very easy to detect.

Many 20th-century maps were originally made in this way and the presence of these little dots does not imply a fake or a reproduction in this case.

62

It is, however, my personal bias that such maps, considered as artifacts, have a lower intrinsic value than those made using earlier techniques. But don't let my bias dissuade you from collecting these maps if you like them. The history of collecting is replete with such predictions and mine will probably, in fifty or a hundred years' time, be proven wrong.

OTHER TECHNIQUES

Cerography or "wax-printing" was developed in the 19th century in the United States. It was used extensively (and profitably) by such publishers as Rand-McNally, and enjoyed a long-lasting popularity in the printing of 19th-century American atlases and school geographies. It was a cheap way to produce textbook illustrations.

Cerography is a form of electrotype. The image is incised in a thin layer of wax. The wax is electroplated with copper, which forms a thin replica of the incisions. The copper is backed by type-metal and the resulting relief plate is used to print directly on a letterpress. While cheap, cerography produced a rather lusterless line.

Relief maps were made, often of plaster of Paris or papier mâché, in an attempt to demonstrate visually, or tactilely, terrain features. A variant of this is the relief map made for the use of the blind. One of the earliest of such maps is illustrated in Figure 4.14. This map was made before the invention of Braille and very few exemplars remain.

STATES AND EDITIONS

The plates used to print maps, be they woodblock, intaglio or lithographic plate, were often modified during their printing lifetime. This resulted, obviously, in different images. Some confusion

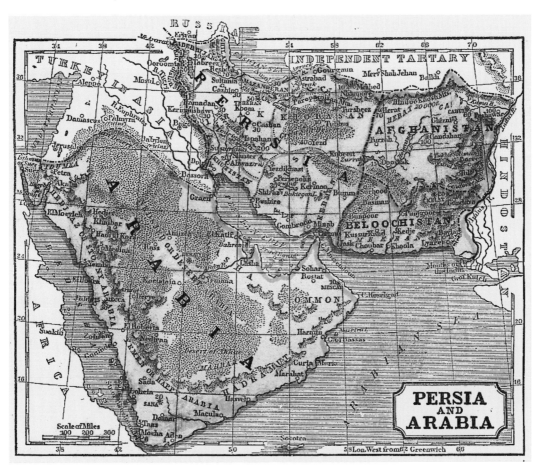

Figure 4.13 This cerographic map, shown full size, is from *A System of Geography for the Use of Schools...* by Sidney E. Morse, published in New York in 1846. The geography book was an early cerographic atlas and the maps were hand colored.

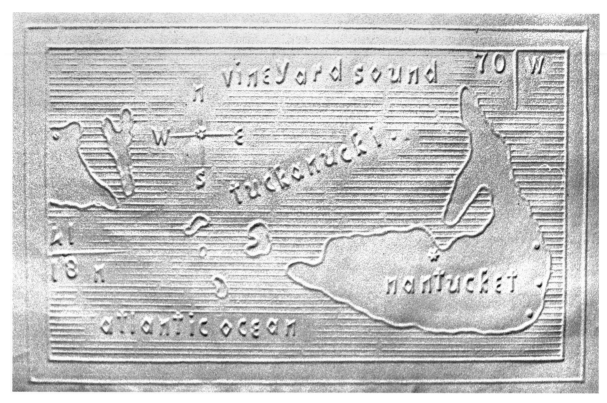

Figure 4.14 *A relief map for the use of the blind.* This map, measuring about 3.5 x 6.25 inches (9x16 cm) was printed by means of a very deep relief impression, resulting in raised lines and letters. Uncolored, it was meant to be examined by fingertips, not eyes. This map predates the Braille system and was printed for the Trustees of the New England Institution for the Education of the Blind, in Boston, 1839. This is an interesting variation of the relief map, which most often is used to demonstrate terrain.

exists about the terms used to indicate the various images produced from plates that were modified over time.

We can borrow the terminology used by our colleagues in the fine art print field, but because maps were produced in much larger quantities (the map trade was probably driven by more commercial interests), we encounter some different problems.

Proof. As an engraver finishes a plate, he may want to follow his progress by pulling a print from the plate. These *pre-publication* images are called *proofs.* We find proofs mainly in the art world; although the map engraver almost certainly pulled proofs, they don't seem to have been kept.

Edition. Every time a batch of impressions are taken, for example, if a publisher is printing another edition of an atlas, we consider all the images in this batch to belong to the same *edition.* The plates may be unchanged from the previous edition. It is important to note, however, that atlases from the "same edition" might have different contents,

further confusing the field. Atlas publishers sometimes replaced some maps in midyear and since the imprint date is the same, they are technically the same edition, but with different maps. Often the only way we can distinguish different editions is by the text on the verso – publishers usually had to reuse the type and it was not left set up from one edition to the next. The edition of a map is generally the same as the edition of the book from which it was removed. For single-sheet maps, the edition is determined by the date of publication.

State. Each significant change made to a plate resulting in an altered image establishes a new *state.* In the case of engraved images, the effort and expense needed to produce a copper plate was large, and publishers tried to get as much use as possible from each plate. If relatively minor changes were needed as new cartographic information became available, they would often burnish out the old lines and re-engrave the region

64

with the new information. Images pulled from the changed plate constitute a new *state*. Thus, depending upon what the publisher did to the plate over time, a map may exist in several states. Alternatively, different editions of an atlas may contain maps of the same state, or a mixture of states.

Let us compare Figure 4.15 to Figure 4.16. This copperplate, *Tierra Nueva*, was used for almost forty years (1561 to 1599) in Ruscelli's translation of Ptolemy, to illustrate the American northeast coast. An early impression from the first edition (1561) is shown in Figure 4.15. Significant changes were not made to the plate until the final edition of 1599, state two, which we see in Figure 4.16. The plate has received significant additions: a boat, a sea monster, several islands and interior mountain ranges. Thus, we say that the plate exists in *two states*.

There is a pragmatic point to be made here. As plates wore during printing, the images they produced showed these wear changes. We see this phenomenon in woodblocks as well as in copper plates. Most often the copper simply wears and the incised lines become weak and pale. Copper *work-hardens* and can become brittle after repetitive printings. Copper plates sometimes crack, especially along lines of stress that are introduced by the incised image. More commonly, they pick up other little nicks and scratches that hold ink and consequently are visible on the paper images pulled from them. All of these normal events in the life of a plate can be used to assemble their printed images into a *temporal sequence*, or *printing chronology*. Thus, we can tell, by carefully comparing different copies of the same map, the sequence in which they were printed from the plate. This is illustrated in Figures 4.17 and 4.18, which are enlarged parts of the map showing such changes with time.

Copper plates were often touched up in order to preserve the quality of the image. The intent of this touching-up process was to restore the depth and clarity of worn lines, but not to alter the informational content of the image nor to change its aesthetic appeal. If we look very closely we can detect truly minor changes resulting from the touch-up process.

For example, in the Ruscelli *Tierra Nueva* details illustrated in Figures 4.17 and 4.18, we can see some of these minor changes. Most discernible are the lines comprising the dark rectangles in the longitude and latitude bars. In the post-1561 editions they have been altered. We also see that the thin parallel lines above the letter "G" now extend farther left than they did in the earlier edition. This suggests that the plate was re-engraved to extend its life. These are deliberate alterations, however we probably should not consider each trivial difference in line to define a new and separate state. Thus, the Ruscelli *Tierra Nueva* map should be considered as simply having two states, one from 1561 to the time the ship and animal and interior detail were added for the 1599 edition.

This map also illustrates the point that *state* and *edition* are independent of each other. The first state of this map was printed in several editions before the second state was created.

Sometimes an entirely new plate or block was made of essentially the same map. We have included in this chapter the little 16th-century woodblock map from Peter Apianus' *Cosmographia* as an example (Figure 4.19). This was a popular book that went through many editions and the wooden blocks for the illustrations were replaced several times. The woodblock for this map was recut several times. Because of the bold difference in the circumferential band and the equator and meridian, we can tell rather easily that the two maps illustrated here were printed from entirely different blocks. One is not always so fortunate and we often have to compare carefully the individual lines. If we do that for these two maps, we note that the lines are similar, but not exactly alike. The fact that the lines in each block are so similar should not be a surprise, since the cutter probably pasted a copy of the old map face down on the new block and carved the wood through the paper, copying the image that was on it. It is worth studying these illustrations carefully, because they demonstrate how to make detailed comparisons between images and what one should look for.

COLOR

With the exception of some early woodblock maps that were printed with both red and black, most Western engraved and woodblock maps were

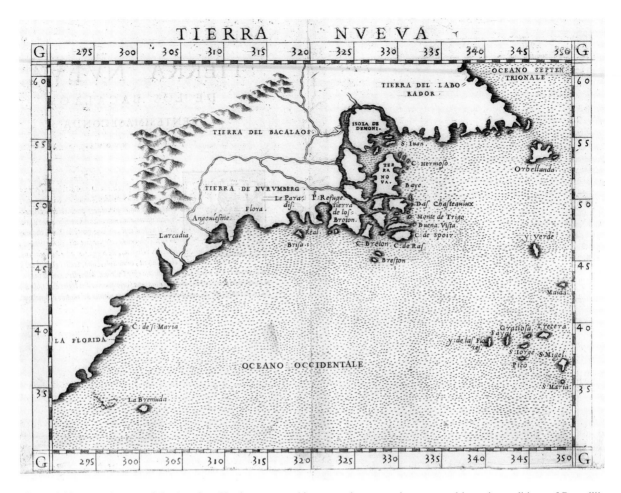

Figure 4.15 An early map of the American Northeast coast, this copperplate engraving was used in various editions of Ruscelli's *Geography* published in Venice between 1561 and 1599. Measuring about 7 x 9.5 inches (18 x 24.5 cm) the map was uncolored in its original form. This map is derived from the slightly smaller version in the very scarce Gastaldi atlas of 1548 and is among the earliest northeast coast maps obtainable. We can identify this as an early impression from the 1561 edition, because the damage to the plate west of Bermuda is minimal and there is very little damage along the bottom border. These regions are shown in detail in Figure 4.17.

printed with a black ink on white or off-white paper. Prior to the 19th century and the invention of lithography, any color that one found on a map was generally applied by hand. Maps made after about the 16th century were sometimes available from the publisher as either uncolored or colored varieties. Obviously, the colored variety was more expensive and consequently a considerable number of maps were sold in an uncolored state. Colorists used brushes to apply watercolors to the newly printed map. If only geopolitical boundaries were colored, we speak of *outline color.* Often, each geopolitical unit received an overall wash of color and we refer to this as *full color,* or *wash.* There are many permutations of this basic scheme. Sometimes the boundaries were accentuated by a

darker color than the wash; sometimes the sea or ocean was colored; often it was not.

Cartouches might be colored or left uncolored. Colored maps by Ortelius, the Blaeus or Mercator, for example, generally had the cartouches colored as well. If the map is colored, but the cartouche is not, we speak of *body color.* Homann or Seutter almost never colored the cartouche and when we see one that is fully colored our first suspicion is that the color is later.

Some maps were colored in a truly sumptuous manner, often for patrons or customers who were royal or clergy. These maps would often have splendid full hand color, and might have borders or highlights in gold leaf.

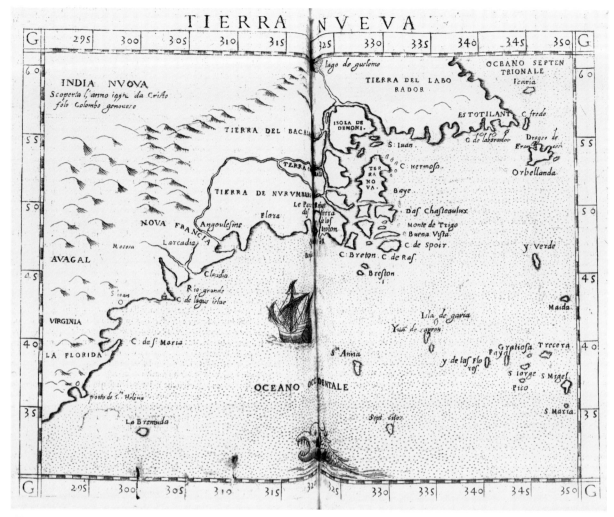

Figure 4.16 Here is the 1599 edition of the Ruscelli *Tierra Nueva* first printed in 1561 (Figure 4.15). This map was photographed in the atlas, hence the deep centerfold. This is also the second state of the map, containing new islands, interior mountain detail, more place names and a boat and sea creature. Notice also the short cracks along the bottom border. The copper plate evidently got brittle and cracked after repeated printings. The cracks held the ink after wiping and printed as though they were incised lines. These areas are shown in more detail in Figure 4.18.

Applying a thin layer of gum arabic to a painted surface was a technique often used to add visual depth to a color. This gives a "wet" look and adds a richness to the image. Such enhancement was often done to colors in the cartouches, less often to the map itself. Rarely do modern colorists use this technique.

Usually *original* or *contemporary color* was applied after printing but before binding into an atlas. These terms therefore refer to color that was done at about the time the map was made. Note that it means contemporary with the manufacture of the map, not contemporary as in "modern." Color might be *later color* that was applied to the map some time after it was printed. If applied in recent times, color is *modern color* or *recent color.*

Modern color is encountered often since the current map market seems to demand colored maps. Many collectors will have their maps colored to order and, indeed, they will appear very attractive. An industry has grown around applying hand color to old maps. Usually an experienced eye can discern modern color, but it is becoming increasingly difficult. Some experienced colorists can apply modern color that is virtually indistinguishable from original color.

The green color found on early maps was often the pigment, *verdegris* (verdi=green, gris=Greek),

67

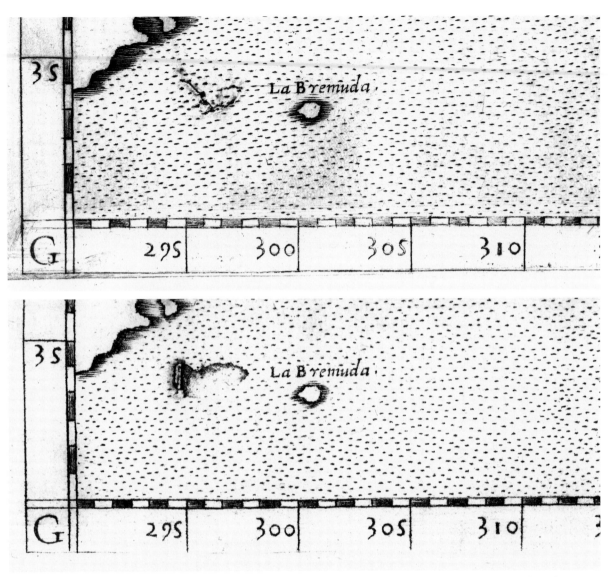

Figure 4.17 Copper plates wear as they are used and the wear is visible in the images they produce. This pair of details is from two different copies of the 1561 (first edition) of Ruscelli's *Tierra Nueva*. The image at the top is from the map shown in Figure 4.15 and represents a very early strike from the plate. There is minor damage west of Bermuda. In an impression taken later in the plate's printing history, but still for the 1561 edition, we see more extensive damage to the region west of Bermuda. This has become a significant defect in the plate and the printers apparently tried to effect some sort of repair. Compare this set of images to the pair we show in Figure 4.18.

literally Greek green. This pigment is copper acetate. It was made by reacting elemental copper with acetic acid fumes, such as those given off by fermenting grape skins. This pigment undergoes characteristic aging, often oxidizing to brown. It is also acidic and reacts with paper, often causing browning to the paper, visible on the verso (Figure C-8). Sometimes this damage is so severe that the paper actually falls out where it was painted green.

These characteristics of (some) early green color are sometimes considered to be diagnostic of early coloring. Alas, this is too easy. Some colorists today use techniques that effectively mimic the appearance of early green. Indeed, the ability of modern colors to mimic the early pigments is remarkable and it is becoming difficult, in some cases, to determine if the color is early or late. Advice to the neophyte: buy only from reputable dealers, and look at and handle as many maps as you can. Learn what style of color is appropriate for a particular map. Only in this way will your own eye become discriminating enough to differentiate between old and young color.

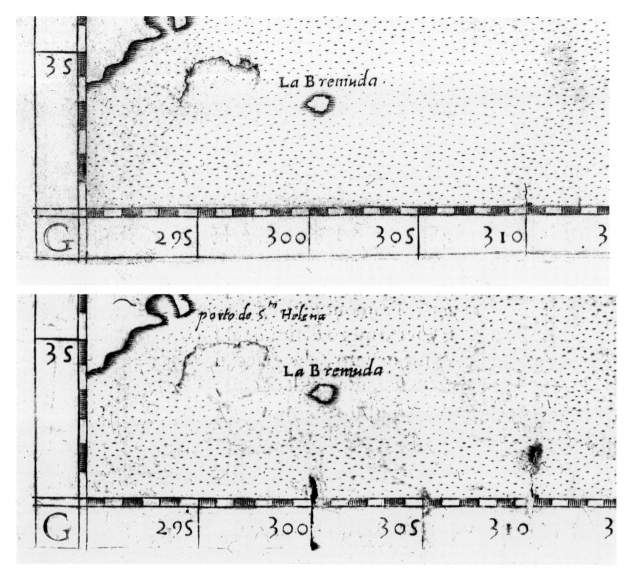

Figure 4.18 This pair of details is also from the Ruscelli *Tierra Nueva* map. The top image is from a 1574 edition. If we compare it to the images in the previous figure, we see that the defect in the copper plate to the west of Bermuda has gotten larger. Also, there is now a crack in the plate extending into the image at "310" along the lower border. In the bottom image, there is yet a further change in the defect west of Bermuda. Moreover, we see that the crack at "310" has enlarged and there is a new one at "300" and another one just starting at "305." Careful comparison of this image to the ones in Figure 4.17 will reveal many minor differences due to wear, inking and wiping, and some retouching of the copper plate. Printing chronologies can be established by such comparisons.

Maps are sometimes colored after receiving other treatment. A dirty or stained map might be washed and bleached to remove soil and then re-sized (see page 85 for discussion) to accept color more evenly. Maps treated in this way sometimes acquire an unnatural whiteness and often feel stiffer and somewhat granular to the touch. Color applied to these surfaces will often appear bright and garish (when compared to original old color). Collectively, these attributes also can provide clues that help one to differentiate between old and new color.

The details I have described and illustrated in this chapter will help the collector learn to read the information contained in the lines, color and surface of a map, and from these to identify the manner in which the map was made. This will be very useful in working through the next chapter.

69

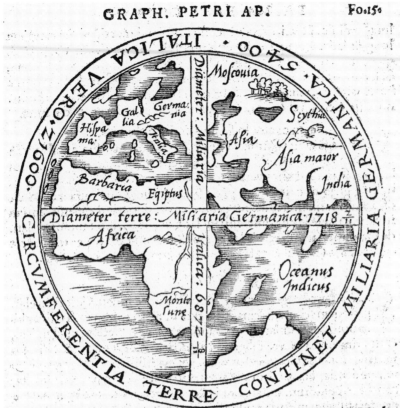

miliaria Alemanica, Sueuica, aut Itala, circuitus terræ certiſſimis Mathe

Figure 4.19 The round map at the left is a woodblock map, shown full size, from the 1564 edition of Peter Apianus' *Cosmographia*. It shows the eastern hemisphere, all of Africa and the apocryphal southern continent. The Earth's circumference is given around the perimeter in both German and Italian miles and its diameter in German miles is printed in the equator while the diameter in Italian miles is shown in the meridian. If we compare this map to the one on the next page, also from Apianus' book but from a later edition (1574), several differences are noticeable. Most apparent is the fact that the equator, meridian and circumference are now black, with the included text appearing as white, i.e. unprinted. This alone is sufficient to tell us that the map is from a different block.

Assume, however, that we have no such dramatic difference. We can still determine that the impressions are from different blocks by comparing more subtle detail.

In the lower left (A), the magnified image is from the 1564 edition; the lower right (B) is from the 1574 edition. Although the lines may

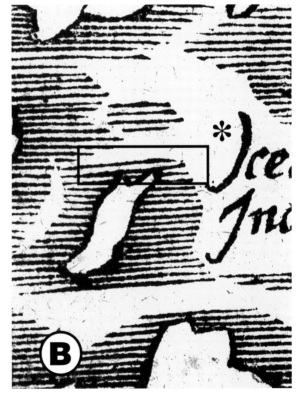

70

superficially resemble engraved lines, we note numerous breaks (such as the one enclosed in the rectangle) that are characteristic of woodblock printed lines. There are other discernible differences in the thickness and distribution of the lines in the two images. In general the lines in the 1574 edition are thicker and longer. Note also that part of the letter "O" in the 1574 image is missing (at the asterisk). This probably broke off, a common occurrence during printing of wooden blocks. More discernible are the differences in the detail below on this page. Compare the lettering style of the 1564 image (below left) to that of the 1574 image (below right). The trees also are quite different and the mountains, particularly in the circled region, have distinctly different lines. None of these differences (with the exception of the broken "O") could be the result of wear *or deliberate alteration* of the earlier block.

miliaria Alemanica, Sueuica, aut Itala, circuitus terræ certiſſimis Mathema

71

Another in Seutter's series of colossal figures. This one depicts a European warrior king and on him is engraved all the names of the kings before him. Full original hand color. Augsburg, c. 1730. Approximately 22.5 x 19.25 inches (57 x 49 cm).

Chapter 5

Facsimiles, Forgeries, and Other Copies

This is not a scare chapter; rather, it is a "good-news" chapter. The antiquarian map world has been remarkably free of deliberate forgeries and they appear infrequently enough not to present a real threat to anyone's peace of mind. I would be remiss, however, to ignore them in this book.

The material in the last chapter will help us identify the process used to produce an image; this will enable us to differentiate most copies from the original. But first, some definitions.

A *reproduction* is a copy of a map made without intent to deceive. It is simply a less-expensive modern version of an old map, often printed in a size very different from the original. Reproductions can be purchased from poster shops, museum gift shops and other places that sell interesting and brightly colored bits of paper, but generally not from dealers in old maps. The person or shop selling the reproduction makes no attempt to fool you, and most reproductions are very easy to detect. Not only will the image be printed by modern techniques (usually offset lithography), but the paper will be modern, often a coated stock, to work better with modern techniques. Sometimes reproduction maps are printed on dreadful, fake age-toned, "parchment paper" that is supposed to mimic old map paper.

A *facsimile,* on the other hand, is made with great attention to mimicking the paper and lines of the original and is designed to be as close to the original as possible, although not made with the intention of deceiving.

A *fake,* or *counterfeit,* or *forgery* is made with the intent to deceive. Unscrupulous individuals may try to pass off a legitimate facsimile as the real thing. In a much publicized recent case, an individual consigned a facsimile globe to Christie's, in London, and it was sold at auction as an original.

The history of art is replete with fakes. Prints seem to be a very tempting target since, superficially, they would appear to be easy to copy exactly. For years the forgers were making Dürer prints, Rembrandt etchings and other big-money items. Fortunately, the map world has been relatively immune to this kind of thing and even the most experienced of us have seen very few true fakes. A few, such as the Gemma Frisius/Apianus facsimile, are so well known that they are not much of a problem. Nonetheless, as we become more involved with our hobby, it is useful to acquire the skills necessary to clarify the occasional ambiguous artifact.

In order to succeed, a fake must be able to mimic or be made with the same technology as the original. As we saw in the previous chapter, lines made by the various printing techniques are, more or less, distinctly different from each other. Thus, we can look at a map and, using the criteria laid out in the last chapter, determine how the lines were made. If, for example, we can determine that the lines were made by lithography, such an image cannot have been made prior to about 1800, when lithography was invented. However fine the lithographic process might have been, the lines simply cannot mimic engraved lines – *the technique is different.* In virtually every instance careful examination reveals the technique by which the lines were made.

With the advent of photography, several new techniques were developed that can make it more difficult to distinguish between the original and the photography-assisted facsimile.

Some photoreproductive techniques, such as *photogravure,* can be uncannily imitative of the original. This process was used to make reproductions of photographic prints that have tonal qualities very similar to the originals. Photogravure was used to make fakes of Rembrandt etchings that were difficult to detect. Photogravure reproductions of Dürer prints were so good that the publisher put a stamp on the verso to identify them as copies.

Woodblock fakes. The woodblock line is easy to copy and copies and can be difficult to detect. The earliest fakes were simply newly carved woodblocks, copied after the original. A woodblock print was pasted face down on a new block of wood, the paper oiled to make it transparent, and a new block was cut. The cutter cut through the paper and tried to copy the lines exactly. This is also the technique legitimate publishers used to cut new blocks when the old ones wore out; it was probably the technique used to cut new blocks for the Apianus maps discussed in the last chapter. However, even with care, the *original cannot be reproduced exactly* and careful line comparison will reveal minor differences. This can tell us if the block was recut. We then need to understand the circumstances under which it was recut in order to determine if the newer image is a later state or a forgery.

In the 19th century, forgers coupled the new technique of photography with lithography to make fakes. A print could now be photographed and a lithographic plate made from the photograph. This process duplicated the detail of the original line exactly, including the woodgrain and little cracks and splits remaining in the block. **However, a lithographed line is planigraphic, and can be distinguished from a relief-printed line by careful examination.** Very careful examination will show that the line is lithographed, not printed from a woodblock, but few buyers carry a microscope to the viewing. I am one of the few who does just that. I carry with me a small pocket microscope that lets me look at the lines under oblique light.

I have seen lithographed versions of several woodblock maps. There are at least two versions of the Münster Americas map made in this manner. They both are very good and appear at auction every now and then.

Sometimes a forger will include some red herrings to throw off suspicion. In the case of one of the fake Münster Americas, I noticed that the map had been colored rather crudely and was backed on the verso with thin Japan tissue. Most viewers would concentrate on the color and the backing and not be overly suspicious that the map itself was a fake. An original map, backed, and with poor color, would still be an eminently salable item, but reduced in price. Another of these maps, also a colored exemplar, has, on the recto, curious figures of natives overprinted on North America. The exemplar I saw had text on the verso.

There is also a Gregor Reisch world map with twelve windheads that has been making the rounds. Several dealers have brought it to my attention and have told me that, when examined very carefully, it appears to be a very well-made facsimile. I have not personally seen this map. The original is a woodblock map published in Freiburg in 1503.

Copperplate engraved maps. The engraved line is very difficult to copy. The most common types of copies are very high-quality photolithographs. In this technique, the original is photographed and then a litho plate is made from the photograph. However, the lines will still be planigraphic and not intaglio, even though they may mimic the shape of the etched line. Even when given hand color, we can see the difference – the whole map will look "flat" as compared to an engraved map. A lithograph most often does not have a genuine platemark, and if a map that is claimed to be an engraving does not have a platemark, then we should immediately suspect that it might be a lithographic reproduction. Of course, if the map is reproduced by halftone (as are many modern copies), then the job is infinitely easier – if we see dots the verdict is clear.

One way of making good facsimiles of engraved maps is to use the technique of *photoetching,* or *photogravure.* In this process, the original is photographed and an intaglio plate is made by etching through a photo-resist. The lines, although they print as intaglio lines, do not have quite the crisp look of the original, especially if they are close together or intersect. The acid tends to etch laterally under the resist, as well as down into the plate, creating a wider, less brilliant line. This, along with the loss of crispness resulting from the photographic process, diminishes the resolution. Whereas in the original, closely engraved and

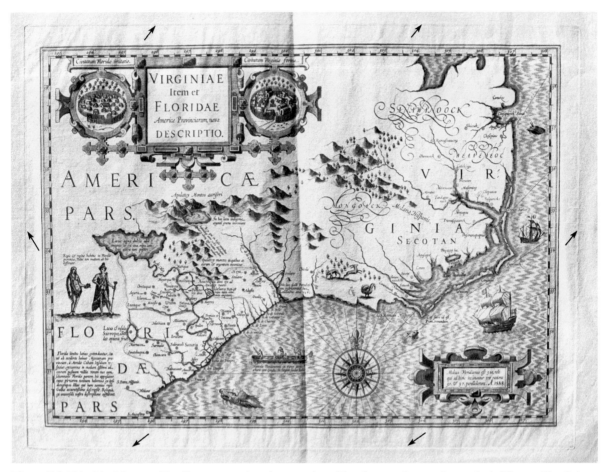

Figure 5.1 This fake Mercator/Hondius map was bought at auction. The photograph was taken under brilliant, raking light to reveal the extent of the platemark around the whole image. The platemark is absurdly wide for a map purportedly of this age. This map is a very skillful forgery made by a photoengraving process and printed on laid paper. Closer microscopic examination revealed problems with the lines, but the diagnosis of a forgery could be made by naked eye observation if one was aware of the platemark. Note that the centerfold is at an angle, and not down the middle, suggesting that the map might never have been included in an atlas.

intersecting lines are distinct and separated clearly, in the fake the lines tend to run into each other, and their intersections are particularly indistinct when compared to the original. Resolution loss occurs also in lithographic fakes and is a powerful way to detect a replica if one cannot do it by the intrinsic nature of the line itself. Nonetheless, detecting a good photogravure copy of an etching is extremely difficult and one sometimes must rely on other hints, such as the type of paper used.

Lithographed maps are probably the easiest to fake. The technology used to produce them is still readily available. Many of these maps are simply not worth the expense of reproduction, but it is quite possible to do so exactly and we must be on our guard. In this respect we should be aware

that lithographic forgers have invaded the arena of modern first edition books. This is, to me, a bizarre market where the price of a book has staggering differences based on trivial condition points. If, for example, a book is "worth" $2,000 in absolutely perfect condition, it might be "worth" only $200 if the dustjacket is torn or has a minor stain. Modern dust jackets can be reproduced exactly by modern printing techniques and with such a premium placed on them it was inevitable that someone would do it. And they did. The same might happen for maps and we should be aware of that possibility.

There is one rather expensive American 19th-century map that I personally know has been reproduced by halftone lithography with intent to deceive. I was, a few years ago, offered a copy of

75

Figure 5.2 The fraudently wide platemark is obvious in this enlargement (above) of part of the map in Figure 5.1. Note that although the quality of the reproduction is very good, the inscription bears the spurious date "1588." Compare the inscription and the platemark to those of the authentic exemplar below. Note also, that in the authentic map, the platemark (arrows) is so close to the heavily colored neatline that it is virtually invisible in the photograph.

76

the Mitchell Texas for a very modest sum. I had been offered the map by a dealer who sent me a color photograph. The photograph fooled me – the map looked perfect. Everything was right, good color, folds, chipped edges and appropriate wear. However, as soon as I opened the package I knew that it wasn't right, even though *the piece of paper looked old and worn*. The original to this map was hand colored and the color on this one looked weak and pale to the naked eye. Low-power optical magnification revealed that the color was lithographed by the halftone process and dots were distinctly visible. It likely began life as an innocent reproduction but somewhere along in its history

someone decided to "age" it and try to pass it as an original.

THE MERCATOR VIRGINIA

Figure 5.1 shows a Mercator *Virginiae Item et Floridae...* map that proved to be a fake. I have seen several of these on the market, the first one for sale by a dealer in Germany. My attention was drawn to this map by its obvious recent color. I think, again, that the color was a red herring used to divert attention from the technical aspect of the image. A potential buyer would be concerned about the color, not the possibility that the whole map might

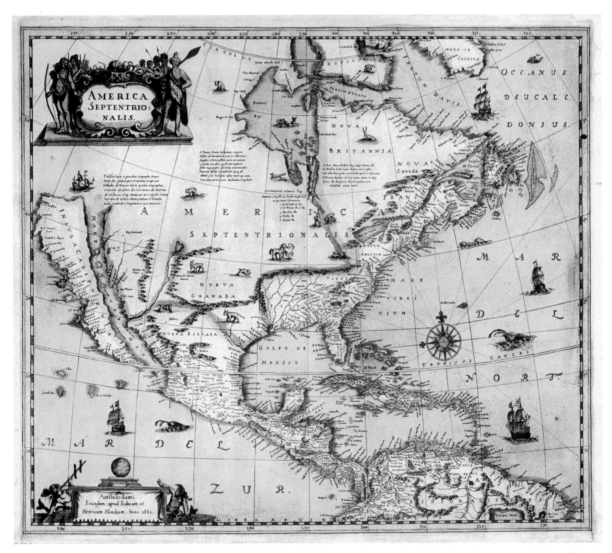

Figure 5.3 This map is a copy of the Jansson North America shown on Page 165. The map is presumed to be derived from a Hondius precursor. The map shown in this photograph is engraved and printed on laid paper, but has all the characteristics of a fake.

Figure 5.4 This detail is from the authentic Jansson map, showing the upper left corner of the map, including part of the cartouche. Note the platemark (arrows). The platemark is narrow on the top and side. Note also the circled regions and compare them to their counterparts in the next figure, which is of the "Hondius" map. Both maps are engravings, printed on laid paper.

be a fake. I wanted to buy it as an example of a fake, but it was priced as though authentic. The map in Figure 5.1 also had very good modern color. It was bought at an American auction and only later did the buyer realize it had an authenticity problem.

A number of things are wrong with this map that we can illustrate and use as a case study. The image size is correct, but the platemark is much wider than found on maps of that period, and much wider than found on authentic exemplars of that particular map. The centerfold, although neat and crisp, was at an angle. It starts out in the correct place at the top, but the map was misfolded by about a quarter of an inch at the bottom. The fold should have come through the right side of the canoe. Had this map been bound into an atlas along this fold, it would not have fit. Further, a spurious date had been placed in the cartouche in the lower right (Figure 5.2). When I examined the lines using a microscope, it was clear that they were not engraved. Nonetheless, this was a very good facsimile, good enough to be bid up at auction.

THE JANSSON NORTH AMERICA

The Jansson *America Septentrioinalis* is a beautiful copperplate engraved map of North America

78

Figure 5.5 This detail is from the putative Hondius map showing the upper left corner of the map, including part of the cartouche, the same region seen in Figure 5.4. Note the top platemark that is much wider than that of the authentic map and the left side platemark that is narrower than that of the original. *This map could not have been printed from the same plate.* The areas circled in the previous figure show some of the most obvious differences. Although the engraver tried to copy the Indians, he achieved a very different picture. Copies, however carefully done, usually lack the surety and sparkle of an original. The copyist cannot permit creativity, but must exercise a self-conscious care that is reflected in the overall qualitative appearance of the copy. This is generally more apparent in "art" than in maps, but we can see the difference if we just look at the Indians in these two examples.

showing California as an island (Map 47, , page 165). There are two known states of the authentic map: one with Jansson's name in the lower left cartouche; the other with the cartouche empty, but bearing slight traces of a previous inscription that was burnished out. It is believed that Jansson bought the plate from Hondius, so it is assumed that an earlier state with Hondius' name might exist. However, there are no known authentic copies of this map bearing Hondius' name in the lower left cartouche. Is this (Figure 5.3) the missing first state, or is it something else?

Even though the two maps look very similar superficially, they are printed from different plates. We can tell this at a glance if we have both in front of us to make a comparison, but I shall use these maps to illustrate the process we can use to differentiate original from copy.

Figure 5.6 The sailing ship above is from the spurious map shown in the previous figure. The ship below, quite different, is from the authentic map.

First, we measure them. The Jansson image (neatline size) is 46 x 55 cm. The putative Hondius is 46 x 54 cm. This is within the normal size range we can expect, based upon normal variation in paper shrinkage and expansion. So we need to proceed further with the exercise of differentiating them.

Let us next compare some of the detail of the cartouche region in the upper left corner (Figures 5.4 and 5.5). First we note the distinctly different platemark characteristics of the two maps. The Jansson map has a much narrower platemark, in keeping with maps of this period.

Qualitatively, the Indian figures are quite different. It might be hard to make this judgment if we were looking at just one map, but with two before us we can tell readily that their expressions are quite different and their poses are not quite alike. If we look at a few regions (marked by circles) where the other differences are most distinct, we can quickly see that the two impressions are not from the same plate. There is no way that one can be just a reworked image constituting another state. *Every line is different.*

If we look at the ship just below Island Californa in the Jansson map we see a broadside view of a ship heading east. In the "Hondius" map the ship is heading northwest and we view it from the stern (see Figure 5.6). Many other differences can be noted in this area.

Despite the fact that both maps are engraved and printed on laid paper, we can decide with certainty that they are from different plates.

The Jansson map is a well-known beautiful map, scarce but not rare. What of the other map? Is it truly a predecessor map done by Hondius? Or is it a fake? I have seen three of these curious items and know of yet another. All are on a heavy laid paper, and two of them are in rather poor condition. None have text on the verso. I know the provenance of the map I used for the pictures in this book and have been able to trace it back to the mid-1930s. Obviously these were not made recently. Some care had to go into them; they are hand colored. The exemplar I have been able to examine most closely is well colored, but it is not a 17th-century color. It also does not appear to be a modern color. All the maps I saw had a centerfold,

80

Figure 5.7 This is a lithographed copy of Blaeu's map of Europe. It is printed on laid paper, has a centerfold and narrow platemarks consistent with the period. I assume the platemarks are artificial. The map shows wear and tear and I photographed it against a black background so that I can show the entire perimeter of the margins. Note the marginal age-toning, the chips, wrinkles and minor tears often found on old maps. Note also the spurious date in the cartouche. This is a very attractive map that *could* fool someone who is careless or trying to snap up a bargain.

but this does not, by itself, mean that they came from atlases.

Was this map a nicely made reproduction or a fake? I do not know why it was made, but although it is good enough to fool some, I think it was not made to deceive.

THE BLAEU EUROPE

Figure 5.7 shows another fake map copied from Blaeu's *Europa recens descripta*. I own this map and bought it, as an example, for very little money from the previous owner, who was also aware that it was not genuine. Very well done, the map has fine hand color. The image is 40.6 x 54.7 cm and the platemark is 41.5 x 55.7 cm. The date in the

cartouche is spurious because the first edition of this map was printed in 1617 and later editions were undated. The map is printed on laid paper that appears to have acquired the wear and tear of long-term careless handling. Unless the "aging" has been done deliberately and very well, this is not a newly made map.

However, even a naked-eye examination reveals that the map appears dull and lifeless. An engraved map, even a later one, will have the dimensionality contributed by raised, engraved lines. So will a photoetched reproduction, since the etched line is also raised. If we look at the lines in this map using oblique light and a hand lens, we can determine that the lines are not intaglio, but planigraphic. The diagnosis is that this map is a

81

well-done photolithographic copy that will only fool those who cannot distinguish between a litho and an engraved line.

A FINAL THOUGHT

Nothing here is meant to frighten the new collector. As you grow into your hobby, if you follow some of the simple guidelines in this book, and learn how to tell different printing techniques and papers apart, you will be fine. Outright fakes in the world of old maps are still, thankfully, uncommon. Nonetheless, until you have acquired the expertise to make your own diagnoses, I still think your best bet is to seek the advice of an experienced dealer.

Curious miscellany – 5 An unusual image relating to maps or mapmaking

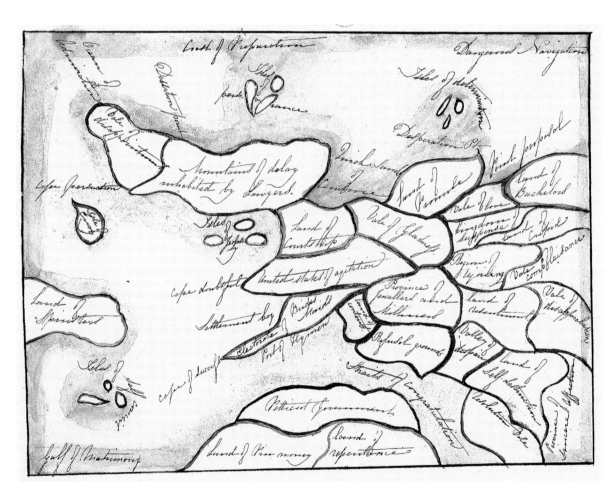

Among some of the curious cartographic traditions is that of making maps showing the "passage through life" or, as in this case, *A map of Matrimony.* This little (5 x 6.75 inches, 13 x 17 cm) English manuscript map on paper is probably a late-19th-century amateur effort. We see countries such as *Land of courtship, Land of bachelor, Land of repentance, Land of parsimony, Isles of self-conceit.* Obviously, the cartographer had a strong view of marriage! Similar maps have, over the years, appeared on postcards and in "humorous" publications.

Chapter 6

Condition and Conservation

Condition is to a map what location is to real estate, or original finish to an antique chair: a major determinant of desirability, hence price. In this chapter we will discuss *condition*, how to recognize problems and restorations and how these factors influence pricing. Let us use a Blaeu Americas map as an example. A copy with original color, some visible wear, and perhaps a few creases, but one that is nonetheless a nice exemplar, is a highly desirable map. A copy with narrow margins, perhaps a tear or two repaired on the verso, and the color either new or touched up, but that is still visually attractive, is also generally desirable, but will sell for less than our first example. Any more problems, say the margins were trimmed to the platemark and have been replaced, or the color is not of the best quality, or there is some spotting or fungal growth or dampstaining, and the map may become difficult to sell, even at a steep discount. Particularly damaging to the perceived desirability of a map are tears *into the image* (as opposed to tears involving the margins only) and *loss of image* (see Figure 6.1).

On the other side of the quality spectrum, a brilliant impression, fresh and clean, with wide, clean margins, fine original color and no visible damage or repairs, can be sold for well above the market price of an average copy. Such price sensitivity to condition is common in the world of maps, where a collector may have a pick of several on the market. A truly rare map, one that exists only in a few known exemplars and that might not appear on the market ever again, is somewhat less price-sensitive to condition for obvious reasons.

Misunderstanding condition is perhaps the most common reason that it is so difficult to buy maps from non-specialists. It takes experience to make a judgment about any given map; one needs to have seen many old maps and be familiar with the exemplar in question. Any general antiques dealer who gets the occasional map into inventory would be well advised to educate himself by examining the maps in a serious collection, either private or public. It is surprising just how good condition can be, and without this kind of knowledge the likelihood of minimizing condition problems is large. One need not eschew a repaired map, or one with condition problems, but one certainly should not pay a top price for it. With this in mind, let me waffle a bit and introduce a caveat. As maps are more and more collected, the overall condition of those remaining available, or appearing on the market, has tended to decline a bit. Longer established collectors and dealers are constantly readjusting their thinking on this subject; the brilliant exemplars of just a decade ago are no longer common.

DESCRIBING CONDITION – I

There are several terms that are commonly used to describe condition in the world of rare books and they are generally transferrable to the map world. I shall list them here so that we have a common language, and return to them later in the chapter when we can discuss them in more detail. I emphasize that although these terms are both useful and widely used, they are obviously subjective and therefore not very precise.

Fine (F) is about as good as a particular map can be. The impression is a good, strong one; there are no defects or flaws to either the paper or impression and there are no stains, discolorations or repairs. The margins are wide and clean. Fine is sometimes called *mint*, a term I dislike since maps are printed, not minted.

Very good (VG) is the next-best grade. A VG map might have some delicate age-toning or very light foxing (see page 91). It might have slightly narrow margins. But it does not have significant repairs or unsightly blemishes. VG is about the lowest grade a serious collector will entertain, *again, except for particularly scarce or unusual maps.*

Good (G or Gd) is less than it sounds. A good map may have a weaker impression, narrower margins, some repairs and evidence of having been cleaned. It may have foxing and abrasion.

ORIGINAL COLOR OR LATE (MODERN) COLOR?

Color that was applied to the map at or about the time it was printed is called *original* or *contemporary* color. Note that, in this sense, contemporary refers to the period of the *map*, not the present. If the color doesn't appear to be original but is clearly not modern, then it might be described as *early color.* Color that has been applied recently is called *modern* or *recent* **color.** **Later color** is a more generic catchword.

A frequently asked question is, "How can I tell the difference between original and later coloring?"

The most obvious answer, yet for the beginner the most vexing, is "through experience." After one has seen and examined literally thousands of old maps, one develops an experienced eye that is uncanny in judging color. Such subjective judgment cannot be dismissed out of hand. Much of art authentication is based on expert opinion and judgment. One notes first whether or not the color is

typical of the period. Certain makers, such as Homann, had very distinct color and if a pale, English-type transparent watercolor was applied to a Homann map it would cry out in a moment. Sometimes later color is more difficult to discern. There are several other clues that can be used. For example, the old green color came from a compound called *verdegris.* Verdegris often (but not always) oxidizes over a long period of time and becomes brownish, sometimes overtly dark brown. It also soaked into the paper and, as it browned, the browning became visible on the verso. This is almost an infallible test; there are very few natural false positives. That is, when a brownish color is seen on the verso to match the green on the recto, it *usually* means the green is old. However, we have seen some very good modern color where the colorist has mimicked this effect by browning the appropriate areas on the verso. We also note that the *absence* of such show-through is not firm

Delin. M.Burghers sculpt. Univ. Oxon.

Figure 6.1 The map in this picture (a double-hemisphere world map by Edward Wells, 1704) is missing part of the lower left image and also has a narrow left margin. The margin and the corner can be replaced, and the missing image supplied in facsimile, but the map's value will not be equal to one not needing repairs.

evidence for modern color, since not all old color exhibits the tendency to migrate through the paper.

Verdegris also has the nasty property of being acidic. Over the years it damages the paper and it is quite a common occurrence to see paper disintegrating where it had been painted with verdegris. Old green regions can literally fall out of the map, leaving holes behind. Green outline color leaves empty space where once it delineated countries. If your map has this kind of damage, then you know the color is original.

If a colored map has a small wormhole, look at the verso to see if a small ring of color has come through from the recto. If it has, then it's pretty good evidence that the color has been applied *after* the worm made the hole. Obviously not original color. The same principle can be applied to other defects in the surface. If a defect, such as an abrasion or tear, is obviously one that occurred after the map had seen some service and if the color is

quite uniform across it, then the color most likely is post-defect. Wherever paper fibers are damaged the paper becomes more absorbent. This often happens along the centerfold as a result of repeated folding and unfolding. If the color is more intense exactly along the line of folding, then it may also suggest that the color was applied after the map had been folded. Original color usually was applied to the map *before* it was folded and put into an atlas. Modern colorists try to avoid this tell-tale line by *sizing* the area well to try to get the paper to accept color more uniformly, but this is not always successful. Sizing is a weak solution of a substance that fills the pores of the paper and provides a non-fleecing surface for the application of color. Often the sizing is a very weak solution of gelatin; more recently it might be a very dilute polyvinylacetate (PVA), the same stuff that is Elmer's glue. Dilute ox-gall was a classic sizing. This was used into the 19th century and can still be obtained for the purpose. Its major disadvantage is that it can react idiosyncratically with the paper and blacken. Many 19th-century colored book illustrations are ruined by the darkening of the ox-gall size.

Sometimes maps are colored to hide defects, restorations, or the signs of washing, bleaching or repair. Mild dampstaining can often be disguised by judicious color. It is worthwhile to look for such things.

Figure 6.2 The map in Figure 6.1 has had the left margin widened and the missing corner replaced by the addition of paper similar to that used for the map itself. The missing image was replaced in facsimile. Repairs such as this are relatively simple but nonetheless can be costly.

Nothing I say here should be interpreted to imply that I think there is anything inherently wrong with modern color on old maps. That is a personal preference for each individual collector.

WASHED AND CLEANED MAPS

Washing and bleaching a map to remove stains and discolorations are perfectly acceptable conservation techniques. Often a simple wash in oft-changed water is adequate to remove the grime of centuries. Sometimes more drastic intervention is required.

Bleaching to remove stains is usually done by soaking the map in a dilute aqueous solution of a bleaching agent such as hypochlorite (common laundry bleach). Some conservators use a solution of Chloramine-T instead of hypochlorite. Bleaching agents are damaging and must be washed out thoroughly or the paper will ultimately deteriorate. Paper made from linen rags can be bleached in a very gentle manner. Wet linen bleaches in bright sunlight. This phenomenon was used by laundresses doing household linens – tablecloths and napkins, for example. It is also used to lighten age-toned maps. They are moistened and set out in the sun. This technique needs no chlorine that might be left behind to cause mischief later on.

Bleaching can leave the paper looking artificially white. If you suspect that a map has been bleached, try smelling it. Unless the conservator was very careful and used an adequate number of rinses, the odor of bleach can sometimes be detected. Often such washed and bleached maps will be toned to hide some of the starkness. In the old days the whiteness was diminished by soaking the paper in a solution of tea. Unfortunately, this undid whatever deacidification steps were taken because tea is acidic.

Washing and bleaching also removes some of the original surface body from the paper and if watercolor is applied to it the color will run, or fleece. This is analogous to trying to watercolor on a blotter or paper towel. In order to overcome this problem colorists first apply a sizing to the bleached paper. Paper that has been sized recently often has a peculiar feel to it, sort of like a newly starched shirt compared to an oft-laundered cotton handkerchief. We have noticed that gelatin size, which is often sprayed onto the paper surface, leaves tiny, shiny flecks on the paper. This material looks like very fine mica and sparkles in oblique light. One has to look very hard for it because it does not detract from the overall appearance of the map but it is indicative of recent sizing, hence recent color work.

REPAIRS

We have all used cellophane tape to repair paper. However, when left on for any length of time tape leaves a residue and causes the paper to be virtually permanently stained. *Self-stick tape must be avoided when dealing with valuable paper items.* (There is one exception: an archival repair tape that is specially made for this purpose.) Any map that has had cellophane tape repairs and has suffered discoloration is considered severely damaged and may have lost significant value.

Other types of repairs must be judged individually. Of course, any tear into the map image reduces its value. If it has been professionally repaired the damage is minimized, but not eliminated. Proper professional repair utilizes an appropriate paper and a reversible (generally wheat) paste that neatly mends the tear without introducing additional new stresses into the paper. A reversible paste is one that can be removed at a later time without damage to the paper.

Increasingly, we are seeing the appearance of heavily repaired or restored maps on the world's markets. This is especially true for world maps. The market for world maps, especially those pre-1700, has been particularly brisk in the past few years and many of the VG or better maps that were in dealer inventories seem to have been collected. Some of the maps now appearing on the market are restored so heavily that few long-time collectors, who are used to maps in better condition, would even consider buying them. Many of them are remargined and they appear to have new color, suggesting that they were originally uncolored or that in the process of repair they were bleached very heavily to remove stains. Large areas can appear in very skillful facsimile and many of them are backed entirely with thin Japan tissue to strengthen them. Most of these extensive restorations appear to be taking place on the Continent. We suspect that large numbers of maps in dreadful condition are languishing in dealers' drawers waiting for market prices and scarcity to increase to the point where restoration becomes economically feasible. Moreover, we think that the scarcity of the better maps means collectors will ultimately buy these repaired ones.

I emphasize that there is nothing intrinsically wrong with a repaired map, but we should know about the repair before buying the map. One way to detect many modern repairs is to examine the map with ultraviolet (UV) light. The UV light excites some molecules, causing them to emit their own light of a different wavelength. This is called *fluorescence.* Among substances that exhibit fluorescence are such things as wheat paste, PVA, modern paper (modern sizes fluoresce) and many other modern adhesives. If we look at an old map with UV, we see nothing – perhaps a few specks of bluish light, but no major fluorescence. A modern repair will generally be brightly fluorescent.

Inexpensive portable battery-powered UV light sources can be purchased specifically for this purpose. They are very useful when map-hunting but remember, **do not look at UV light with the naked eye.**

BACKING AND LAYING DOWN

I have mentioned the process called *backing.* In this procedure a very thin sheet of (usually) Japan tissue is glued to the back of a map. This is largely done as a conservation measure and is designed to give strength to an otherwise fragile sheet of paper. Sometimes it is difficult to tell that such a sheet of tissue has been attached; it is only by noting that the printed line or word is slightly dulled that we can be certain that such a backing was applied. Backing was sometimes done using thicker paper, but that practice has largely been abandoned for several reasons, not the least of which is that it is very unattractive.

A backed map may be less desirable than one unbacked. They are sometimes hard to sell (unless very scarce in any condition) and most collectors shy away from them. Backing is done for a reason and a map that was sick enough to require backing generally sells for a significant discount.

We sometimes find maps that have been backed with heavy cardboard or a similar material. We emphasize that it is a serious defect for a map to be

laid down in this fashion. This was a common practice among framers until about twenty years ago. Large prints can be difficult to hold flat when framed, and it was a lot easier to glue (lay) them to a stiff backing. This practice often ruins the map, or necessitates repair that is too expensive to justify. A map that has been laid down has lost a significant amount of value; a common map might even become unsalable and a scarce map salable at a vastly reduced price. *For this reason, we never buy a map that is framed; one must always be able to see the verso.*

It might be worth trying to remove the backing from an unusually good or expensive map, but the conservator's charges will be high and the map will probably always show signs of its previous mistreatment. There are several problems here. Old backings were often highly acidic and the map may have been damaged by the acid. If the framer, in his attempt to keep a map absolutely flat, used a water-soluble mounting medium, then it is possible, in principle, to soak the map free from the backing. This may be all right if the map is uncolored, but for colored maps the long soaking required to loosen aged glue may cause damage to the color. Not all traces of glue may be removable and the back of the map may always look unsightly.

In more recent times some framers have used *dry mounting tissue* to lay down maps. These materials are allegedly archival in that they are non-acidic and actually can form a barrier to acid that might be lurking in the backing. However, some dry mount material is equally damaging to the map and can be virtually impossible to remove at reasonable expense. Some are soluble in (non-polar) organic solvents, but their removal becomes an expensive and difficult procedure. Our advice is to avoid buying any laid down map unless you're a gambler. If you have one for sale, realize it's got a serious problem and price it accordingly, usually a mere fraction of what it would fetch otherwise. And be thankful if someone buys it.

Let me emphasize that some maps were backed normally. These include not only the linen-backed folding maps and wall maps, but many kinds of sea charts. Some sea charts (such as the blue-backs) had a second sheet of paper glued to their verso to make them stronger in use. This is perfectly normal.

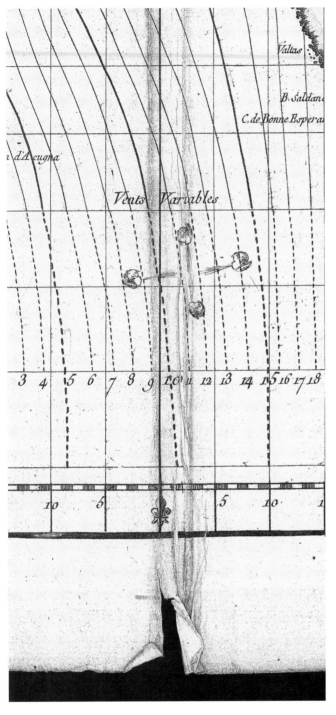

Figure 6.3 Typical centerfold damage. The map is Bellin's *Carte Des Variations De La Boussole...* of 1765. A large map (21.5 x 34.25 inches; 54.5 x 87 cm), it was subjected to many misfoldings and creasing along the original centerfold. There is a split visible near the very bottom but no paper has been lost. Consequently this split can be repaired easily by applying a patch on the verso. There is some minor soiling along the creases, but a conservator can remove much of this. Large maps are more susceptible to this kind of damage, especially at the bottom.

Maps were often laid down on substances other than paper. We have seen them glued to masonite and to wood. One of the worst atrocities we have seen was a Blaeu world map decoupaged to a serving tray!

THE CENTERFOLD

The split is the most common of centerfold problems. Splits generally occur in the bottom part of the centerfold (Figure 6.3), since atlas pages are turned from the bottom and this is where the stress is introduced. If the split has not entered the image and is confined to the margin only, then it is not overly important. However, if it has entered the image then it should receive a professional repair, not cellophane tape.

Abrasion also occurs most often along the centerfold. There is no easy fix to this other than having the color (if the map is colored) touched up. Misfolding often occurred, especially in large folio maps, and we frequently note that there are several "centerfolds", not just one, all resulting from careless closing of the atlas over the years. It is very difficult to "unfold" paper without wetting it and such misfolding is best left alone. Nonetheless, all these defects should be noted and they all, individually and collectively, have a greater or lesser effect on price.

Centerfolds are often darker than the surrounding paper. The chief reason for this is the glue used to attach the ***binding stub*** (or guard) to the back of the map. A binding stub is a strip of paper that is glued to the fold region of a map so that the binder can sew through the stub when putting the map into a book and not put stitch holes in the map itself. Such stubs should be left on; removal can tear the map. However, old glues may become brown with age and darken the paper. We note this commonly in maps by Zatta. Again, mild age-darkening of the centerfold region is not a major thing, but if it is highly visible and consequently objectionable it reduces the market for the map considerably.

Figure 6.4 This photograph shows part of a re-margined Blaeu Africa map. Parts of the original margins remain and the side margin has been largely replaced by gluing on a separate strip of paper. The neatline was trimmed off in places and the paper match is not very good. Nonetheless, this map is now frameable.

Courtesy: T. Suárez

Some maps that were bound into books without the use of binders guards exhibit **stitch holes** in the centerfold. If this is a normal condition for the map (as it is in the Schedel world map of 1493), then stitch holes are of no great consequence.

PROBLEMS WITH MARGINS

Over the years, atlas bindings wear out and often the contents are rebound. It used to be common for binders to plane the edges of the newly bound volume to make them neat and smooth. Unfortunately, each time the volume was rebound some of the margins got trimmed away by the plane, producing *trimmed maps*. Finally, after several rebinds, depending upon how much each binder liked his plane, the margin would disappear and there would be actual loss to the image (see Figure 6.4). This is a very serious defect, since although one can replace margins and create a facsimile map surface (see Figure 6.2), it's much like adding new legs to our Windsor chair and has about the same effect on value. Obviously, as noted already, marginal loss and repair is a relatively small matter compared to image loss.

Figure 6.5 (right) Another narrow margin on a severely trimmed Wells map is seen here. The remaining margin (the light line between the image and the black background) is barely adequate for matting.

Maps were often trimmed to fit into existing frames. Often such narrow margins are concealed behind mats. This is another reason among many why dealers do not generally buy framed maps. If margins are just a bit narrow it's not a big problem, nor does it reduce the value of the map significantly. Some maps, such as the Renard coastal maps (charts) and the van Keulen charts, are notorious for having very narrow margins. If this is a common characteristic of the particular map then so be it; most exemplars will be like that. However, in the case of other maps, if the margin was trimmed for some reason then it may reduce the map's value. Certainly if it is cut back to the platemark or the neatline then the map is sorely harmed and may have lost considerable value. In Figure 6.5 we see the right side of part of Edward Wells' double-hemisphere world map. Note how closely it has been trimmed, probably when the atlas was last rebound in the early 19th century (as judged from the binding style). The margin has been trimmed to the plate-mark, reducing the value of the map. Fortunately there has been no loss to the image and the margin can be easily extended to facilitate matting and framing.

Remargining can be done in several different

89

ways. Most common is by pasting strips of similar paper to the verso of the map. Although the least desirable from a cosmetic standpoint, this is the cheapest and easiest method and, if done properly, is not overly intrusive. More difficult is the process of *leaf casting,* a technique that employs a liquid slurry of paper pulp to actually make a piece of paper *in situ* to replace a missing piece, i.e. a margin. This is a very sophisticated type of repair and sometimes is virtually undetectable, since the slurry can be made from scraps of original old paper. I have seen maps remargined in this way where the restorer then created a fake platemark in the newly cast margins. If any of the image requires replacement, it might be done by adding a *manuscript facsimile,* or even by "welding" a printed replica of the missing portion into the paper of the original.

Marginal stains are less important, obviously, than stains that involve the printed surface of the map. If they are very unsightly then value is compromised. Look for the type of stain, too. A uniform browning with a sharp demarcation line may indicate *mat burn*. This results from using a mat made of cheap acidic material. Over the years an acidic mat damages and browns the paper it contacts. Always examine a matted map for this problem. If it is present, it might be assumed that the rest of the material in contact with the map was acidic and that the whole map may need treatment by a conservator. The treatment process of deacidification will inhibit further damage to the map, but can be expensive.

CONSEQUENCES OF IMPROPER FRAMING AND GLAZING

Improper framing and glazing create one of the most damaging environments to which a map can be subjected. Sometimes maps can be found in severely damaged condition because they have spent the past fifty years or so pressed against glass. It used to be quite common practice to simply put a map into a frame (sometimes even trim the margins to make the map fit the frame), recto against the glass, and put either corrugated cardboard or thin slabs of pine behind it to hold it pressed tightly to the glass. Two things happen to maps in this environment. Condensation on the inner surface of the glass, a normal occurrence as the humidity and temperature change, begins to damage the map. The paper becomes damp and may develop foxing or mildew. It can, over the years, also get dirty, since the moisture helps the paper absorb stain from the frame or backing. The poor-quality backing is often highly acidic. The acid from the backing attacks the damp paper and does severe damage, hydrolyzing the cellulose fibers and making the paper very brittle. Many such maps are beyond any reasonable repair and are a total loss. Even when this process is just beginning, the map's value is severely compromised. For this reason dealers will generally not buy maps that are in frames. Inspection of the verso is essential.

How to judge brittleness? If you are afraid to roll a map into, say, a three-inch-diameter tube for fear of cracking it, then you're dealing with a very brittle map. Such a map needs heavy and expert restoration work and is much reduced in value, possibly to 25% or less of a map in good condition. I add my usual caveat here, and emphasize that in the case of very rare maps, condition defects may become irrelevant. Also, there are some maps, such as those printed on thick paper or some nineteenth-century maps, that are intrinsically brittle and the brittleness does not imply pathology.

The mat surrounding a properly framed map is not just decorative. Its chief function is to hold the map away from the glass, preventing condensation damage. If it is necessary to frame a map without using a mat, then special spacers are available that serve to maintain separation between the map and the glass.

OFFSETTING

Offsetting refers to the transfer of image from one printed sheet to another. Offsetting was more common during the early days of printing than now. Ink takes a long time to dry and if a book or atlas is bound before the ink is completely dry, part of the image can be transferred from the printed surface to the paper pressed against it. Modern work is less susceptible to this problem because modern presses dry the ink as the sheets leave the press. Sometimes offsetting occurs if the paper becomes damp and then color from one sheet can be seen,

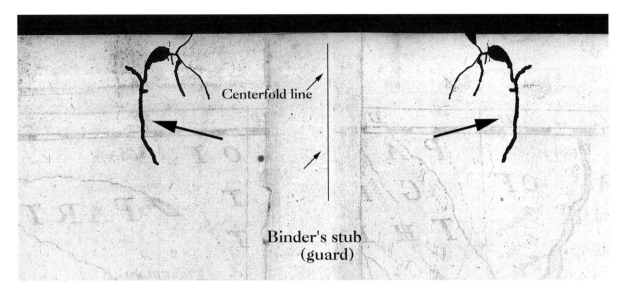

Centerfold line

Binder's stub
(guard)

Figure 6.6. In this photograph of the verso of a map, taken against a black background, we see the bilaterally symmetrical wormholes (bold arrows) which show the path of the worm through the book when the map was folded along the centerfold (thin black vertical line), resulting in the Rorschach-like pattern. Most often we see simple holes as evidence of worm feeding. Note the binder's stub, or guard, still on the map and the show-through of the image on the recto, slightly visible from the verso.

mirror image, on the other. A curious kind of offsetting is seen when a sheet of paper is kept against a surface that has been colored by highly acidic color such as verdegris (see Figure C.8 in the color section). This map has a mirror-image offset on either side of the centerfold. In this case, the acidic verdegris has damaged the opposing sheet. Such acidic damage could have been prevented by using a sheet of barrier paper, but once the damage is done it can be arrested but not reversed. Mild offsetting that is barely visible does not have a major impact on a map's value, but if it is severe, such as seen in Figure C.8, then it reduces the value. In general offsetting is unsightly and anything unsightly reduces the market for an object that is sold largely because it is attractive.

MISCELLANEOUS DEFECTS

Wormholes are small holes that are the result of a larva (worm) eating its way through a book or stack of paper. If the worm was traveling normal (perpendicular) to the plane of the paper only a small hole will result. More damage is done when the worm changes direction and begins to travel parallel to the plane of the paper (Figure 6.6). A few wormholes do not matter much if they do not interfere with the image, or if the map does not look like Aunt Maud's lace tablecloth. Wormholes

can readily be repaired from the back and a few wormhole repairs come with the territory.

Dampstaining could be serious. At its mildest it merely discolors the paper. However, if paper is damp for a long period of time it begins to deteriorate and lose "body." It assumes the characteristics of facial tissue because the long cellulose fibers have been broken into shorter lengths. In the worst-case scenario dampness is combined with an acidic mat or backing. Then the paper breaks down very quickly and can literally come apart in ones hands. Such defects reduce the price of a map significantly. Mild dampstaining can be removed by washing the map and the body can be restored to some extent by re-sizing the paper.

Foxing refers to rust-colored spots (that bear a fanciful resemblance to the color of a fox) that appear in patches. Usually it occurs in paper that has been damp or stored at high humidity for a period of time. Foxing represents an iron-based stain resulting from fungal growth. Left unchecked, it will ultimately weaken the paper and result in disintegration. A foxed map can be improved cosmetically by washing and bleaching, but then we run into the risk of damaging the color as discussed. Any map so treated is more difficult to sell, hence not as valuable. Many of the maps we see in antiques shops often have a significant

91

amount of foxing or other damage. Some papers, such as those made by the English firm, Whatman, in the first few decades of the 19th century, are particularly susceptible to foxing.

Mildew is another form of growth that depends upon moisture. It can seriously weaken paper and, moreover, result in ugly black stains that are very difficult, and sometimes impossible, to remove.

Remember that both foxing and mildew are *infections*. They are contagious, spread by spores which, if the conditions of temperature and humidity are right, can infect other paper. This is not to say that one should become paralyzed by the fear of spreading an epidemic of foxing throughout one's collection, but a word of caution is worthwhile. Don't bring a large quantity of heavily foxed or mildewed old books or paper into your library. If you must have them, it might be worth consulting a conservator about having them first disinfected by thymol vapor.

Maps bearing *accession* or *library stamps* often appear on the market. These marks are the equivalent of stamps in, or numbers on the spine library books, and indicate that the item was once part of an institutional collection. Stamps are of two general kinds. The type that raise the paper, permanently deforming the fibers, are called *blindstamps.* These are similar to the impressions made by notary and corporate seals and involve squeezing the paper in a metal die. More common are simple rubber stamps that apply an inked ownership statement. Sometimes the stamp is located in the margin, in which case it does not detract materially from the item. Rubber stamps were often applied to the verso, and this is not a problem if the ink has not bled through to the recto. Unfortunately, a significant number of librarians have, over the years, applied stamps to the image itself (Figure 6.7). As material becomes scarcer, we can learn to live with these, but we note that they do affect the image and may detract somewhat from the value of the map.

Institutional stamps do not invariably mean that the map is stolen. Over the years institutions and libraries sell unwanted material on the open market. Such *deaccessioned* material is often marked with library stamps. The writer knows personally of many libraries that have sold part of

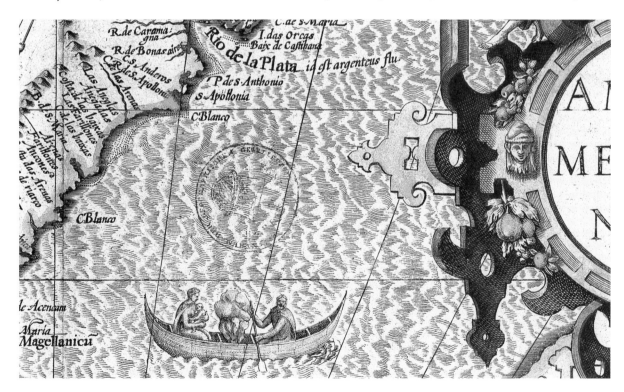

Figure 6.7 Located just above the boat is an accession stamp that was applied directly to the image of this Mercator/Hondius map of South America. The stamp is not particularly noticeable in the original and does not detract from the overall visual appeal of the image. This detail is reproduced actual size.

92

Figure 6.8 This is an actual-size detail of part of the city view of Heidelburg by Sebastian Münster (Basle, 1555). It is a woodblock print and there is letterpress on the verso which shows through in places. The large letter "A" is an example of this show-through. In the lower part we see additional text show-through.

Show-through is a problem most often in maps that have large blank areas, such as sea charts, or views that have a lot of low-contrast sky area. Show-through is visually distracting. I am particularly fond of the Blaeu and Jansson maps of the eastern coast of North America, but they are unfortunately plagued by show-through and it is difficult to find copies that are not so afflicted. It is often possible to minimize show-through by placing a black backing behind the map when it is framed.

Age-toning is a polite way of saying the paper has browned. Often associated with poor-quality paper, particularly 19th-century and later machine-made paper and those containing wood pulp, age-toning is often accompanied by brittleness. If you examine the Johnson map, Figure C.5 in the color section, you will see an example of uneven age-toning, a characteristic of these maps. In this case, the age-toning is darker around the edges.

Age-toning of good, hand-made paper usually suggests that the map was stored improperly or may suggest some other pathology. Simple age-toning should not be confused with the darkening resulting from contact with acidic paper or wooden backing.

Spotting defects. The easiest and most practical way to see many of the defects I discussed in this chapter is to hold the map up to a bright light, or a window, and examine it visually. This will reveal many repairs, especially older ones, and a variety of defects. Obviously, you can't do this if the map is matted and framed.

DERELICT MAPS

Sometimes truly dreadful-looking maps appear on the market. Seemingly everything is wrong with them. Such *derelict maps* often can be restored, but the question that must be asked is if the project is worth the effort and expense. If the map is of great importance and/or value, such a project should be considered. Of course, from a purely commercial standpoint, it must be determined that the *repaired*

their collection bearing such stamps. Nonetheless, it is always prudent to wonder if such maps may have been stolen and we should take care to determine that they have a cleanly transferrable title. On several occasions we have declined to purchase maps bearing such marks. Recently, following the breakup of the Soviet Union, we have noticed maps that bear impressions from old Soviet institutions. Since it is almost impossible to trace the ownership trail of these maps, we decline to purchase them. This problem is not restricted to the map trade. Old Master prints are now surfacing that also bear such stamps. On some occasions, it has been noticed that they also bear the stamps of German institutions, and they have been determined to be part of the loot that the Soviets took after WWII.

If a map has text or other material printed on its verso, it is often visible from the front, or recto (see Figure 6.8). This is called *show-through.*

Figure 6.9 One of the early Tennessee and Kentucky maps printed in America, this exemplar has suffered greatly. The image on the recto (above) is still strong, but the paper is weak, badly stained and torn. Crude repairs have been made to the verso (below). Should such a map be restored?

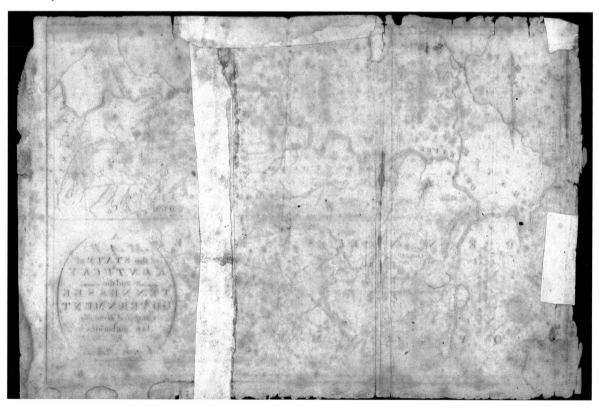

94

map will have enough value to justify the work. Bear in mind that a repaired map usually has a lower value than one in good original condition.

As an example of such decision making, let us consider the map shown in Figure 6.9. I bought this derelict Doolittle map specifically to use as an illustration in this book. It demonstrates well the reasoning that one has to put into the decision to restore, and also demonstrates typical defects that *can* be dealt with. When they *all* occur in one map, though, then watch out!

This map, published by the American, Amos Doolittle, is an important little map measuring 7.6 x 11.5 inches (19.4 x 29.2 cm). We can date it by knowing that Kentucky was made a state in 1792; Tennessee in 1796. The map is therefore most likely a 1795 imprint, and although not rare, is uncommon enough to warrant consideration of restoration. The paper has suffered from chronic damp and we notice overt water stains along the right margin. The paper *feels* bad – it is soft and mushy, it has lost the firmness of good laid paper and has more the texture of facial tissue. There are several multiple folds, the margins are narrow and chipped, and there are tears extending into the image. Here and there we see that small amounts of image are actually missing (top and left). When we examine the verso, we see old paper strips glued on to try to reinforce the edges and folds. There is show-through, but that is not an aesthetic problem since it appears on the verso. Both sides of the map are soiled, and the paper is heavily mottled.

A conservator would begin to treat this map by washing it in a water bath to remove the dirt and loosen the glue under the paper strips. The jagged margin edges would be trimmed to prevent further damage and a similar paper would be sought to use as added-on margins. It would be possible to *leaf cast* new margins and also fill in the chips entering the image. In leaf casting, a paper slurry is made (often using old paper) and new, replacement paper is cast *in situ*, to replace missing parts. However, this map is simply not valuable enough to warrant such measures. Instead, the margins would be replaced using strips of old paper and any holes in the map would simply be backed with similar pieces of paper. Before the map was removed from the water bath, it would be deacidified (probably by adding a carbonate buffer to the water) and then

sized to restore strength to the paper and give it back its "body." However, the paper in this map is *really* bad. Not only is it soft and lacking its original tensile strength, but the folds are very weak and the paper is in danger of falling apart at the folds. Accordingly, the map would have to be *backed* by gluing a sheet of thin, high-tensile-strength paper to the verso, covering the entire back surface of the map. Finally, when the repaired map was dry, the missing detail could be added in manuscript facsimile.

Now, all this is obviously a lot of work to be done by an experienced, highly trained professional restorer. It is not the sort of thing one would want to try using a pan of water in the kitchen sink.

Such a restoration could cost well over a hundred dollars. The map, in original fine condition, commands only a few hundred dollars. From a purely financial standpoint such a restoration does not make sense, especially if one considers that a highly restored map does not command the same price as one in good original condition.

Sadly, we would have to conclude that a map in this price range, of this scarcity, does not warrant such extensive restoration. Unless, of course, it is associated with wonderful personal memories, or if Cyrus Harris, the mapmaker, was your ancestor.

Let us look at another example. This time, I show a map (see Figures C.6 and C.7 in the color section) with some serious problems, but it is a valuable and important map. This map, Laurent Fries' map showing part of the New World, was published first in Strasbourg in 1522, with subsequent editions in 1525, 1535 (Lyon) and 1541 (Vienne). The last two editions were published by Michael Servetus, who was burned by the Protestants for heresy. Calvin, his persecutor, ordered copies of the atlas destroyed, hence it is somewhat scarce. The map I illustrate here is from the 1541 edition. Its size is 11 x 14.6 inches (28.2 x 37.3 cm) and has adequate margins.

This is an important map showing America, and has original hand color. The exemplar shown had been improperly matted and framed. It also had some abrasion with loss of surface near the centerfold. There was dampstaining, and other stains were seen on the recto (Figure C.6). The verso was darker than the recto, probably because at some time the map had been against acidic material. The map had also been taped to the back of an acidic mat using masking tape and some strips of self-adhesive black plastic tape, (see figure C.6) possibly two of the worst ways to affix a map to a mat.

This map required extensive work by an experienced conservator to remove the tape and residual adhesive; to clean the surface and remove as much of the staining as commensurate with good conservation practice. Obviously, because of the original color, it was not possible to wash and bleach the entire map as would be done for an uncolored specimen. For cosmetic reasons, we wanted the missing centerfold regions to be replaced. Moreover, since this is an important map, we did not want the restoration to be so "good" that the map appeared artificially new. We wanted the conservator to perform an "honest" repair; one in which the restoration would be unobtrusive cosmetically, and result in an appearance commensurate with the map. It was also important that the map be treated to remove harmful stains and adhesives and be left in a stable condition so that further deterioration would not occur.

The "before and after" pictures show the outcome. Compare Figures C.6 and C.7. You will note that not all stains were removed; it was deemed more important to preserve as much of the original color as possible. Also, there are still some tape stains visible, indicating just how difficult these are to remove. The show-through on the verso is much reduced (a minor point) but still present. The holes in the centerfold were filled with paper and made to appear unobtrusive by judicious toning. Miscellaneous imbedded grime was washed out and the map deacidified. The bill for this work was high, but the result is a map that is worth many thousands of dollars and that will, if properly cared for, not need attention again for a hundred years or more.

A PHILOSOPHICAL POINT ABOUT RESTORATION

The maps we deal with in this book are, for the most part, printed on degradable materials: paper, cloth or vellum. As such, they will inevitably deteriorate. In the process of deterioration, they acquire age-related changes. Some of these, such as the toning of paper or the oxidative changes to pigments, are of little consequence, and we have come to appreciate some of them as attributes. This is similar to our appreciation of the patina on a piece of early furniture. These are products of age; they are not part of the creation of the original item. *And we have come to perceive some of these age changes as being an integral part of the artifact.* Thus, if we remove the patina from old furniture, if we remove the craquelure from an Old Master painting, something is lost. To our eyes *they will look wrong,* even though the craquelure or the patina were not part of the item when originally made. All this is to say that the process of aging, and its products, is an inherent part of any antique.

What, then, is the function of a restorer? The restorer should fix defects, remove dirt, strengthen weak areas and chemically stabilize the piece. It is my firm belief that restoration should not be carried out to the extent that a map looks "new."

This is not a plea, or an apology, for soiled or defective maps. It is a recognition that overzealous restoration can change the nature of the artifact, and change it into something not entirely desirable.

DESCRIBING CONDITION – II

At this point we have looked at several typical defects often found on old maps, and have tried to put them into perspective. With this understanding we can go back and reconsider the descriptors we apply to maps. As we gain more experience and learn about the condition of the *overall population* of maps, we can be more precise and confident that our descriptions will convey an accurate picture of the map to someone else.

We might, for example, describe a map as "having a small centerfold split, no loss of image, extending an inch into the lower image and repaired on the verso, but everything else VG." Such a description will tell the reader that the centerfold split is the only thing wrong with the map. It is always wise to err on the conservative side of a description.

SOME PRACTICAL TIPS:

How to find a conservator. This problem is akin to finding a good doctor. You can consult the telephone directory but you are better advised to ask for a referral. Contact your local map society, your local library or a knowledgeable collector who has had experience. Very often the conservation department of local museums and libraries permit their conservators to take on small amounts of private work.

Be certain that you discuss the problem with the conservator before the work is started and that you provide *written instructions*. It is, to say the least, upsetting (and possibly expensive) to get your map back with work you didn't order done to it. I recall getting a map back from a conservator to whom I sent it for a simple centerfold repair. She, for some reason, washed it and the old color ran. Her explanation? "I thought you wanted it washed." The map lost a thousand dollars in value and I had no recourse since I hadn't given written instructions.

In all likelihood you will have to ship your material to the conservator unless you live in a large city. A little care in shipping is worth the effort.

How to ship maps. There are two ways to pack maps: flat or rolled. I prefer, whenever possible, to roll the map, wrap it in tissue and ship it in a stout tube. *Do not use thin-walled stationary store shipping tubes.* The tube should be at least three inches (about 7.5 cm) inside diameter and have a wall thickness of at least 3/16 or 1/4 inch (approximately 0.5 cm). Alternatively, some people have successfully used plastic drain pipe available from plumbing supply stores. Whatever you use, leave at least an inch or so of padded space at each end of the tube (ends often get crushed in transit) and cover the ends with something rigid, be it fitted plastic or metal plugs or corrugated cardboard cut to size and taped in place.

If a map is brittle, don't roll it tightly. Use a bigger tube or ship it flat. Again, wrap the map in tissue to prevent abrasion and pack it between several layers of corrugated cardboard. Use at least three layers of corrugated board on each side of the package and alternate the direction of the corrugations for added strength. This should be adequate for sizes up to about 20 x 30 inches; thereafter it is best to use a thin sheet of plywood to provide added strength and puncture-resistance.

Curious miscellany – 6

An unusual image relating to maps or mapmaking

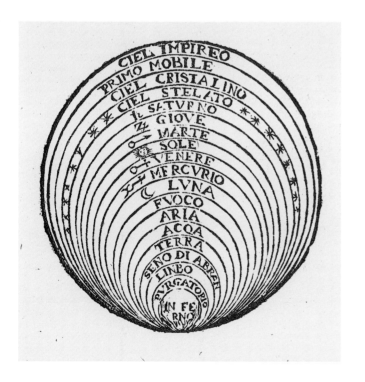

This little woodblock diagram (shown full size) of all of creation, from hell to beyond the celestial sphere, appeared in Gioseppe Rosacchio's *Mondo Elementare, Et Celeste* published in 1604. Although it is a map of sorts, it is more of a late medieval philosophical construct.

Another in Seutter's series of colossal figures. This one depicts a Holy Roman Empire Elector. Full original hand color. Augsburg, c. 1730. Approximately 22.5 x 19.25 inches (57 x 49 cm).

Chapter 7

On Building a Map Collection

What constitutes a collection? What do collectors collect?

We can generalize by observing that most serious collectors specialize. An area or period interests them and they usually restrict their collecting to that subject. Beginning collectors are usually more catholic in their interests but if they remain collectors, they generally focus in and become more specific. However, note also the truism that collectors collect what can be collected. It does little good to lust after a Gutenberg Bible. One simply cannot collect them. The same is happening in some parts of the map world. As older trophy maps either disappear from the market or become too expensive for most collectors, their interest shifts to the more available maps.

Beyond these kinds of blanket statements, we observe that there is almost an infinite variety of collections and collectors. We know people who simply want beautiful maps for decorative purposes; others have a scholarly interest, and others a mixture of interests. We will explore some of these possible collecting motives in this chapter.

Many people collect maps that show places where they have traveled. One of our clients has a wonderful collection of 16th-century city plans of the various European cities she has visited; a businessman buys an antique map showing every place he has been on company business.

Americans are fascinated by their roots and many collect maps of places from which their ancestors have come. It is, for example, possible to build a beautiful and fascinating collection of maps of Ireland, including Irish counties, that go back to the 16th century. This could be accomplished for relatively little money, since even the great maps of Ireland by Blaeu and Mercator cost relatively little compared to, say, their counterparts of North America. Many beautiful Ireland maps of the 18th

century cost only a few hundred dollars or so, and the maps from the 19th century, the time of the great wave of emigration, are often priced even more modestly. The same holds true for most of the countries that sent America its citizens.

Africa is a continent that maps particularly well from an aesthetic standpoint (see Figure C.1). Early maps of Africa, from the 16th century, are still priced reasonably and available from most good map dealers. A visually splendid and culturally important Africa map collection can be assembled for a modest amount of money.

Some collections, such as those of maps of the Holy Land, can start off modestly but become very expensive rather quickly. The great maps in this area are very rare and very much sought after, hence very expensive. This is even more true for maps of North America. In a collecting area such as this where some maps can cost more than a hundred thousand dollars, and even then appear on the market only once or twice in a lifetime, collecting can be very expensive indeed. However, even in this area, it is quite possible to build very interesting collections for relatively little money. Many collectors seek maps showing California as an island, a curious cartographic perversion of the 17th century. Many of the California-as-an-island maps are very rare and very expensive, but some are quite affordable, even to beginners (see Figure C.4.) 18th- and19th-century America maps are still relatively plentiful and can be inexpensive. With few exceptions (Florida, Texas, the southwest, California) 19th-century local America maps are a most affordable subject.

18th- and 19th-century maps of Europe and Africa are pitifully cheap (in my opinion). Fascinating collections around a theme, for example the Napoleonic era, can be assembled for little money but a lot of effort.

If your interests are 20th century, you're lucky, for these maps are generally quite inexpensive. There is a bit of a problem in collecting 20th-century maps: not many dealers stock them. They really are "too recent" to attract the interest of many map dealers, but often are found in secondhand bookshops, flea markets, or in the inventory of ephemera dealers. In the 20th century, although there have been no significant boundary changes in the United States, we have had a rapid and continuous change in roads, railroads and, especially, civil and military aviation installations. All of these are documented on maps that are readily collectable, and that are truly cheap.

Maps of military campaigns are a popular subject. Contemporary maps of the American Revolution are expensive. But if you are interested in other campaigns, especially many of the seemingly perpetual European wars, then prices are much lower. American Civil War maps (see Figure C.5) are a distinct specialty. These maps can be very expensive if one is collecting contemporary manuscript battle maps, for example, but very inexpensive for many of the government-issued lithographic war maps.

Fashion dictates, to some extent, what is collected and what is not. I have always been puzzled why maps of France appear to be undercollected. It is quite possible to assemble a magnificent collection of maps of France (or Portugal, or Persia) for comparatively little money. I think a dominant reason is that the cartography of these regions was well known by the time the making of the kinds of maps we consider in this book was begun. Maps of these areas, even very old maps, do not chronicle significant new information or discoveries. This theory does not explain, however, why so many maps of South America, of the north polar regions, or, in particular, maps of Africa in the 19th century, are also inexpensive. All these maps show emerging cartographic recording and discovery, yet are, for some reason, undercollected. It also does not explain why Holy Land maps *are* expensive.

Maps of continents can form the subject of a collection, as can maps of countries, states, rivers, cities, and lakes (the Great Lakes of the U.S. are among some of the most difficult areas to collect). I noted earlier that trophy maps of the continents by some of the great mapmakers, such as Blaeu

DOCTOR SYNTAX & BOOKSELLER.

Figure 7.1 With some luck and a lot of knowledge one may find splendid old maps in used bookstores, such as the one shown in the Rowlandson etching of 1818. We advise a less chaotic approach to the search!

(see Figure C.3) or Ortelius, are quite expensive. For example, a good Blaeu Africa, Asia or America map is a major purchase. But many other continent maps are available at lower prices. Do not eschew the maps by De Wit, Lotter, Homann, and similar mapmakers.

Sets of maps by the same maker showing the world and continents are often collected. Such matching sets, if made by important figures such as Blaeu, and in fine condition with original color, can be quite expensive.

Maps of the American West explorations, African interior exploration, Himalayan expeditions, East Indies, West Indies, Germany, Swiss Cantons, and European castle towns all form subjects for collections. So do nautical charts, star maps and moon maps.

City plans and views are perhaps one of the more interesting specialties. Maps or plans of many European cities, large and small, were printed from the 15th century on, and many are priced at but a few hundred dollars. Of course, many are in the thousands, but we need not dwell on this. *The important point is that you can build a satisfying collection for a reasonable amount of money.*

BUYING MAPS

How does one buy maps? Where does one find maps to buy? These are important questions and there are no simple answers to them. Much depends upon your pocketbook, knowledge, and oddly enough, where you live. If you live in the middle of Nebraska, for example, there will be no major map dealers near you. You will probably not be able to find a dealer in the entire state who has a good in-depth selection of pre-19th century maps.

Map fairs are a wonderful venue where you can look at literally thousands of maps and meet the dealers personally. Each winter there is a map fair in Miami. Just a few years old, this fair has begun to attract serious dealers and serious collectors. Each spring there is an international map fair in London under the auspices of the International Map Collectors' Society (IMCoS). Usually in mid-June, it is worth the trip just to attend this fair. Also in London, there are monthly map fairs, mostly local events, at the Bonnington Hotel in Bloomsbury. Located just a block or two from the British Museum, this venerable fair has a following of regular exhibitors and buyers.

Book fairs, especially the better ones, such as those sponsored by the major international rare book trade organizations, generally will have several major map dealers exhibiting. This is a network of national associations (in the USA it is the Antiquarian Booksellers Assn. of America, Inc. or ABAA, located in New York City; in England it is the Antiquarian Booksellers Assn., or ABA, located in London. See Appendix G for more addresses and details); all are associated with the International League of Antiquarian Booksellers (ILAB). Member dealers display the logos of these organizations with pride. They are difficult to join. More important to the consumer, all members must subscribe to a code of ethics which includes willingness to take returns.

All is not skittles and beer, however. It would be disingenuous to imply that every dealer is a pinnacle of propriety. In this respect, it is true that one is generally better off dealing with a specialist. Anyone who has invested a lot of time and effort in learning and mastering a very technical discipline has too much invested to be other than strictly ethical. Develop a discerning eye. Make certain the dealer with whom you work has the experience to help you with the *judgments* necessary to do serious collecting. Pick your dealer on the basis of knowledge and inventory. Make sure he/she is someone with whom you feel comfortable. If there is no inventory ("I can get you any map to examine, just tell me what you want and I can have one for you to look at within a week") that "dealer" might be riding on someone else's coattails and expertise.

Do not forget the local used book fairs and antiques fairs. Although you are less likely to encounter specialist map dealers at these events, you may find the less expensive local maps. Second-hand bookshops and antiques shops are also a good source for these (Figure 7.1). Note however that *you should never buy a map that is already framed and glazed without removing it from the mount to inspect its condition.*

BUYING AT A DISTANCE

Dealers in rare maps and atlases are, of course, the foundation of the map trade. Collectively their inventories are huge and you should be able to find many maps of interest available from stock. If you cannot travel to where the maps are, then have the maps travel to you. *Mail-order is a perfectly safe way to buy old maps.* You will also find that once you have established a relationship with a dealer, you can expect that dealer to contact you when he acquires a map that might interest you.

Most dealers issue catalogues or lists. Buying maps from these are really almost risk-free. Some dealer catalogues are splendid illustrated publications, in full color with heavy glossy covers and good, scholarly map descriptions. Others are barely legible and have, at best, poor illustrations of maps. Most dealers will be pleased to send you complimentary copies of their catalogues. It is important to read carefully the conditions of sale, which lays out the dealer's terms, guarantees and the way business is done.

A good working relationship with a few dealers is the best insurance a novice collector can have. Make your own judgments about the dealer. Would you conduct other business with him? Do you trust him as a person, do you trust his judgment, and, quite important, do you share the same aesthetic values? In other words, do you and he mean the same thing when you say a map is pretty? Do you mean the same thing when you assign a "Very Good" condition to a map? Does the word "important" mean the same thing? Recognize the fact that dealers deal in different levels of the map market. Some seem to specialize in "the low end," that is, maps under about $200. Others specialize in maps that might be $3,000 and up. It is almost useless to ask the latter to find you a map that should sell for under a hundred dollars or so, for example an 1890s atlas map of Ohio. It's simply a different market. Also, although most dealers are happy to share their knowledge and expertise with collectors, particularly beginners, realize that it is a business and not (entirely) a public service and that unlimited time is not available. Map dealers' offices are not surrogate museums or libraries. Much of the fun of the hobby comes from digging out information, as well as maps, on one's own.

The dealer's location is immaterial. It used to be that English and Continental dealers were less expensive than American dealers. Often this is no longer the case and, indeed, many European dealers are buying their maps in America. We find also that many of the maps coming from Continental inventories are heavily restored and have modern color.

Don't make the mistake of shopping on the basis of price alone. If you are buying important maps, there are no "bargains." These maps have international prices that occupy a rather narrow trading range. The market in these maps is a mature, differentiated market and one should not expect to find steep discounts when buying such commodities any more than when buying shares of IBM. There are, indeed, some dealers who generally have cheaper maps but these often have been heavily repaired or restored. This is, of course, another reason why you should get to know your dealer well. There is nothing wrong with buying a repaired map (see Chapter 6), or one with modern color added, but the dealer should be willing to tell you what was done, and it should be priced accordingly. The chapter on "Condition and Conservation" will help you decide what you should look for.

Every customer should be able to get a decent guarantee with a map. This is not a so-called "certificate of authenticity." Most of these certificates I have seen are non-informative and virtually useless. What is really needed is an explicit statement by the dealer, including descriptions of any repairs, the dealer's opinion of the age of the color, and a refund/return policy.

COLLECTOR'S RESPONSIBILITY

Up to now I have been discussing mainly the dealers' obligations to the collector. There is a reverse obligation as well. Dealers trust collectors with their maps. We send out, on approval, thousands of dollars worth of maps to people we may never have met. Collectors must behave honorably and pay promptly or return promptly those maps they don't want. These must be returned in the same condition as received, packed properly to avoid damage in transit. It is the *sender's* responsibility to insure and pack properly.

BUYING MAPS AT AUCTION

Auctions are everyone's fantasy. Here is where the free market is at its best and fair market prices will

Figure 7.2 Auction action at a London sale is depicted in this 1887 wood engraving. Later hand color.

attain. Right? Only possibly! Auctions certainly have an important place in the market and they, each year, seem to take more and more business away from private dealers. Perhaps this is a trend that cannot be reversed, but certainly buying and selling at auction is a two-edged sword. Buying at auction successfully requires a high level of knowledge.

There are some specialty "mail-order" map auctions that generally sell less-expensive maps. Large auction houses, such as Sotheby's and Christie's, will often have specialized sales of maps and atlases. These sales usually have medium-to high-priced maps and atlases. Some of the world's great maps and atlases have passed through these houses. So have defective maps, improperly described or improperly attributed maps, and heavily repaired maps.

If you are sophisticated and can travel to the auction rooms (Figure 7.2), then you can often make good use of their services. If you cannot view the maps personally it is worthwhile paying a

knowledgeable dealer to represent you and to examine them for you. If you consistently rely solely on auction catalogues you will eventually be disappointed.

Are auction prices lower than dealers' prices? Sometimes, but often not. Many dealers, myself included, use auctions to sell off the inventory we don't want. We do this because mediocre maps often fetch far more at auction than they would if we catalogued and sold them. Thus, many of the maps consigned to auction come from dealers who cannot (or will not) sell them to private clients. They may have something wrong with them: poor color, repairs, hidden flaws, or whatever. Often unsuspecting buyers assume that the fact that they are in the auctioneer's catalogue makes them good value. On top of all that, there is the buyers' premium, now generally 15% or more. If you buy a map for a thousand dollars at auction, the actual bill from the house will be $1150. It will be even higher if you've bought it in a VAT (value-added-tax) country, such as a European one, where VAT is added to the buyers' premium and not refunded even if the item is removed from the country. Finally, we note that auction catalogues are not cheap. Subscriptions to the map catalogues from just the major houses will cost several hundred dollars a year.

Just as I was writing this chapter, I read that a major London auction house admitted to using phony telephone bidders to "run up" prices. Thus, when bidding on the floor was not going high enough to satisfy the house, they had their own people drive up prices by telephone! Those of us who have spent a lot of time on the auction floor can often spot the bids the auctioneer pulls off the chandelier or the back wall, but the telephone bid scam is particularly difficult to detect. Indeed, telephone bidding at German auctions was frowned upon, if not illegal, until relatively recently.

Ultra-rare and ultra-expensive maps may appear at auction and be more accessible than if a private dealer had them. Here, it makes sense to consider buying at auction. Again, if you are inexperienced you should have a knowledgeable dealer represent you. Dealers will act on your behalf at auctions and give you the benefit of their experience. Their fees are usually modest and well worth it. Our firm uses such services regularly when we cannot personally go to a sale to examine the material.

Perhaps the biggest drawback to buying at auction is that, practically speaking, you cannot return anything. It's yours when the hammer falls. If you buy from a reputable dealer you won't ever have this problem. This is not a hypothetical issue. A while ago my firm bought a book at a large European auction house, based upon their catalogue description. We paid about US$1,000 for it. When it arrived I discovered it was missing a leaf. The catalogue description did not mention this and the auction house refused to accept the return of the book. I had a choice of eating the loss or suing them in their country.

Read the fine print. What happens if title to the goods is contested? Suppose the atlas or map you bought at auction proves to be a stolen item? What recourse do you have if the goods are defective or not as described in the catalogue? Or a downright fake, as happened in 1997 when a famous London auction house sold a replica globe as the real thing. (The *London Times* gave this event

extensive coverage, but I never saw it in the American press.) One of the maps described and illustrated in this book (see Figure 5.1) was bought from an American auction house. It proved to be a fake. Again, a reading of the fine print in the house's "Conditions of sale" section is most illuminating.

Do this little exercise. Read through the pages and pages of fine print governing the conditions of sale in the catalogues of major auction houses. Then look at the catalogue of any major map dealer. Generally it's a simple statement such as: *Everything is guaranteed and everything is returnable.*

If you choose to sell your maps through an auction house, do comparison shopping. Remember, if your consignment is good enough, the terms are probably negotiable despite what the fine print may say. You may not need to pay for storage, insurance, advertising, cataloguing, reserves or illustrations. All of these are "standard" charges auction houses try to load on consignors. If you have good enough material, you may even be able to negotiate commissions.

Auctions are discussed in another context in the next chapter.

STORAGE AND DISPLAY

We all like to display our collections. Maps lend themselves particularly well to this and it is perfectly safe to mat and frame them as long as some precautions are followed.

1. Every framed map should be matted with an acid-free mat of adequate thickness to keep the map from touching the glass surface. If a map (or any paper item) is simply put into a frame and pressed against the glass it will deteriorate rapidly. This deterioration is the result of condensation that collects on the inner surface of the glass. The chief function of the mat (or, in England, the mount) is to keep glass and map separate. If you don't want a mat, you still must keep the map away from the glass. Framers have plastic inserts that fit under the rabbet in the frame. Such spacers maintain an air gap between glass and map.

2. Both the mat and the backing board must be acid-free. If you live in an area where acid-free material is not available, it's better not to frame your maps. Keep them in a folder until you can get the proper materials. The map must be attached to the mat at the top by means of linen or acid-free paper hinges, and be suspended freely from them. It must never be taped down along the sides. The paper must be free to undergo normal dimensional changes without stress.

3. Self-adhesive materials, such as cellophane or masking tape, must *never* be used in the framing job. These will ruin any paper they come in contact with. Indeed, any map that has cellophane tape on it for any reason is seriously reduced in value. There are self-adhesive, archival-quality, translucent tapes (such as Filmoplast P™) that can be used safely and are not to be confused with cellophane tape.

4. Protect maps from sunlight and physical damage. We are much impressed by the newer ultra violet-light-absorbing Plexiglas that can be used instead of glass for glazing. It is much lighter than glass, is shatterproof and will not damage the map if the whole framed assembly is dropped. Most important, it prevents ultraviolet light from damaging the paper or fading the color. This material is well worth the small added expense. Maps should never be hung in direct sunlight. Paper will be damaged and color may fade. Light from ordinary fluorescent lamps is equally harmful. If you must hang your maps in a room with fluorescent lighting, you should acquire special sleeves to place over the lamps to block their harmful emissions.

Harm done by light is a function of its intensity and duration (in addition to its wavelength). This is why museums will display their maps under low levels of illumination. For the average collector, normal room illumination should pose no long-term threat. The occasional strobe-illuminated photo-graph or photocopy can be tolerated (high intensity but short duration) but one would not want to shine a bright floodlight on a map for very long.

At some time in one's collecting history one will run out of wall space. Many collectors solve that problem by

using flat plan chests, or blueprint cabinets, to store their maps. Most museums seem to prefer steel plan chests, perhaps because of the slight possibility that wood might impart acidity to the contents over the years. (I actually suspect the real reason is that steel drawers open and close easier than wooden ones.) Whatever the merits of this argument, I personally prefer wooden plan chests so that my office or library avoids that institutional appearance imparted by steel cases. Reasonable care, not fanaticism, seems to make a hobby or a business more enjoyable! Storing maps in individual acid-free folders should protect them from any possible damage from their wooden environment. These folders are available from a number of suppliers or can be made simply by folding a large sheet of acid-free paper. Somewhat nicer than plain folders are the ones with a Mylar sheet (Figure 7.3). The map is placed under the Mylar sheet and the top of the folder then comes down, affording even more protection. I use these and like them because even when the folder is open, the Mylar protects the map.

Mylar sleeves are very good protectors. They are acid-free, completely clear and rugged. On the

Figure 7.3 Made of heavy acid-free paper, this folder has a Mylar sheet glued along one edge. The map is perfectly safe under the mylar and the top paper cover affords extra protection.

downside, they are heavy and expensive. I never use polyethylene bags of the sort that some ephemera dealers use for their postcards and other paper items. Although some consider them to be safe, I am concerned about residual monomers that might damage inks and paper.

THE MORALITY OF RIPPING UP BOOKS

We do not need to mince words. Most maps are taken from atlases, which are books. (This process is called *breaking;* those who do it are called *breakers.* An atlas considered suitable for breaking is also known as a breaker.)

Whatever it is called, the act of removing the maps from the book destroys the book. Map collectors drive the equation in this direction. In many ways it is much like buying pocketbooks or shoes made from the skins of endangered animals. If nobody buys the shoes people will stop killing the animals. If we don't buy the maps, people will stop ripping up books.

There is no easy answer to this moral dilemma. In reality, most of the damage has already been done, and not only by the breakers. Most dealers or collectors do not take apart intact atlases. We see many atlases that already are missing either a significant number of maps (probably removed by some early owner) or that have suffered extensive damage otherwise. We frequently buy atlases that have parts ruined by water or smoke. Mice and silverfish take their toll, also. If these atlases cannot be repaired, then it makes sense to take them apart and restore and salvage the remaining maps.

The recent sharp price rise in atlases has made the problem of breaking somewhat self-limiting. Unlike animals that reproduce, atlases do not. As the atlas population declines, individual atlas values increase and it becomes less profitable to break them and sell their parts. Thus, it is no longer economically feasible to buy a Blaeu or Ortelius atlas and take it apart for the individual maps. If the prices of individual maps were to double (which they well might) then atlases, at today's prices, would again become fair game for the breakers. However, atlas and map prices have become linked. They seem to be marching together and it is very unlikely that they will get out of register enough to ever again warrant the wholesale taking apart of atlases for their individual maps.

Indeed, we know of instances where it has been economically possible to make an otherwise intact atlas whole by finding the missing maps and reassembling the book. This is much like a chemical reaction that can go in both directions depending upon energy. In the case of maps and atlases the energy is money and the reaction (atlases to individual maps, or individual maps to atlases) will go in the direction that creates the greatest value. When there is no net gain from taking apart atlases the practice will stop.

I think that the case for the 19th century is a still a bit different. There is a tremendous demand for 19th-century maps, especially those of American states. And, there are still a tremendous number of 19th-century atlases out there. Many more atlases were made in the 19th century than in earlier centuries because they were cheaper (lithography vs. copperplate engraving) and they were, at least in part, machine made. The demand for single maps is large enough so that it is still profitable to take apart some of these atlases, but usually the ones taken apart already have other defects, such as missing maps or water damage. Intact, complete 19th-century atlases are becoming sufficiently expensive so that the profit derived from taking them apart is shrinking. Unless something drastic happens to the prices of individual maps independent of the price of atlases, it is unlikely that a significant number of additional complete atlases will be destroyed for their parts.

SOME TIPS

•If you have the opportunity to buy two maps, one in better condition than the other (and consequently more expensive) buy the better one.

•If you decide not to buy a map that a dealer has sent you, return it to the dealer *right away.* Pack it carefully, preferably in the original packing, and insure it. Getting it back safely is *your* responsibility.

•Catalogues cost dealers money to produce and mail. Auction houses charge for their catalogues, dealers don't. It is a kindness to dealers to have them remove your name from their list if you are not interested in their maps.

Chapter 8

Prices – The Markets Speak

SOME THOUGHTS ON PRICES

This is a difficult subject. Map prices often seem confusing and illogical to the novice. That is because the price of a map is only one part of a greater picture and is intimately related to many other factors that also need to be understood. In this chapter I will share my thoughts about prices and pricing and explore some of those interrelated factors that together determine the desirability and influence the price of a map. Among them are:

1. Scarcity. Is it unique? Very scarce? Common?
2. Does it contain significant new knowledge?
3. Does it contain information of cultural importance, to local peoples or to humanity as a whole?

Additionally, the nature of the market strongly influences map prices, and we need to know about the market. This market comprises dealers, serious collectors, casual collectors, institutions, and a larger number of people who buy maps as decorative items. Every buyer creates a distinct but transient market. Since there can be large price differences for the same map in different markets, we need to know something about each if we are going to understand the overall map market.

I will make the argument that, in practice, *a map is worth what the seller can get for it.* This sounds simplistic, but the consequences of this argument are important. There are some maps that trade in a fairly narrow price range that really is a ***consensus price.*** These are maps that have wide appeal and which are bought and sold frequently enough to create an almost public market. For the scarcer, or more unusual, or little-known stuff, a specific seller and buyer *create a market* for each transaction. I don't mean to liken this to a transaction at a bazaar, but rather I emphasize the uniqueness of such transactions.

If I, as a dealer, cannot identify or understand the map and its context, then I cannot create a reasonable market. Dealers sell not only bits of paper, but knowledge, experience and the reputation that collectively make those bits of paper meaningful. The dealer who knows more than I and who has worked harder and/or invested more time, money and energy in building his knowledge and reputation can justify his price better than I can justify mine. The point I'm making here is that the price of an individual map is based on many things, most importantly knowledge *about the map and its place in cartographic history.* Obviously, my credibility as well as my knowledge contributes to my ability to make a cogent argument about the value of a map, hence create the market price.

It might be heresy to say this, but although knowing earlier prices for similar maps might be interesting, such knowledge should not overly affect decisions about prices today. If you don't believe this, just consider what influence three-year-old price records of shares traded on the New York Stock Exchange have on what you are willing to pay for any specific equity today. Most emphatically, just because a book lists what someone in the past has (allegedly) paid for a specific map, *it doesn't mean that you, I, or anyone else can get a similar map for the same price, or for that matter, sell one for that price.* Also, it should be remembered that such lists, by and large, ignore or minimize details of condition and frequently do not mention edition or state. This makes it even more difficult to ascribe meaning to the prices in the guides.

This raises another point. I believe that there is no real difference between a price of, for example, $2,000, and $2,200. Antiques price guides make a great fetish of trying to make such distinctions, but what we really need to know is a general price range for an item *and the reasons for the price.* There are no exact prices because there are no exact values. No two maps are exactly alike and no two dealers or collectors place a map in exactly the same cultural context, which, as we have seen, determines part of the map's value. Some collectors may feel cheated if they pay more than last year's auction price for a map they want. Invariably, over the long run their collection suffers. It matters little if a map was listed in a catalogue for $1,000 two years ago, but the fellow selling one today is asking $2,000 for a copy. The only relevant question is "How was this price arrived at?" (Of course, one also has to consider the question, "How badly do I want the map?")

It also doesn't matter what the map cost the seller. *The price the seller paid for a map has no influence whatever on the value of the map to me or subsequent buyers. It affects only the seller.* If he bought it for ten bucks, good for him! If he paid a thousand dollars for a map that no one else can sell for five hundred, then it's his loss. *Overpaying for a map does not make it worth more.*

Factors that determine price have long-term consequences for someone building a collection. Over the years I have had the privilege of appraising some very fine map collections. I have discovered that, in general, those collections with the most important maps in the finest condition were usually made with the intimate help of serious dealers. These collectors were also willing to pay fair prices for the maps when they bought them, and were willing to pay a premium for condition. Invariably, their collections became worth substantially more than they paid for them. The lesson is that quality counts.

AN IMPORTANT NOTE ABOUT PRICE PHILOSOPHY IN THIS BOOK

This book explores map collecting, buying and selling as an integrated activity. Determining prices, obviously, is a part of this activity. Throughout this book I emphasize that *there are no fixed, exact prices for maps.* Many people want exactness and will be attracted to price guides that may try to give the *impression* of exactness. I do not try to do this because *it cannot be done.* It is somewhat useful to know what other similar maps have sold for in the past, it might be more useful to know what the price

range is of maps currently on the market, *but it is most useful to know why some maps are more desirable, hence more expensive than others.* If there is a secret to all this activity, this is it. A list of prices past can cause more trouble than it solves. We know of collectors who have turned down maps offered to them because the price was higher than they saw in some guide. Years later they were still looking for these maps, always hoping to buy at two-year-old prices! These same people would not expect to do this when buying stock in a bull market, yet for some reason they expect it to happen when buying maps. Inexperienced dealers will often try to justify map prices by showing what others have charged for (presumably) similar maps without actually knowing *why* such prices prevailed. Given all the circumstances of issue, condition, color, etc. a map listed in a guide for $3,000 might be a bargain at $5,000. Conversely, a map listed in a guide at $5,000 may not be a bargain at $3,000.

SOME DETERMINANTS OF PRICE

Throughout this book, I have alluded to price and pricing in several different chapters and in several different contexts. This is because *there is no single determinant of a map's price.* The price is an amalgam of many factors, all of which have their own contexts.

Every map is to some extent subjective, reflecting the knowledge upon which the image was based. This knowledge, of course, reflects the culture of its makers, the intellectual biases of the age in which it was made and the reason for which the map was made. We must also know a bit about the knowledge of geography at the time a map was made. All these factors contribute to the perceived value and importance, hence price, of a map *in our culture at this time in history.*

In general, maps that represented the forefront of geographical knowledge are important. Maps that just reiterated this knowledge are generally considered less important, and those that were issued with grossly outdated concepts are often among the least desirable. But there are exceptions. For example, the Münster Americas map (see Map 25 in Part II) showed outdated coastlines of the Americas when it was printed. This map, however was *very early* and also was the first separate map of the American continents. Even if "outdated," if a map is an example of an important *map type,* such as a T-O map, or Macrobius' map, then it will be important to collectors.

Rarity breeds desirability in some maps. Even though the scarcest states may not be the "best" cartographically, some collectors will only want the "scarce" versions.

Scarcity can work the other way. There is an adage, "collectors collect what they can collect." Some items are so rare that it is not practical, or even possible, to collect them. In the map world, it would be nearly impossible to collect all of the Sotzman maps of individual American states, even though they are late 18th-century maps. Lafreri-type maps (Map 23, page 142) are sufficiently scarce that many collectors have never seen one. Average collectors often tend to shy away from the very scarce *because they are unfamiliar.* And this is how some great collecting opportunities are lost.

Against the background of some of the less tangible price determinants, there are, of course, specific points.

The following considerations are important in determining a map's market price.

1. Mapmaker. Some makers are more important than others. Generally great mapmakers tried to be as accurate as possible, using only the most reliable and up-to-date information at their disposal. They also often had a great sense of beauty; their works are usually well thought out, well laid out and well executed. Some mapmakers are more derivative and simply repeated much that was on others' maps. I think that Robert de Vaugondy (1686-1766), for example, falls into this latter class of mapmaker and most of his work is priced much lower accordingly.

2. Condition. This cannot be overemphasized. Obviously, collectors should seek out the best condition. Even though these maps are the most expensive, they appreciate the most and are the easiest to sell should the collector ever decide to do so.

Europeans are often less obsessed with condition than American or British collectors, and we have seen very high prices realized by maps in rather poor condition on the Continent.

3. State. This is an important consideration in some maps. Some states are exceedingly scarce, others just more desirable for one reason or another.

4. Region. Some areas seem to be collected more widely than others. For example, maps of France seem particularly undervalued, which means that if you're interested in France you can still form a truly great map collection for relatively little money. At the present time, county maps of France, local maps of European regions (especially Eastern Europe), South America, and Portugal are among the least in demand. Quite attractive maps of these regions by major mapmakers, in good condition and ranging from the 1590s through the 18th century, can be bought at modest prices. Other regions, such as Africa, have had their collecting fortunes wax and wane. Africa seems much underpriced at this time. Asia, particularly Japan, has seen some very large price increases as the Westernization of the well-educated, affluent population increases. The same holds true for the Middle East. Years ago it was difficult to sell maps of Anatolia or Arabia, but these have developed a nice market as a middle class emerged in that region. There is now, for example, a small but thriving map trade in Istanbul.

5. Period. In general, maps that were made at the time a region was being explored and mapped are the most sought after, hence expensive. As the Pacific Northwest was being mapped, for example, a great number of maps were produced showing what proved to be erroneous geography, or incomplete details based on partial knowledge. These maps are more important than the post-discovery maps that followed.

Maps showing the western areas of North America in the 1830s and 1840s are in great demand, since they represent the early phase of cartography leading to statehood. Maps of the eastern states of that same period are generally less in demand; they were already mapped, surveyed and divided up. Maps of Europe, on the other hand, do not show much difference over a very long period. Except for decorative differences and political outlines, Europe maps from the 1650s are quite similar to those from the 1750s and 1850s. This is one reason why the price of a map may not be related all that closely to its age. Beginners are

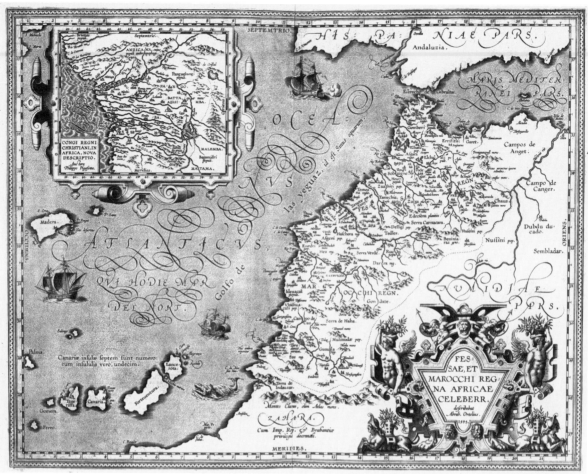

Figure 8.1 *Fessae, Et Marocchi Regna Africae Celeberr.* This map, by Ortelius (dated 1595, edition of 1598), is a beautiful map of a very undercollected region and a good example of a splendid map by a major mapmaker that can be had inexpensively. 15.3 x 20 inches (39 x 50.5 cm). This exemplar is uncolored.

always surprised to learn that there are a large number of maps printed in the 1500s that are priced under a hundred dollars!

Although some specialist collectors are interested in political boundary changes, in general maps that solely reflect this kind of change as opposed to landform discovery changes are often not considered as "important." As always, there are exceptions. For example, Texas as a Republic was an important phase in North American history and maps showing the Republic of Texas are in high demand, especially in Texas.

6. Visual appeal. This is an important, **but** *very subjective,* factor. An attractive map of an avidly collected region is generally more salable, hence costly, than a plain one.

There are maps from the Golden Age of cartography that are very beautiful yet, in my opinion, vastly underpriced. Regardless of how attractive or how good their condition, they seem to sell at chronically low prices. This is because their regions are not collected widely and the demand for them is low. I list and illustrate a few that are particularly appealing to me. Others may disagree or would add to this list. Decide yourself!

•The Ortelius Fez and Morocco (Figure 8.1) is beautiful but the map is very inexpensive.

•The Ortelius map of the Azores is one of the most attractive maps ever made (in my opinion).

•Jansson or Blaeu's Moluccan Islands (Figure 8.2).

110

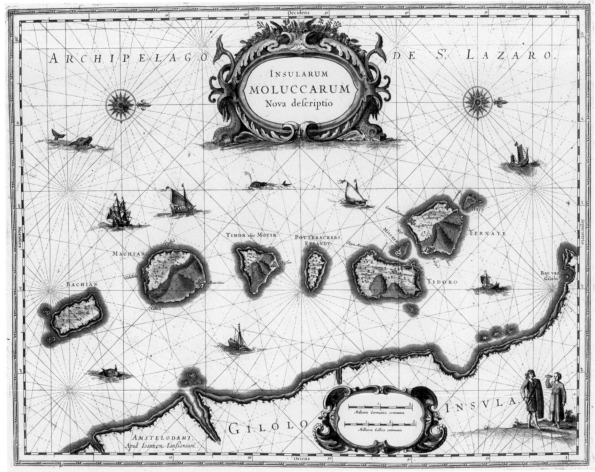

Figure 8.2 *Insularum Moluccarum Nova descriptio*. Jansson's first edition of this map (Koeman Me35) 15 x 19.5 inches (38 x 49.8 cm). This exemplar has original color.

•Blaeu's Scottish counties. Some of Blaeu's finest cartouches are found on these maps. In original color they are spectacular.

•Blaeu or Jansson's Venezuela, Guiana, Paraguay (Figure 8.3).

•The various Straits of Magellan maps by Blaeu, (Figure 8.4) or Jansson. These are spectacular maps with sea creatures and sailing ships, yet are priced very low.

All the above maps are highly decorative yet any can be bought in fine condition and in fine original color for surprisingly modest amounts. They provide an opportunity for collectors of modest means to acquire beautiful maps by some of the world's great mapmakers.

7. Color. In today's market, colored maps seem to sell better than uncolored maps, and many colorists are diligently applying color to uncolored

maps to increase their salability. We wonder what imprecations future collectors will hurl our way when, years from now, fashion will again turn (as it always does) and the uncolored maps will be the more prized!

I am generally neutral on the issue of coloring old maps. I see nothing wrong with it if it is done well and one does not apply color inconsistent with the map's period. But I am uncomfortable when I see a very scarce, important old map colored in a blind, thoughtless attempt to make it sell more easily. Truly important and scarce maps should not need color to sell. They should never be colored or altered, any more than one would change the color of Mona Lisa's hair because contemporary fashion prefers blonde.

From a decorative standpoint it might make sense to add color to a map and, indeed, the interior

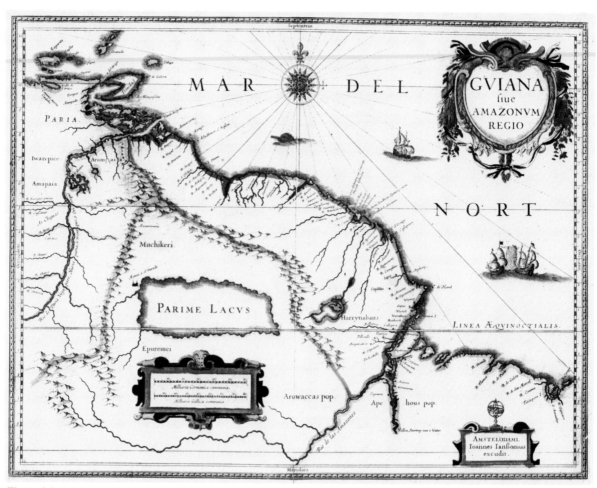

Figure 8.3 *Guiana sive Amazonum Regio*, by Jansson, is very similar to Blaeu's. This copy has fine, original outline color. The map measures 16 x 21 inches (41.2 x 53.5 cm).

designer market almost always insists on colored maps. However, should one pay a premium for a map just because it has been colored recently? If the color is poor, the effect is equal to putting a coat of red enamel on a Windsor chair, and, in my opinion, the map should sell for less. If the color is good and a prospective buyer thinks it enhances the visual appeal of the map, then well and good. Color does not always add very much to the price of a map (unless it is original color); but it does tend to make it easier to sell.

THE DECORATIVE MARKET

One of the largest parts of the overall map market is the market that supplies decorative maps to interior designers, or people who do not collect maps *per se* but want one just the right color and the right size to go over the sofa. There seems to be a different pricing structure in this market – the final consumer, or owner, does not seem to care as much about the cost of the map as about the visual appearance it adds to an interior. It is also difficult to assess true cost in this circumstance, since the map is generally bought as a package that includes matting, glazing and an expensive frame.

TROPHY MAPS

There are, in the map world, a group of maps that are universally sought after. These are the maps that serious collectors with deep pockets try to acquire in the best possible condition, and money is no object. They are pivotal maps in the history of cartography: important, often beautiful, and scarce, rare, or very rare. Sometimes all of the above.

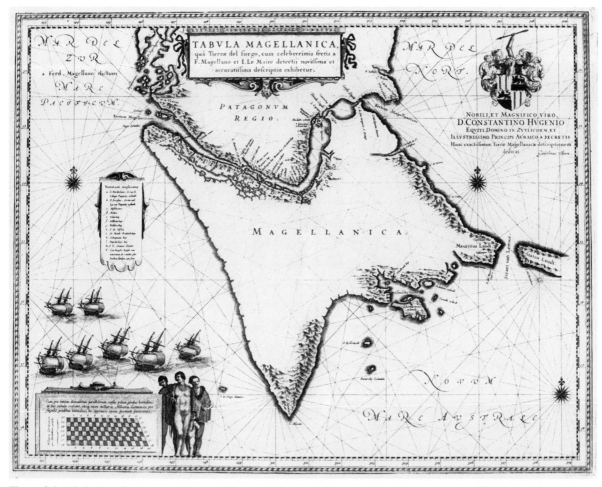

Figure 8.4. *Tabula Magellanicaa, quâ Tierræ del fuego....* This map by Blaeu, published in Amsterdam in 1640, has original outline color. A fleet of sailing vessels is in the lower left above the group of standing natives. There is a dedication in the upper right. The map measures 16 x 21 inches (41.2 x 53.5 cm).

The market for "trophy maps" is not for the faint of heart; it involves serious money and requires difficult decisions. However, there are literally thousands of beautiful, important (or at least significant) maps that span a five-hundred-year range of our history and that are available for collecting. You can build a beautiful and indeed very important collection centered around these maps. But if you decide to go after the trophy map, beware! Our hobby can become very expensive if the cheapest map we collect costs $5,000. And the occasional ultra-rarity costs orders of magnitude more. And we have to wait years to find one.

DEALERS AND PRICES

As in most things, there is a hierarchy in the map world. Serious collectors who buy maps costing many thousands of dollars do not, in our experience, get the bulk of their maps by wandering the countryside looking in small antiques shops in the hopes of finding a "deal." They generally buy at auction and from established dealers who have, over the years, built up a trade in their type of map. These dealers are known for the quality of map they carry as well as their knowledge; they are likely to have an international clientele and are known in the map trade worldwide. Such dealers do not sell maps on price alone, but rather provide a greater service. Major dealers take great pride in dealing in an honorable fashion. Nothing will harm a dealer's reputation faster than misdescribing (and consequently mispricing) a map.

Likewise, there are dealers who specialize in less expensive maps. It shouldn't be too surprising

113

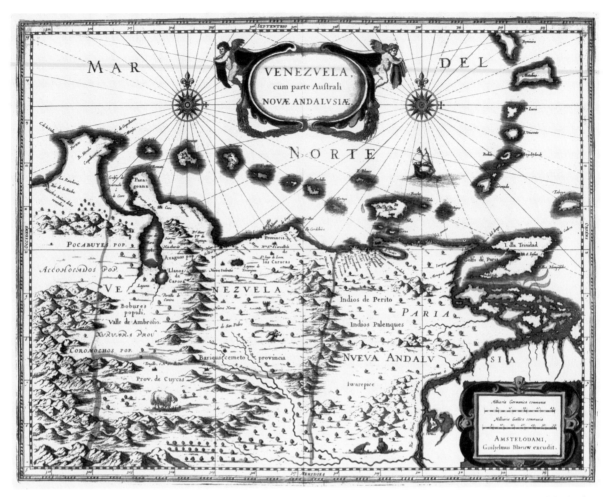

Figure 8.5. *Venezuela, cum parte Australi Novæ Andalusiæ...* This map, by Blaeu, published in Amsterdam in 1668, has original color. It measures 14.75 x 19 inches (37.5 x 48.5 cm). This map often has a darkened centerfold.

that this is a distinctly different market from the very expensive map market. There are equally great differences between the general used book shop and the specialist dealer in very rare books. They are basically different businesses, with different goods, and different patrons. I do not mean to imply that one is better than the other, simply that they are different.

Inexperienced dealers may try to establish a price based upon their cost for the map. This sometimes works for the more routine items, but misses the point entirely when dealing with more serious items and violates our axiom that purchase price does not determine "value" or selling price. If a clear and cogent argument can be made for a price of, say, $500 for an item, then that should be the price regardless of what it cost the dealer. The markup might be 20% or 2000%; it doesn't matter.

There are certainly times when a price can be negotiated, even with vigor. None of us who have been rare map dealers for any length of time object to discussing the basis of our prices. However, we all understand that there is little enough good material available and dealers incur significant expenses finding and acquiring it.

Dealers and collectors should have a symbiotic relationship. The specialist dealer is not just a middleman, moving merchandise and taking his markup. There is true value added in finding the map in the first place, then recognizing it, authenticating and researching it and finally making it available to the ultimate collector.

Finally, a general observation that map collectors should keep in mind. The map business is a very capital-intensive business. Many dealers,

114

especially those selling higher-priced maps, operate on surprisingly small margins.

In all my years in business, I have often heard, "Can you give me a discount on this map? I saw it somewhere else for less." I have never heard, "Let me give you an extra 20% for this map. I just saw it elsewhere for a lot more!"

COLLECTORS AND PRICES

Collectors should expect to pay fair market prices for their maps. The issue from a collector's standpoint boils down to the question, "What constitutes fair market?"

Serious map collectors are very knowledgeable, often knowing more about their special area than any generalist dealer. They also should be able to judge condition and understand how it affects price. The latter is particularly important. A map of the Americas, by Blaeu, in fine condition, that might have an enthusiastic market at a substantial premium, might command a substantial discount if it has centerfold problems. A 19th-century Mitchell map of Missouri that might fetch under a hundred dollars in fine condition might not be salable at any price if it has a tear. The ability to make the kinds of condition judgments that have this effect on prices comes with experience, and many collectors have lots of that.

Obviously, if a map has a long and regular track record at major auctions, or in reputable dealers' catalogues, there will exist a public record of its trading range. To go back to our Blaeu Americas example, we have sold copies of this map privately for much less and for much more than its auction track record. This means that we have, in each case, *established our own market price.* We were able to do this because of a number of factors, including our knowledge and reputation. In each case, we were able to make a cogent argument for the correctness of our valuation. Still, there is a broad trading range outside of which a map will not sell. No matter how nice it is, a Blaeu Americas *should* not sell if priced at some truly absurd number. At least not today.

Most knowledgeable collectors will deal with a relatively small group of dealers. This world is really quite small and most serious collectors know most serious dealers (and vice versa). There is usually a buoyant equilibrium that is reached with a collector becoming comfortable with a group of map dealers whose inventories complement the collector's growing collection and the collector's interests. Rarely does the same dealer have a large selection of low-priced maps (say under $100) and also a large selection of high-priced maps (say over $3,000). Consequently, a collector in one range or another does not usually deal much with a dealer at the other range. This situation prevails in other areas as well, for example antiques, jewelry, automobiles or clothing.

BUYING AT AUCTION

Many collectors share a particularly appealing fantasy regarding auctions. They believe that the auction house is the best place to buy and also the best place to sell. This argument implies that we somehow can buy at "wholesale" when buying at auction, but, miraculously, the price becomes high retail when we sell! Somehow, it doesn't work out that way.

Virtually all auction houses have a "buyer's premium," which means that the buyer pays an additional 10 to 17% (or even more, in some cases) *on top of the hammer price.* In Europe you will also have to pay VAT (usually around 17%) on both the goods and on the buyer's premium. On top of that, of course, are the annual expenses of subscribing to the glossy catalogues.

Most auctions are not "absolute." They generally set a price, below which the item will not be sold. This price is called a *reserve.* For example, if a map is estimated to fetch $1,000 then it might have a reserve of $800. If bidding does not exceed the reserve, the map will not be sold but will be "bought in," that is, go back to the consignor (who generally has to pay a fee anyway).

There seems to be a general notion that somehow the price one winds up paying at auction is the "right" price. After all, someone was willing to pay just one increment lower! This does not take into account "auction fever," that sudden onset of bidding frenzy that overtakes beginners. Far more than price is at stake in the auction room. Ego, manhood, prestige and a display of power: these are always at work in the auction rooms and those who are not conscious of them will be trapped.

There are other dangers.

Do auction houses "run up" bids? Emphatically "yes," according to recent reports in the British press. One of the world's largest auction firms, located in London, admitted that it used fictitious bids, sometimes via fake telephone bidders, to raise the prices of lots that would not otherwise go above the reserve.

Nonetheless, the auction rooms are where many of the world's great maps and atlases change hands, and if we want to play in that league, this is one of the ballparks.

My advice: if you plan to make serious purchases at auction, and are inexperienced or cannot view the sale, have a dealer represent you in person. When my firm cannot be present at a sale, I pay a colleague to represent us. I negotiate a fee for services, usually a percentage of the hammer price, and find that this is very cheap insurance, indeed. I would advise, however, to engage the services of an experienced dealer.

HOW TO SELL YOUR MAPS: SELLING TO DEALERS

The nuts and bolts of selling one, or several, maps has not really been dealt with extensively in any book, to our knowledge. There are, as in any transaction, two main factors to be considered. Both the buyer and seller need to feel comfortable about the transaction. This is especially true if one is selling maps from the estate of a deceased relative. There is often a lot of emotion involved in these transactions.

Antiques dealers usually come upon maps in rather small quantities at a time, generally framed. Map dealers do not like to buy framed maps since they cannot examine them for condition. Many maps that have been in a frame for a long time need attention (see Chapter 6), and a dealer needs to know this before making an intelligent offer. Yes, having your maps matted and framed was expensive, but that cost does not accrue to the map itself – *an expensive mat and frame does not make the map inside more expensive.*

When you finally decide to sell your maps, have the confidence of your convictions, but be absolutely honest with yourself about condition, color and state – all those things we talk about in this book. With a bit of experience and a lot of self-

control you can determine the general value of many maps. You should certainly be able to decide if a map is in the hundreds or thousands. Remember also that there is *no exact value for any map.* Each map is unique in its own way – these are not shares of General Motors or IBM! For that matter, try ascertaining the "value" of a share of IBM and then relate that to its selling price!

It is probably most efficient to contact only the few dealers who seem to deal in your type of material and see what happens. Offer your map in a letter, preferably with photographs (Polaroids are just fine) and a good description. You may wish to consider a consignment sale, wherein the dealer will sell your map on a commission basis rather than buy it outright. You might first ask the dealer for a copy of his latest catalogue. Not only to check prices, but to get an idea of the class of the operation. Someone who puts out a good, illustrated catalogue is not fly-by-night, and you should be able to readily trust such dealers to examine your maps on approval and send them back if they don't want them. You can be pretty certain that someone who has just issued illustrated catalogue number thirty or forty isn't going to steal your map.

Dealers will buy maps at a price that permits them to make a decent profit. In other words, they buy below the price at which they think they can establish a market. In general, if one is selling into the trade, the best prices will be realized by selling to the appropriate specialist dealer. This situation is roughly similar to that found in the antiques trade or in the used book marketplace. A dealer specializing in inexpensive maps may not be able to create a market for a high-priced map and consequently cannot pay as much. Such a dealer may have to use a secondary market (usually another dealer) to move the map. A dealer is most likely to pay more for a map that he thinks he can sell quickly. And he certainly will pay much less (if he buys it at all) for a map he thinks he might still have in inventory when he dies.

It is a matter of ethics that a private individual coming to a dealer with a map for sale be treated fairly. It is wrong, and in many states illegal, for a dealer to "lowball" a price. A dealer cannot, in some states, claim ignorance, since by placing himself before the public as a dealer there is implied a degree of expert knowledge. It is probably fair and equitable to offer a price that approximates prices one has paid for comparable material in the recent past.

Clearly, if you want to maximize your price when you sell into the trade, you should first research the map as best you can, try to establish its place in the overall map pricing structure, and be honest with yourself about the condition.

SELLING AT AUCTION

Auctions serve a valuable function in the world of old maps. They bring together buyer, seller and maps in an atmosphere where prices can be established quickly by mutual consent in what may appear to be an open market. However, all is not peaches and cream.

If you decide to sell your maps at auction, you may need to wait a long time before the deal is completed. As a consignor, you should try to negotiate the terms. If the value of your consignment is high enough you may be able to negotiate a very favorable consignment contract: perhaps a flat 10% fee, no cataloguing charges, no illustration charges, the item is to be covered by the auction house's insurance at no cost to you, no buy-in fee (if it doesn't sell), and return shipping at the house's expense. In other words, if the auction house sells your maps, it gets 10%; if it doesn't, it gets nothing. Payment in full should be expected no later than 30 or 45 days after the sale. If you don't negotiate terms, you might find that you are being charged 20% or more of the selling price for commission, plus a charge for insurance, plus a charge for cataloguing and illustration. Should your consignment not sell, you might be liable for a buy-in charge, an insurance charge, and a packing and shipping charge to return your property to you. If auction houses don't want your business, you can always find a dealer who does, and get paid up front right away – no surprises!

I suggest you read the auction house's fine print very carefully. Some auction houses will not pay you unless they receive payment for *all* your items they sold. In other words, if you consign ten maps and they are knocked down to ten different buyers, the auction may not pay you for any of them, unless all ten buyers have settled with the house. And that could take a long time, if there's one major deadbeat in the group.

We advise avoiding the smaller, local auctions that are really extensions of someone's used book business. While they may exude a nice, folksy charm, the economics of that kind of business are such that it can invite problems for the consignors.

Finally, it is important to realize that an auction house's estimates should be regarded as advertising, not as appraisals.

APPRAISALS

If a seller wants an appraisal on a map, it should be so stated and a fair appraisal price agreed upon. Every dealer gets a steady stream of telephone calls or visits asking, "How much will you give me for this map?" when in reality the question is, "What is the fair appraisal value of this map?" If you want to sell a map, be prepared to sell it; if you want to determine its value, be prepared to pay for this information. You can, using this book, understand some of the *bases* of value. But if you want an *appraisal,* nothing substitutes for the experience of a dealer who has seen many thousands of maps in his career and whose livelihood depends upon his ability to determine their price. "Independent" appraisers rely heavily on price guides and catalogue prices to determine "value." This, for the reasons discussed already, is no substitute for the hands-on expertise of a practicing dealer.

In Britain the custom in the antiques trade is to charge an appraisal fee based upon a percentage of the total appraised value. In the United States, this practice is frowned upon and in most instances, a fee-for-services basis is required. Good appraisals are difficult to do and are not cheap. It has been my experience that general antiques dealers or bookdealers are not usually qualified to appraise rare maps and atlases. They tend to rely upon price guides and such appraisals may not only be woefully incorrect, but, moreover, may not hold up in court.

I always make it very clear to clients that we will not buy a map or collection that we have appraised (unless many years have gone by), in order to avoid any hint of conflict of interest.

Appraisals for insurance or charitable gift purposes should attempt to determine the actual replacement value of the maps (what it would cost you to buy them today). Appraisals for inheritance

117

purposes should reflect their current cash value if you were to sell them into the marketplace all at once. The fact that these valuations, each perfectly valid, would be different underscores the argument that there is no one true "value" for any map.

MAPS AS INVESTMENT

Finally, we deal with this question: Are maps good investments?

A map collection pays dividends in the form of intangible benefits. Enjoyment, satisfaction, pleasure, beauty. Collectively, this is a very nice return on any invested capital.

Maps have appreciated in recent years and they are worth much more than they cost, say, ten years ago. We have never appraised a *good* collection that was worth, in its entirety, less than its cost. Most collections, built over a ten- or twenty-year period, have appreciated substantially. Certainly part of this may result from the occasional "find" that every serious collector comes across from time to time, but most of the appreciation has resulted from the actual market price increase of the maps themselves. Does this mean that maps make good

financial instruments? Our personal opinion is "probably not." Don't buy maps for this reason. Buy them because you like them; buy them because you enjoy them, and buy good ones. Your ultimate financial profit will be an added dividend. Never forget that you are buying at retail and, if you have to sell, you will be selling at wholesale. You need to be in the market for a long time to offset this effect.

Historically, good collections that have had time to "ripen" (say, ten years or so) have almost always sold for more than they cost. But to note the disclaimer so often used in the financial world, "past performance is no guarantee of future profits." Nonetheless, the track record is pretty good, especially for the higher-quality maps. Buy the best you can afford: it is always easier to sell high quality at high price than poor quality at low price. Do not, however, expect to sell a map or collection in three or four years for a profit, except in rare cases.

But, most of all, buy what you like, buy what you enjoy, and the best profit of all will be yours.

Pleasure.

Curious miscellany – 8

An unusual image relating to maps or mapmaking

Monsters are part of many early maps. This is a detail from a woodblock plate depicting terrestrial and (mostly) aquatic monsters that appeared in Münster's *Cosmographia* published in the 16th century. The detail is shown full size and the entire plate measures 10.25 x 13.5 inches (26 x 34.4 cm).

Part II

The maps collected

Chapter 9
An Introductory Survey of the Diversity of Printed Maps

The specific purpose of this section is to illustrate the changing appearance of printed maps over their 500-year history. The maps shown here were selected to provide visual demonstration of the way the overall appearance of maps has changed through this time.

This section is not intended to chronicle the mapping of any area or areas: others have already done this and done it well. Most map books, however, emphasize the rare, very important and very costly at the expense of the vast numbers of other maps that are available, important, modestly priced, and that fall into the "neat map" category. Thus, it comes as a surprise to some beginners to map collecting that many maps from the dawn of printing are not only available, but are plentiful.

Past emphasis on "trophy maps" implicitly demeans the rest of the printed maps and tacitly suggests that they are less worthy of being collected. I disagree with this and aim here to present a more egalitarian view of mapdom. It is true that some of the most important very early maps are quite scarce and expensive and some, indeed, are known in only a few copies; nonetheless collections of early maps can still be assembled at reasonable cost. In my own business I often get telephone calls from beginners asking about the availability of "old maps." Frequently these callers think of early-20th-century maps as being "old" and are intrigued when they realize that for about the same price as an ordinary 19th-century American state map, they can have a 1560s map. True, this does little good if the individual

wants to collect maps of Ohio, but nonetheless, it illustrates the point that very old maps are available at affordable prices. Many of these specimens from the early days of Western printing have an intrinsic interest simply because they appeared so early.

I have tried to include only those maps that are available, with some degree of regularity, on the world's markets.

Obviously no claim to completeness can be made, but I have tried to assemble a sequence of images that present the salient points of the visual evolution of printed maps.

Some experienced collectors will find fault with the brevity of this section. Others will surely disagree with my choice of maps and, as a corollary, those I have omitted. Choices have been made deliberately. I originally wanted to use only one hundred maps. Reason intervened and I agreed to more.

Some of the descriptions will contain a reference, indicating where you can find a more detailed discussion of the map. These references, such as Burden 64, refer to the reference book (in this case Philip Burden's *The Mapping of North America)* and the map number or page number. In those instances where an author has more than one book, I have given a short-title for clarity. All the reference books used for this purpose are listed in Appendix B.

Join me, now, in this illustrated, guided tour through 500 years of mapmaking. I hope you will find new friends here! Savor the flavors and how they change during several hundred years of mapmaking.

1482—*An Early Printed World Map*

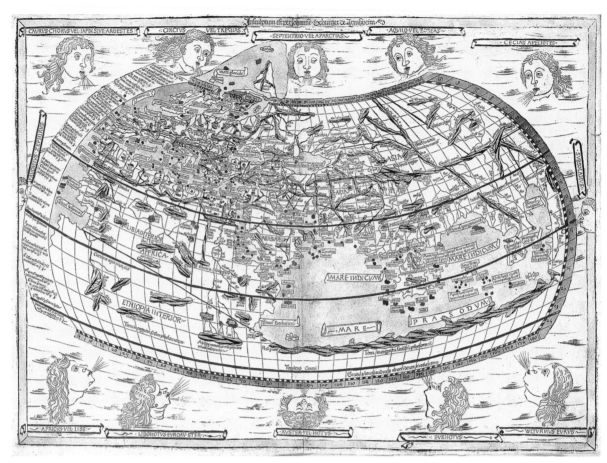

Nº 1   **[Ptolemy, Claudius.]** Ulm: 1482. Woodblock world map. Image is 15.75 x 21.75 inches (40 x 55 cm). Colored.

This is a splendid, rare, very early world map, based on Ptolemy's geography. It is from a work that is apparently the first edition of Ptolemy printed with woodblocks, and the first to be printed north of the Alps. The block is signed by the cutter, Johann. Some maps, printed from another block, lack this signature. There are only a very few known exemplars and this "unsigned" map is even rarer.

Shirley, *World* 10, 11.

122

1493 – *The Schedel World Map*

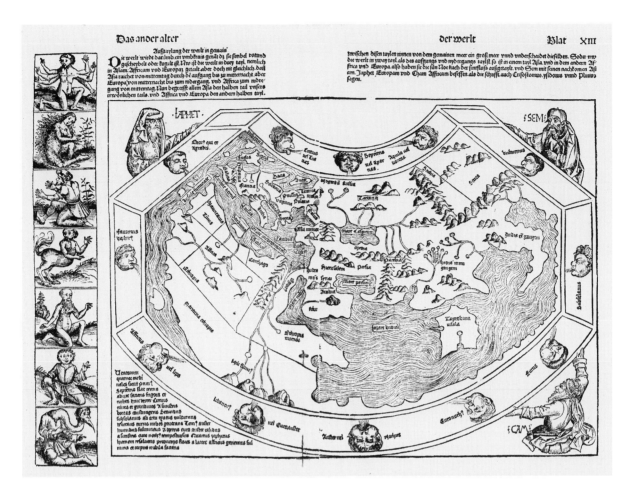

Nº 2   **Schedel, Hartman.** Nuremberg: 1493. Woodblock world map from the Nuremberg Chronicle, German edition. Image is 14.5 x 20 inches (37 x 51 cm). Uncolored. Overall sheet is 17.5 x 24 inches (43 x 61 cm).

Verso has German text and woodcut illustrations, most of which are of grotesque and malformed people. This map shows the world known to Columbus and is one of the earliest collectable world maps. It is also one of the great intellectual constructs linking the medieval world with that of the age of discovery. Jerusalem is at the center and we note the monsters, medieval inventions, arrayed vertically along the left border. The map is surrounded by windheads and woodblock images of Japhet, Shem and Ham, the sons of Noah, representing a religious iconography relating to the post-diluvian state of the world. The cartography itself reflects the total known world just prior to the discoveries of Columbus and Diaz. We note the heavy reliance on Ptolemaic maps and also the acceptance of unproven theoretical landmasses, such as the connection between Asia and Africa that creates a landlocked Indian Ocean. The map is without a scale, latitude or longitude. Although often found with repairs and centerfold defects, this map should be considered by anyone building a collection of world maps.

Shirley, *World* 19.

1493 — *The Schedel Map of Europe*

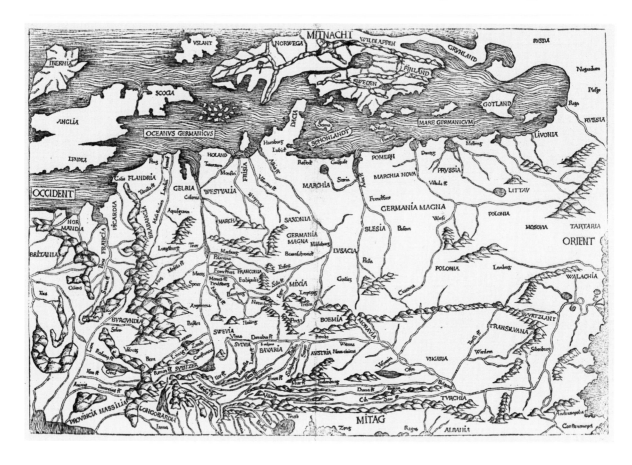

Nº 3

Schedel, Hartman. Nuremberg: 1493. Woodblock map of Europe from the Nuremberg Chronicle, Latin edition. Image is 15.3 x 23 inches (39 x 58 cm). Uncolored.

Since this map was bound at the end of the book it is usually found with much more damage than the world map. It is the last great Europe map made before the world was changed forever by Columbus. In an almost atavistic woodblock manner, Northern and Central Europe, and their mountains and rivers are shown in a schematic fashion. Constantinopel (sic) is in the far lower right; Ireland (Ibernia) in the far upper left. Bits of the Mediterranean intrude along the lower margin but none of the southern countries are included (or the western reaches of France). We see, perhaps, the view of Europe held by the Nuremberg publishers, perhaps reminding us of the famous *New Yorker* magazine cover depicting a New Yorker's view of the United States.

124

1493 – *Schedel City Views*

Nº 4,5   **Schedel, Hartman.** Nuremberg: 1493. Woodblock city views from the Nuremberg Chronicle. Top: *Constantinopel*. Image is 9.25 x 20.75 inches (23.5x53 cm). Bottom: *Roma*. 9 x 21 inches (23 x 53.5 cm). Both are double-page views and share the page with text (not shown here).

Although many of the city views in the Nuremberg Chronicle are "generic" views that were used several times with different identifying captions, the large ones such as those shown here are representational. Views of the major cities are much sought after; views of the lesser cities are priced surprisingly low.

Views with original color are very scarce. The verso of these views sometimes contain interesting woodblock prints, for example, the Rome view has a large woodblock view of Genoa on its verso. They often have repairs to the centerfold.

MAPS AFTER 1500 – A TIME OF TRANSITION

The 16th century began with portrayals of new lands, new maps of old lands, and a continuation of the tradition of showing old geography juxtaposed to new. In the latter part of the century, when map printing was a hundred years or so old, it had matured to the point where many different types of maps were being produced. Some of these maps are now are now extremely rare, some are staggeringly expensive; most are neither. We can recognize the germs of later maps in maps of this period, and they express well the intellectual continuity of the science. Indeed, many of the 16th-century maps retain an archaic appearance. The woodblock technique waxed and then waned and copperplate reigned by the time the century was passing.

Many of the maps of the 16th century are Ptolemaic maps, but we see an increasing emphasis on "modern" maps. Of course, we begin to see the appearance of maps showing the New World, and it is always surprising to see how rapidly the Americas were mapped. Asia and the East Indies were explored and mapped and, despite their distance from Europe both in miles and in time, maps of these places kept pace with discovery.

There are many early-16th-century mapmakers whose names may not be very familiar to beginners at map collecting, but who, nonetheless, were giants of the period who produced some of the germinal maps of the world.

By the time the century was running out, two of the world's great mapmakers appeared on the scene. Both were Dutch. Ortelius, working in Antwerp, and Mercator, working in Amsterdam, produced some of the most important (as well as beautiful) maps and their names are engraved in the history of cartography as indelibly as the lines on their copperplates. Maps by both makers are attractive and much sought after, and their names are virtually household words, among even neophyte collectors. Their maps of the world and continents are among the most collected and recognized trophy maps. Not all of their maps are expensive, however. Prices drop rapidly as one looks at the less-collected areas of the world, and many truly beautiful maps are readily available from most dealers for modest sums. The collector is urged to look at some of the less widely collected maps from this period – for they represent some of the most interesting bargains in the world of old maps.

I have included many maps from this period since it is a foundational century of mapmaking.

1517 – *The Universe*

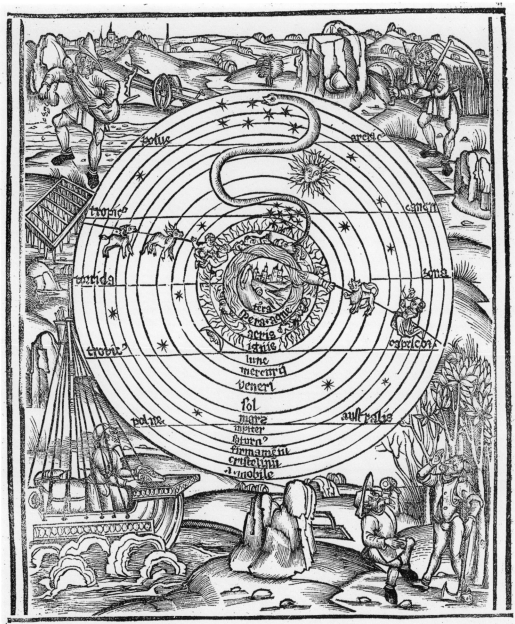

Nº 6   Lyons: 1517. Woodblock map of the universe. 7.25 x 6.1 inches (18.5 x 15.5 cm). This diagram of the universe appeared in an edition of the works of Virgil printed in 1517.

This is a complex map of the geocentric pre-Copernican universe. The concentric rings represent the orbits of the planets, moon and sun and various celestial spheres. The map has the ecliptic superimposed. In addition, it shows the tropics. There is a rich iconography in the corners.

Such maps of the universe, representing a Western model combining observation and belief, were rather common. For example, the Nuremberg Chronicle has a series of such diagrams covering all the days of Creation and becoming sequentially more complex, or "complete." This one was used in the early 16th century, appearing in several editions of Virgil's *Aeneid*.

1528 – *Bordone*

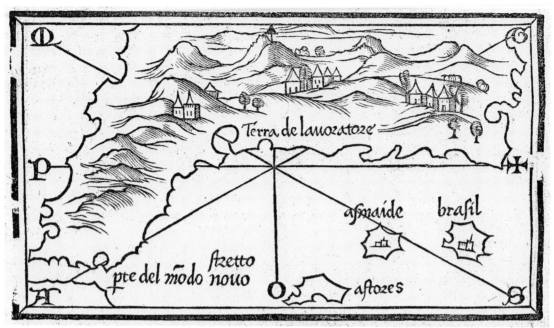

N° 7, 8   **Bordone, Benedetto.** Venice: [1528]. Woodblock map of America (above) and Japan (below). Illustrated full size. Uncolored. From *Libro di Benedetto Bordone...de tutte l'Isole del mondo...*

Although it is difficult to reconcile Bordone's depiction of the coast of North America with current maps, this is an important image. The mythical islands of Brasil and Asmaide are shown, as are the Azores.

The Japan map is the first printed separate map of Japan to appear in a Western book. Note the show-through from the text on the verso. Both maps, although scarce, are still obtainable.

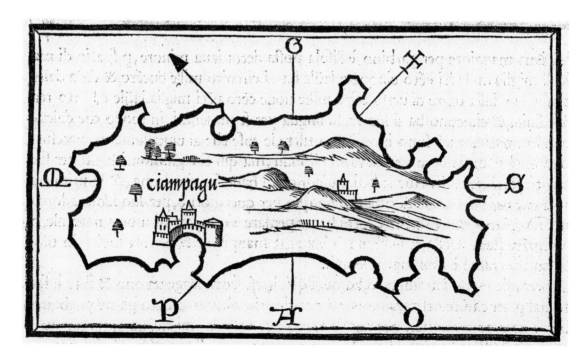

128

1522-1541 – *Fries' World*

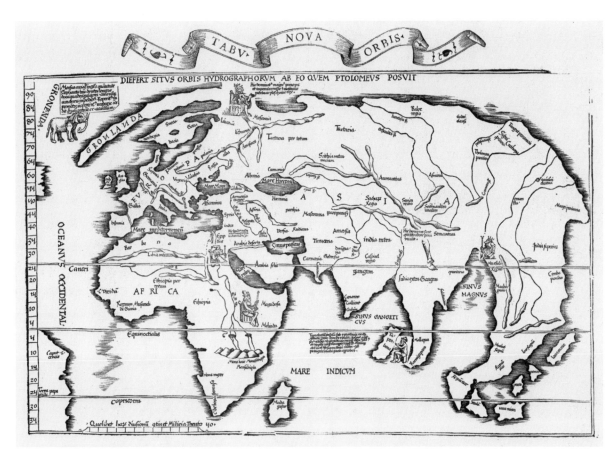

Courtesy: T. Suárez

Nº 9   **Fries, Laurent.** Strasbourg: [1522]. Woodblock world map. 11.25 x 17.75 inches (28.5 x 45.5 cm). Uncolored. From *Orbis Typus Universalis...*

This map is based on the Waldseemüller map of 1513. Laurent Fries published an edition of Ptolemy in Strasburg and included maps based on Waldseemüller's earlier work, but somewhat reduced in size.

In the far western reaches, we note small parts of America.

This popular map is one of the earliest world maps showing America that is still accessible to the present-day collector. It went through several editions through 1541 and

they can be distinguished by the way the title is presented above the map. In our illustration, the title is in a banner and we note part of the neatline in the upper right missing. This identifies this map as the 1535 issue. The 1522 was titled *Tabu. Gran Russie;* the 1525 had no title and the top right neatline is missing for the first time. The 1541 issue has no banner; the title is *Tabula nova totius.*

Shirley, *World* 49.

1535 – *Fries' South Africa*

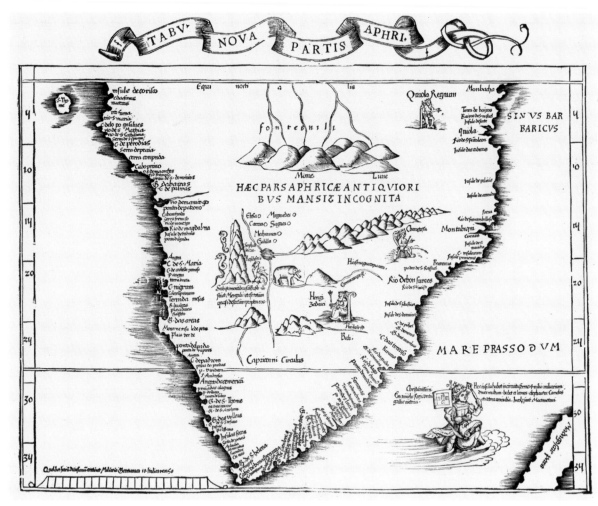

Nº 10 **Fries, Laurent.** Lyons: 1535. *Tabu. Nova Partis Aphri.* Woodblock map. 11.75 x 16.5 inches (30 x 42 cm). Uncolored.

This map of southern Africa is based on the Waldseemüller map of 1513. As with the rest of Laurent Fries' maps, they were basically reduced versions of the earlier issue of Ptolemy. This map, of course, is a "modern" map, since the south of Africa was unknown to Ptolemy.

The interior is still very rudimentary.

The King of Portugal is riding a bridled sea monster.

This map illustrates the point that very early maps of many parts of the world are quite available to today's collectors.

As in the preceding map, there are several editions. Norwich, *Africa* 150.

1548 — *Macrobius' World Map.*

N° 11   **Macrobius.** Lyons: 1548. Woodblock world map. 3 x 3 inches (8 x 8 cm). In: *Macrobii Ambrosii Aurelii Theodosii, Viri Consularis, & illustris, In Somnius Scipionis, Lib II. Saturnaliorum, Lib. VII.*

This world map is based on the concepts of Ambrosius Theodosius Macrobius (fl. 399-423 A.D.), a Roman geographer. The map is part of Macrobius' commentary on Cicero's *Dream of Scipio.*

The map represents the concept that the earth is divided by two ocean girdles encircling the globe.

Although the map illustrates only the known world, it does show the hypothetical antarctic continent. It also shows a circumnavigable Africa and an open Indian Ocean, concepts that were later to prove very important in the history of exploration.

The map was reproduced many times. Here, it is illustrated in a 1548 edition of Scipio's *Dream* and is shown full size, still within the book.

ca. 1548 – *Münster's World*

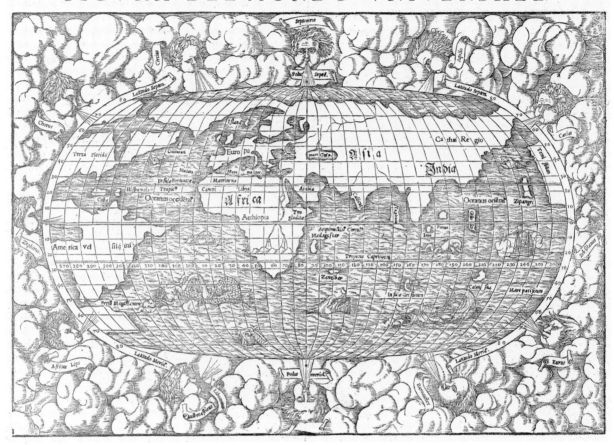

Nº 12   **Münster, Sebastian.** Basle: c. 1550 or later. *Figura Del Mondo Universale.* Woodblock world map. 10.25 x 15 inches (26 x 38 cm). Uncolored.

The woodblock world maps that appeared in Sebastian Münster's *Cosmographia...* are still readily available, but their prices have been going up and the quality of the available exemplars has been gradually declining.

There are two Münster "modern" world maps, made from different woodblocks. The one shown above is the second, appearing after 1550. It is on a oval projection surrounded by windheads and clouds. The woodcutter's initials "DK" are seen in the lower left.

Münster's *Cosmographia* was a popular book and appeared in many editions and languages. Some of them were printed on rather inferior paper.

Most of these maps were not colored at the time they were printed and many of the colored specimens we have seen on the market have been colored later.

Shirley, *World* 92.

1548 – *Münster's Europe*

MODERNA DISCRIZZIONE DELL' EVROPA

Nº 13　**Münster, Sebastian.** Basle: c. 1550. *Moderna Discrizzione Dell' Europa*. Woodblock map. 10 x 13.25 inches (25 x 34 cm). Uncolored.

Sebastian Münster's *Cosmographia...* contained numerous double-page woodblock maps. This dramatic map of Europe shows the continent with south at the top, demonstrating the more fluid approach early mapmakers had to orientation. The "north at the top" tradition had not yet become fixed. The *Cosmographia* appeared in many editions and in different languages. The blocks show varying amounts of wear and several were recut as the century progressed.

This map of Europe shows mountain ranges in the old "molehill" style. Forests are suggested by little groups of trees and rivers by pairs of lines. There is a splendid period sailing ship in the Atlantic off the coast of France. The dark spot on the ship's hull appears to be old candle wax that had dripped on the map in the distant past.

1548 – *Gastaldi's South America*

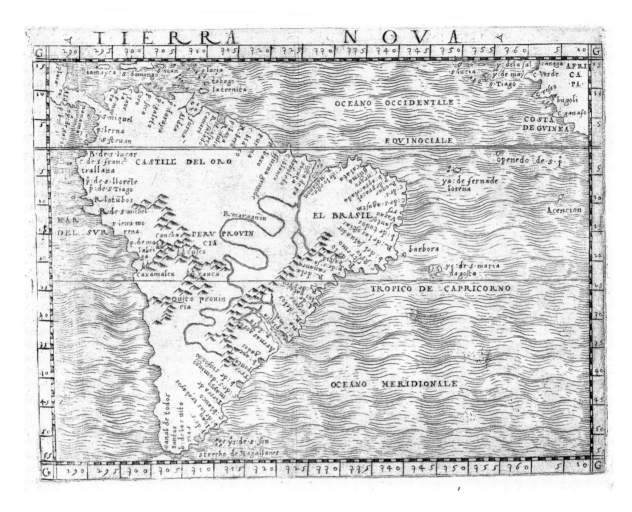

N° 14   **Gastaldi, Giacomo.** Venice: 1548. *Tierra Nova.* Copperplate engraved map of South America. Image is 5 x 6.75 inches (12.5x 17 cm). Uncolored.

The little Gastaldi atlas is quite scarce. It is considered by many to be one of the first true atlases. The maps are quite attractive, with typical wavy lines for the sea. They were not printed with any degree of uniformity and one often finds the impressions uneven. The plates were not uniformly inked. Within the same atlas, some plates can be very dark while others are very light.

This is the first printed map showing a separate South America that can be collected.

Gastaldi was one of the 16th century's great mapmakers, responsible for this little volume that broke definitively

with the Ptolemaic tradition. Despite paying lip-service to the tradition of reconciling "modern" geography with Ptolemaic geography, these maps clearly deviated from the classical pattern and may be considered an intellectual watershed in cartography.

There are many "firsts" that can be claimed for this atlas. It is the first 16th-century atlas to use the copperplate technique, the first atlas in Italian, and the first "pocket" atlas.

1548 – *Gastaldi's Mariner's Map*

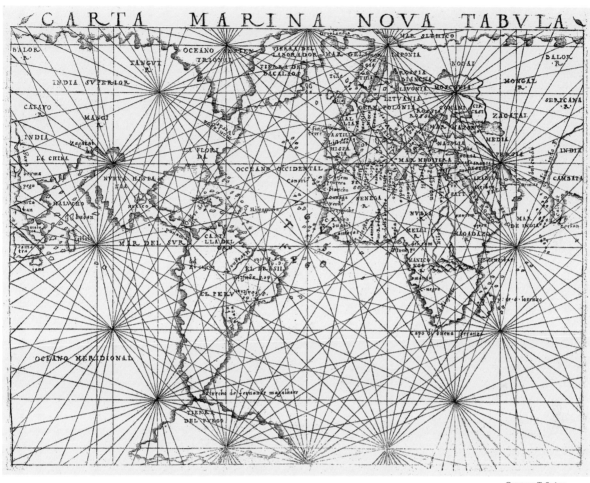

CARTA MARINA NOVA TABVLA

Courtesy: T. Suárez

Nº 15   **Gastaldi, Giacomo.** Venice: 1548. *Carta Marina Nova Tabula.* Copperplate engraved map.
Image is 5 x 6.75 inches (12.5 x 17 cm). Uncolored.

Known as the "Mariner's Map," this map is on a scale much too small to be a meaningful or practical sea chart. It is nonetheless crisscrossed with rhumb lines, creating a seaworthy appearance through a spiderweb of nautical images. The Mariner's world map is particularly interesting. All the continents are linked by means of connections between Asia and North America: North America with Greenland; Greenland with Europe; and Europe with Asia.

California is a peninsula. This map predates the discovery of Hudson's Bay and the Great Lakes, and the permanent Northeast American settlements. Nonetheless, the area of New England contains *Montagna Verde*, where today's Green Mountains are!

A slightly larger version of this map appeared in all of the editions of Ruscelli, from 1561 to 1599.

c. 1550 – *Münster's City Views*

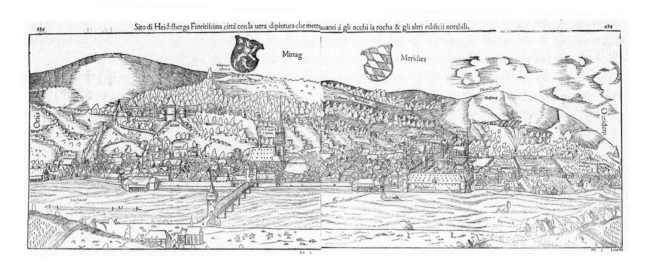

N.º 16, 17

Münster, Sebastian. Basle: c. 1550. Woodblock city view. Top: *Heidelburg*, 10 x 28.5 inches (25.5 x 72.5 cm). Below: *Colmar*, from a Latin edition of Münster, measures 9 x 13.5 inches (23 x 34 cm). Both views are uncolored.

Münster's *Cosmographia* had numerous city views. Some were relatively small text woodcuts, others were long panoramic fold-out views. The larger views are well-executed woodblock prints, often with titles in banners flying high above the cities. Most double-page views measure about 23 x 35 cm (9 x 14 inches) but some are only half as high with letterpress text below. Several are multiple sheet fold-outs, such as Vienna, 9 x 30.75 inches (22.5 x 78 cm). Single-page views are generally about 13 x 16 cm (5 x 6 inches). Some editions had a small view of Mexico City.

Most views were issued uncolored and it is likely that most colored copies have later color. It is

important to note that the title, generally printed above the woodblocks, was letterpress and appeared in the language of the edition. Thus, similar views appear with titles in different languages. The *Cosmographia*, a large book, was in print well over fifty years. The number of copies produced must have been staggering. Doubtless, in an effort to keep the cost down, many were printed on very poor paper. This paper is often coarse and of uneven thickness. Consequently, some of the images are rather poor impressions. It is unfortunate that many people equate some of the Münster images with poor quality. Exemplars printed on better paper are simply wonderful. Centerfold darkening is common, as is show-through from the text on the verso. Prices depend more on the city and condition than on the date of issue.

c. 1550 – *Daniel's Dream*

Courtesy: T. Suárez

N° 18   **Solis, Virgil.** c. 1560. *[Daniel's Dream].* Woodblock world map. 3 x 4.5 inches (7.6 x 11.5 cm), map only. Early color.

Bibles often contained maps, especially illustrating the Holy Land and the world. This little woodblock map illustrates Daniel's dream of the four kingdoms (*Daniel* 7), and was designed by Virgil Solis. It appeared first in a bible published in Frankfurt in 1560. The four beasts, symbolizing the earthly kingdoms, are shown, not arising from the sea, but rather on continents. The map is restricted to the Old World and the continents are much distorted. There is a ship in the left and we note that the windheads are all at the top.

Many of the Daniel's Dream maps are very similar and it can be difficult to distinguish between them.

1561 – *Zeno Map*

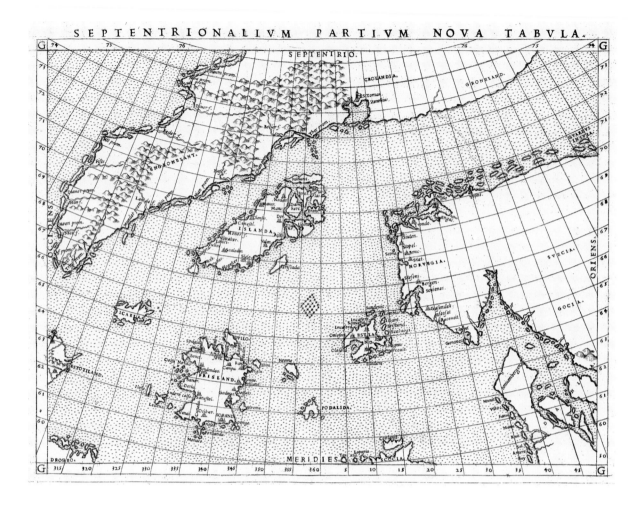

Nº 19 **Ruscelli, Girolamo.** Venice: 1561. *Septentrionalium Partium Nova Tabula.* Copperplate engraved world map from *La Geografia de Claudio Tolomeo...* (editions 1561-1599), 7 x 9.5 inches (17.7 x 24.2 cm). Uncolored.

In this map, the North Atlantic is littered with mythical islands, some traceable to the Zeno legend. In 1558 Marcolino published a book by Nicolò Zeno in which the "discovery" of several of these islands was noted. The book was allegedly a compilation of northern voyages by Zeno's ancestor made over a hundred years earlier. Zeno himself edited this map for Ruscelli's geography. The information on this map was used later by Ortelius in his map of the northern regions of the Atlantic.

Among the mythical islands shown, Frisland in particular had a long cartographic life. Many later maps continued to show it, some with exquisite detail and even locating its capitol city.

This map shows some famous cartographical misconceptions. Along with maps showing California as an island, the Prester John maps of Africa, the wonderful Schlarraffenland and perhaps the Kircher map of Atlantis, this map should be part of the backbone of a collection of maps of mythical geography.

1561 – *Ruscelli's World*

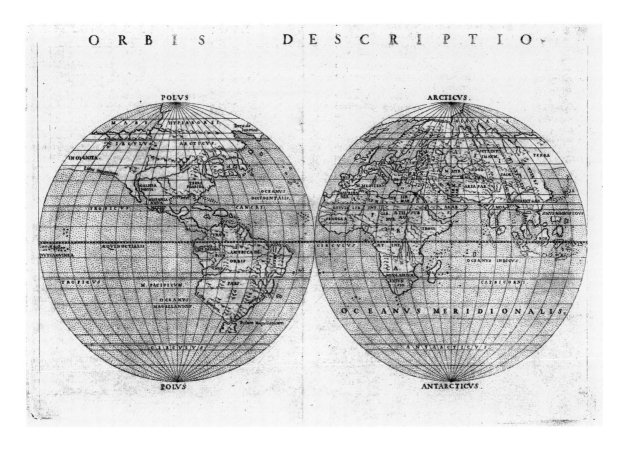

N° 20   **Ruscelli, Girolamo.** Venice: 1561. *Orbis Descriptio.* Copperplate engraved world map from *La Geografia de Claudio Tolomeo...* (editions 1561-1599). Each hemisphere 5 inches diameter (12.5 cm). Uncolored.

Ruscelli's editions of Ptolemy included a Ptolemaic world map and a double-hemisphere "modern" world map. This is the first printed double-hemisphere world map to appear in an atlas. Ruscelli translated the Ptolemies into Italian and had a new set of maps prepared. Various editions, from 1561 to 1599, contain many of the same maps with some alteration and additions in the final editions.

The double-hemsiphere map shown here is the first state, from the first (1561) edition. Later states include a New Guinea and a large antarctic continent. There are other minor changes, particularly among the islands.

Ruscelli's work also included a rectangular "modern" world map, the so-called Mariners' Map, illustrated in this section.

Shirley, *World* 110

1563 – *Ramusio's Africa*

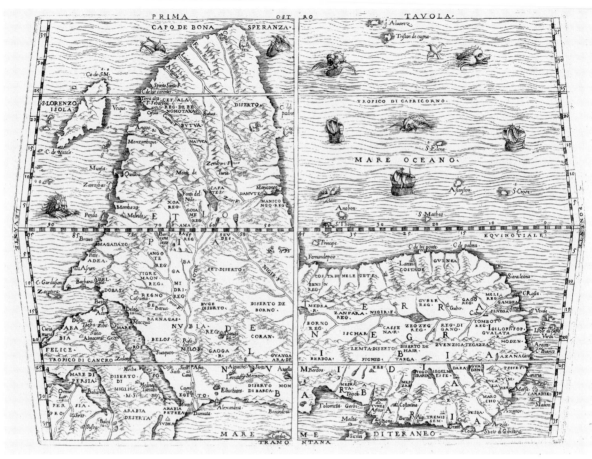

Nº 21   Ramusio, Giovanni Battista.   Venice: 1563. [Africa] *Prima Tavola*. Copperplate engraved map. 10.75 x 15 inches (27.5 x 38.2 cm). Uncolored.

There is an earlier woodblock version of this map, first published in 1554, but the block was lost in a fire. This is a copperplate engraved copy of that original map and first appeared in the third (1563) edition of Ramusio's *Delle Navigationi et Viaggi*.

Showing many of the early misconceptions about Africa, such as the origin of the Nile, this is a germinal map for any

Africa collector. It is based on information from two Arab geographers (Edrisi and Leo Africanus) and retains some of the Ptolemaic ideas. Despite appearing in Ramusio's book, the map is actually by Gastaldi.

One of the more dramatic "south-at-the-top" maps, it has great visual appeal.

Norwich, *Africa* 6.

1565 – *Ramusio's Hemisphere*

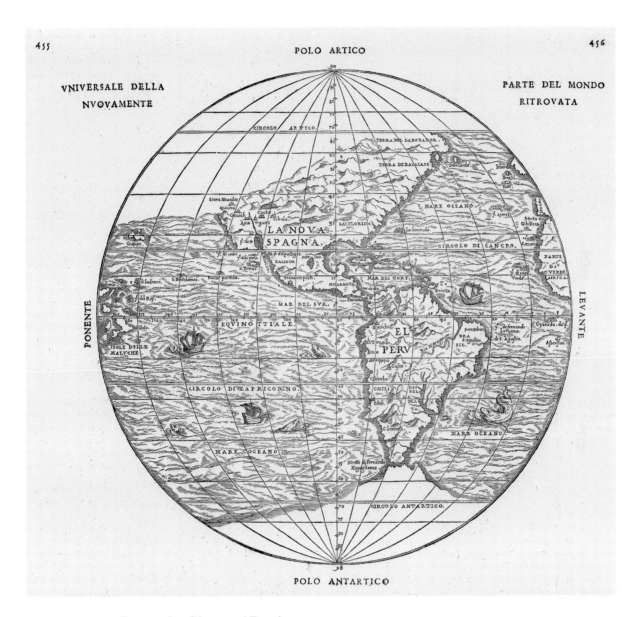

POLO ARTICO

VNIVERSALE DELLA
NVOVAMENTE

PARTE DEL MONDO
RITROVATA

PONENTE

LEVANTE

POLO ANTARTICO

Nº 22   **Ramusio, Giovanni Battista.** Venice: 1565. *Universale Della Parte Del Mondo Nuovamente Ritrovata*. Woodblock map. Second state. Diameter of hemisphere is 10.5 inches (26.7 cm).

A wonderfully up-to-date map, this hemisphere was perhaps the best available map of the period, although the name America does not appear (North America is called "La Nova Spagna" and South America "El Peru"). Ramusio has removed the False Sea of Verrazzano that is so prominent on the Münster map of the same period. California is a peninsula and there is good detail in the region. The map shows the Sierra Nevada mountains discovered only a few years earlier. We note the rather good depiction and location of Japan. Indeed, among the firsts for this map appears to be the use of the name Giapam for Japan: compare this with Münster's map of the Americas (Map 25). Finally, we note that this map is an early example of a map wherein the purported continuity of the American and Asian coasts is doubted. The definitive break will occur a few years later. A germinal map. Burden 34.

141

1567 – *Gastaldi's Italy*

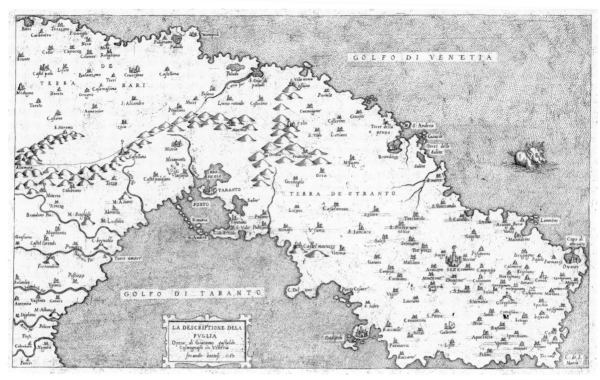

N° 23   **Gastaldi, Giacomo. (Bertelli).** Venice: 1567. *La Descriptione Dela Puglia.* Copperplate engraved map. 8.25 x14 inches (21 x 35.5 cm). Uncolored.

After Gastaldi's death (c. 1565) his beautiful map of the heel of Italy's boot was published by Bertelli. A very detailed map with many place names including Taranto and Brindisi *(Brandizzo)*, which was the ancient Roman port.

This is a splendid example of the Italian Lafreri-type map.

A delightful little creature lives in the sea and is shown twice actual size in the picture at the right.

1572 – *Benito Arias Montanus' World*

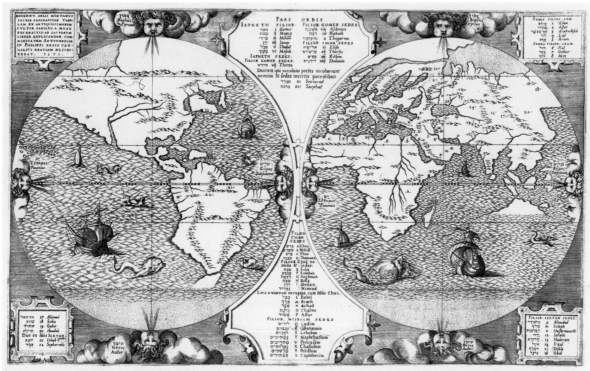

N⁰ 24 **Montanus, Benito Arias.** Antwerp: 1572. *Benedict. Arias Montanus Sacrae Geographiae Tabulam...* Copperplate engraved world map. 12.5 x 20.75 inches (32 x 53 cm). Each hemisphere is 10.25 inches (26 cm) diameter. Uncolored.

This map is an example of a bible map. It is from an eight-volume polyglot bible edited by Benito Arias, who was also called Montano because of the place of his birth.

The first edition was printed in 1571, but virtually all of the copies were lost when the ship transporting them to Spain went down. The map shown here is from the second edition and is also the second state of the map, with the word "gentes" added below the word "Iektan" in the lower left panel.

Nicely engraved, the map was made to illustrate theological issues, especially the dispersion of the tribes of Israel. It has received much attention because of the landmass shown in Australia's location, well before European knowledge of Australia. In all likelihood, this was simply one of the engraver's quirks and does not imply a much earlier knowledge of Australia.

The map is very attractive, with nicely engraved oceans filled with monsters and ships. Well-executed windheads surround the hemispheres. Text is in Hebrew and Latin.

c. 1574 – *Münster's Americas*

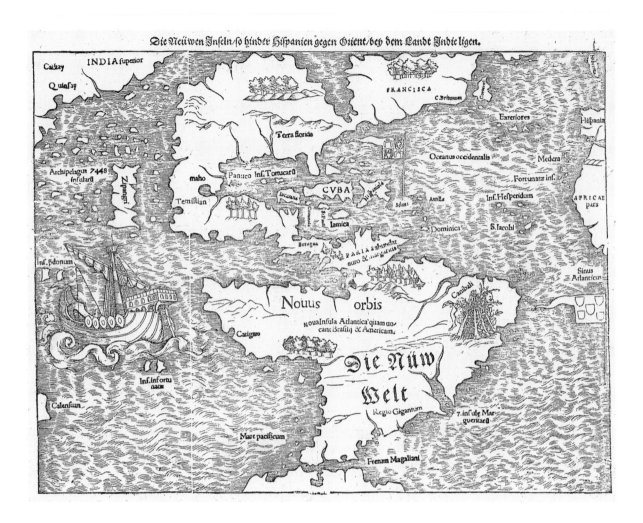

Die Neüwen Inseln/ so hinder Hispanien gegen Orient/bey dem Landt Indie ligen.

N° 25   **Münster, Sebastian.** Basle: c. 1550. *Die Neüwen Inseln/ so hinder...* Woodblock map. 10 x 13.25 inches (25.4 x 34 cm). Uncolored.

Perhaps the high point of all the woodblock maps by Münster is the Americas map. From 1540, editions of Münster's *Geographia Universalis* and later, until 1580, his *Cosmographia* contained this famous New World map. Showing both North and South America, this is apparently the first printed map to show the Americas as separate continents. North America has a very narrow appearance because the false Sea of Verrazzano pokes in from the Pacific, almost bisecting the continent. Verrazzano thought that the waters of the Chesapeake Bay were the Pacific and this error was continued on maps for some time. The map has several other wonderful features. The ship in the lower left represents Magellan's. Japan, here called Zipangri, is close to the Pacific coast of America.

The first state of this map lacks the German "Die Nüw Welt" that appears on the map above. Prices for this map have risen steadily, with the demand for the first state justifying a higher price as compared to the later states. The text on this map was printed from moveable type, and the map had different titles in different editions. There is a single edition (Basle: 1552) that has latitude and longitude grids. Original color is very rare on these maps and most of the color one encounters is later. Münster's *Cosmographia* had about forty editions and there are many states to this map, most determined by the different words printed on the map by means of moveable type.

Goss, *North America* 6; Schwartz & Ehrenberg 18; Kershaw 1; Burden 12.

144

c. 1575 — *Münster's Spain*

N° 26   **Münster, Sebastian.**   Cologne: c. 1575. *Della Region Spagnola Nuova Discrizzione.*
Woodblock map. 10 x 13.25 inches (25 x 34 cm). Uncolored.

A nice early map of the Iberian Peninsula. Mountains
are shown using the archaic "molehill" symbols. Coastal
detail is exaggerated. The map has latitude marks but omits
longitude. The panel in the lower right has descriptive text.

The mate to this map is the Ptolemaic version with
typical truncated Ptolemaic projection.

c. 1580 – *Guicciardini's City Views*

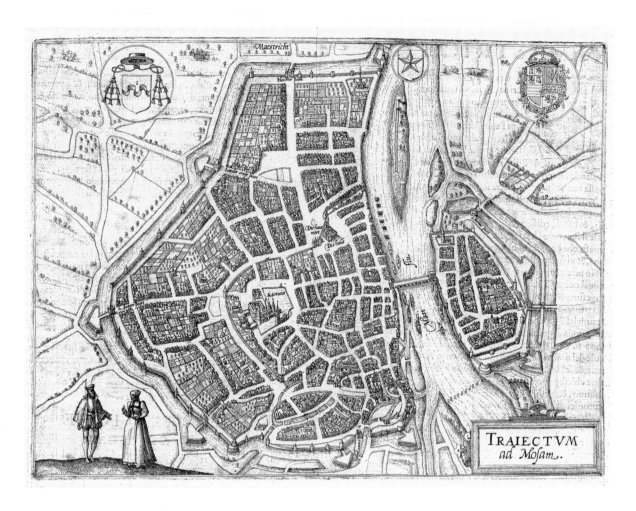

N° 27   **Guicciardini, Ludovico.** Antwerp: c. 1580. *Traiectum ad Mosam.* Copperplate engraved city view. 9.5 x 12.5 inches (23.5 x 32 cm). Italian text on verso. Uncolored.

Lodovico Guicciardini produced a series of copperplate engraved town plans of the Low Countries that appeared in *Descrittione di tutti i Paesi Bassi.* This was a popular book and there are editions from 1567 to 1660. It is difficult to determine from which edition any given individual sheet was removed. Many of these are beautiful examples of copperplate engraving and printing. We often find them printed from plates that were not fully "wiped" giving them a beautiful, soft background. They are usually uncolored and most of the colored examples acquired their color recently. These maps are not as widely collected as the views by Braun and Hogenberg and they generally sell for much less. Indeed, some of the Braun and Hogenberg images appear to be derived from the Guicciardini views.

The early editions were published by Plantin in Antwerp, the later ones by Jansson in Amsterdam.

146

c. 1580 – *Braun & Hogenberg's City Views*

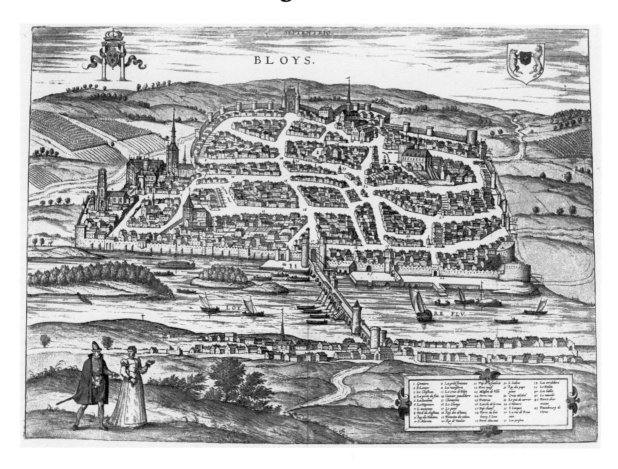

N° 28   **Braun, Georg & Hogenberg, Frans.** Cologne: c. 1580. *Bloys.* Copperplate engraved city view. 12 x 17.25 inches (30.5 x 44 cm). Uncolored.

Some of the earliest printed city views were part of Schedel's Nuremberg Chronicle. Many other city views appeared later, especially in the 16th century. The city plans and views by Georg Braun and Frans Hogenberg are perhaps the most sought after and famous of all 16th-century city plans. These were part of the great illustrated work *Civitates Orbis Terrarum* that was first published in Cologne in 1572. The complete work was in six volumes with publication dates of 1572; 1575; 1581; 1588; 1598 and 1617. Many editions followed, and some of the plans were issued as late as 1710. Not all countries were represented equally in this work. There are 118 German views but only 2 American views (Cuzco and Mexico city on one plate). There were 49 Iberian, 20 British, 6 Asian, and 18 African views. Belgium, France and Holland had 48, 39 and 55 views respectively. In all, there were slightly under 550 views and plans. As with most such images, the Braun and Hogenberg views of well-known cities (London, Paris, Amsterdam, Rome, Frankfurt, Vienna, etc.) are most in demand, having the largest global appeal, and consequently are quite

c. 1580 – *Braun & Hogenberg's City Views*

N⁰ 29 **Braun, Georg & Hogenberg, Frans.** Cologne: c. 1580. *Santander.* Copperplate engraved city view. 12.5 x 14 inches (31.7 x 35.5 cm). Uncolored.

expensive. Views of the smaller, less well-known towns have a specialty market and sell quite well in Europe, but at lower prices.

We note that many of the views of lesser-known cities also find a market in America where, because of their European ancestry, many individuals buy images of the more obscure European places.

It is also difficult to identify the edition from which a single sheet might have come. In general the plates from the Cologne editions have smaller margins than those from later editions. Some of the Braun & Hogenberg plates were engraved by Hufnagel, the engraver who did many of the Ortelius maps. These often carry a price premium.

1580s – *Ortelius' Miniature Maps*

Nº 30

Ortelius, Abraham [after]. Plantin, Antwerp: 1588. *Gallia.* Copperplate engraved map. Full size. Uncolored.

Miniature atlases proved to be very popular and, possibly because they were less expensive than large folio atlases, sold well. In particular, pocket versions of Ortelius' *Theatrum*, usually issued uncolored, were very well known and are collected avidly today. The maps in the first versions of the *Epitome* were engraved by Philip Galle and printed by Plantin. Individual maps from these atlases are charming and are reasonably priced. There were many editions in different languages and the publishing history of these maps is complex.

Nº 31

Ortelius, Abraham [after]. Plantin, Antwerp: 1589. *Typus Orbis Terrarum.* Copperplate engraved world map. Full size. The plate shows significant pitting and scratching compared to the plate used for the France map above. Uncolored

King p56

c. 1581 – *Bünting's World*

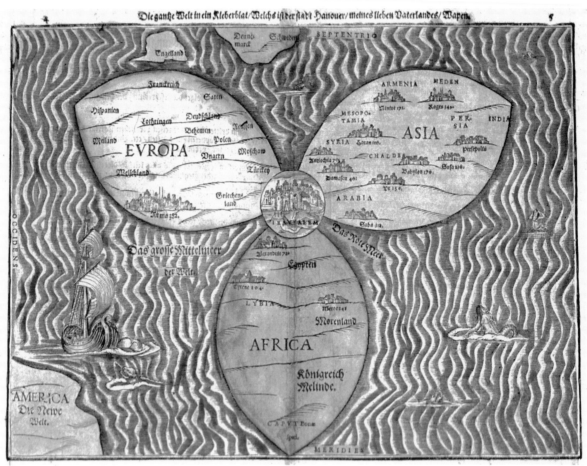

N⁰ 32 **Bünting, Heinrich.** Magdeburg: c. 1581. *Die Gantze Welt in ein Kleberblat...* Woodblock world map. 10.5 x 14.5 inches (26 x 36 cm). Colored.

This woodblock map is in the shape of a clover leaf, representing the Trinity and the tripartite nature of the ancient continents (Europe/Asia/Africa) which were divided among Noah's sons. Jerusalem is at the center.

The new continent of America is not part of the cloverleaf; rather it appears separately in the lower left. England and Scandinavia are also not part of the cloverleaf. It has been suggested that the cloverleaf design not only represents the Trinity, but represents the arms of Bünting's native Hanover.

This map is scarce and is a most curious world map. *Itinerarium Sacrae Scriptura...* in which this map appears was a popular book and was reprinted many times. Since the type was set into the woodblock separately and not part of the block's integral carving, variations are noted.

Shirley, *World* 142

1587 – *Ortelius' Europe*

N⁰ 33 **Ortelius, Abraham.** Antwerp: 1587. *Europae.* Copperplate engraved map. 13.25 x 18.25 inches (33.5 x 46.4 cm). Full original hand color.

The Ortelius Europe is a double-page atlas map with a centerfold running vertically.

If we compare it to the Mercator Europe of about the same age, we see different distortions in the landmasses and also see that the Ortelius map was more definitively a *Europe* map, with detail stopping rather abruptly at the Europe-Asia boundary. In the Mercator map detail runs on into Asia.

N° 34   **Ortelius, Abraham.** Antwerp: c. 1587. *Angliae Regni Florentissimi Nova Descriptio, Auctore Humfredo Lhuyd Denbygiense.* Copperplate engraved map. 14.75 x 18.5 inches (37.5 x 46.8 cm). Full original hand color.

This is a very attractive map with England yellow and the bit of Scotland shown a pale pink. Ireland, largely behind the cartouche, is green. Cities and towns are highlighted in red; forests in green. The elaborate cartouche is fully colored, as is the distance scale in the upper right.

There are five splendid sailing ships and a couple of sea monsters.

This copy has French text on the verso and is the second state.

Shirley, *British Isles* 98.

1590 – *Porcacchi*

N° 35

Porcacchi. Venice: 1590. *Mondo Nuovo.* Copperplate map. 4 x 5.6 inches (10.2 x 14 cm). On full sheet of Italian letterpress text measuring approximately 12 x 8 inches. Uncolored.

Porcacchi's work *L'isole piu famose de mondo*, published in Venice in the latter 16th century and early 17th century, was a true *Isolario* or book of islands. Earlier in the century we had Bordone's woodblock *Isolario* (see Maps 7 and 8) and the century ends with Porcacchi's copperplate work.

In addition to islands, Porcacchi's work contained a modern world map and this nicely engraved North America. The map is shown full size above, and the inset at the right illustrates the entire page of map containing letterpress text and engraved headpiece.

Burden 42.

1595 — *Theodore De Bry*

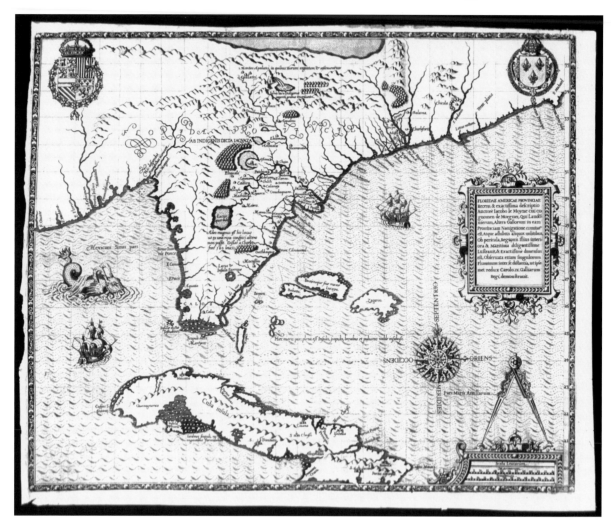

N° 36 — **Le Moyne/Theodore de Bry.** Frankfort am Main: [1595]. *Floridae Americae Provinciae Recens & exactissima descriptio Auctore Iacobo LeMoyne...* Copperplate. 14.25 x 17.75 inches (36.2 x 45.2 cm). Uncolored.

Theodor de Bry, in series of books, published several copperplate maps near the end of the 16th century. In addition to a world map, he had several showing American detail. These maps are usually found with some defects, often being trimmed close as a result of successive rebinding. The plate for this map was a bit large for the paper on which it was printed and many of the copies have been trimmed into the image. The map used for this illustration was photographed against a black background to illustrate the full extent of the margins, which are unusually large for this map. Original color when present is deep and brilliant, making the maps most attractive. Unfortunately, they are scarce and expensive, especially when the condition is good.

A voluminous literature exists on this map and we can but touch on some of the high points. Often considered one of the most attractive and important maps of North America, this map was produced by Theodore de Bry after the original manuscript of Jacques Le Moyne, who had accompanied the colonizing expedition of Ribaut and de Laudonnière to Florida (1562-1565). Issued as part of de Bry's Great Voyages *(Brevis Narratio eorum quae in Florida...)*, the map is notable for showing Port Royal for the first time on a printed map; it also shows Fort Caroline and a number of the major Florida lakes. Most of the attention is focused on the mainland. This is also a major early Cuba and Bahamas map.

1595 – *Theodore De Bry*

N° 37   **LeMoyne/Theodore de Bry.** Frankfort am Main: [1590-1607]. *(The Town of Pomeiooc.)*
Copperplate engraved town view. 11.5 x 8.6 inches (29.2 x 22 cm). Uncolored.

A fascinating view of a 16th-century native Virginia village. This is one of the few very early accurate depictions of American town life at the time the Europeans arrived. "They are in a manner like those which are in Florida, yet they are not so strong nor preserved with such great care..." The temple is at "A" and the king's lodging at "B." The town's water supply is from the pond at "C."

There is only one known state.

c. 1597 – *Quad's World*

N° 38   **Quad, Matthais.** Frankfurt: 1597. *Typus Orbis Terrarum, Ad Imitationem Universalis Gerhardi Mercatoris.* Copperplate engraved map. 8.5 x 12.25 inches (21.5 x 31 cm). Uncolored.

An unusual world map that includes the mythical Atlantic islands: Groclant, Thule, Frischlant and S. Brendam. The western bulge of South America is retained and there is an extensive antarctic continent. Greenland is an island but California is not. A scarce world map. First state.

Shirley, *World* 197.

1598 – *Ortelius' Modern World*

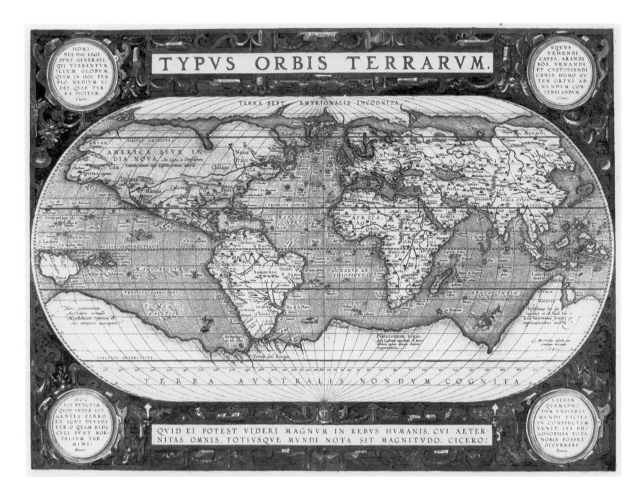

N° 39   **Ortelius, Abraham.** Antwerp: [1587] 1598. *Typus Orbis Terrarum*. Copperplate engraved world map. 14 x 19.25 inches (35.5 x 49 cm). Full (later) hand color.

An unusually attractive world map, this is Ortelius' third (and last) world map, and was used for all editions of his *Theatrum* until the 1612 edition. Although the map has a 1587 date it was not used in an atlas until 1592. There is a second state, very scarce, showing a change in the southern tip of South America.

The corner medallions contain classical text; Cicero and Seneca are represented equally. The Solomon Islands are so marked for the first time on this world map.

Shirley reports that there are some forgeries of this map but I have never seen one.

Shirley, *World* 158.

c. 1603 — *Bayer's Celestials*

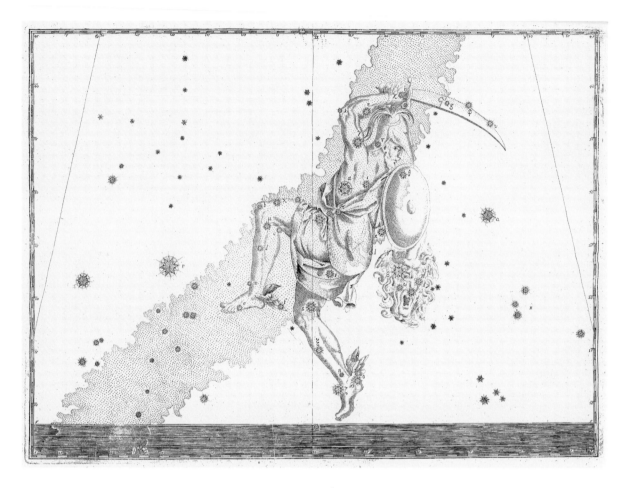

N° 40   **Bayer, Johann.** Augsburg: 1603 or later. *[Perseus.]* Copperplate engraved celestial map. 11 x 15 inches (27.8 x 38 cm). Uncolored.

Bayer's star charts, comprising his *Uranometria, Omnium Asterismorum Continens Schemata...* had a lasting influence. They introduced the use of letters of the Greek alphabet to denote, in descending brightness, the major stars in each constellation. This nomenclature is still in use today.

The charts were derived from a number of star catalogues, including Tycho Brahe's. The plates were republished many times, the last edition being in 1689. A few copies were sold with original or early hand color; these are quite spectacular. We have seen one atlas with early color in which the brighter stars were highlighted with gold leaf. Some editions have text on the verso which is prone to show-through.

Warner p18.

c. **1610** – *Mercator's Europe*

N⁰ 41   **Mercator, Rumold.** Amsterdam: 1610. *Europa, ad magnæ Europæ Gerardi Mercatoris...*
Copperplate engraved miniature map. 6 x 7.75 inches (15 x 19.5 cm). Full original color.

This map was printed in 1610 by Rumold Mercator from the original plate of 1595. Note the typical Mercator moire pattern in the ocean.

As noted in the description for the Ortelius Europe map, the detail in this map continues into Asia, making it more than strictly a Europe map.

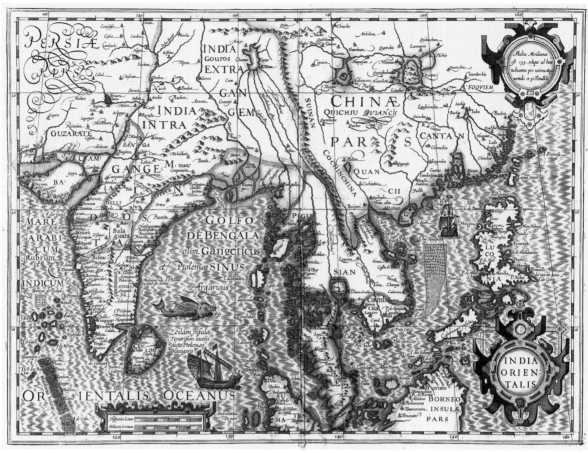

Courtesy: Martayan Lan, New York

Nº 42

Hondius, Jodocus. Amsterdam: 1613. *India Orientalis*. Copperplate engraved map. 14 x 19 inches (35.5 x 48 cm). Full original color.

A beautiful early map of the East Indies. Remarkably well charted, this region was, of course, of enormous importance to the Dutch. This map combines cartographic importance and a high aesthetic sense, typical of many of the maps of this period by the great mapmakers.

1616 – *Hondius' Miniature Celestial*

N° 43   **Hondius, Jodocus.** Amsterdam: (dated) 1616. *Globus Coelestris.* Copperplate engraved celestial map. 3.75 x 5.5 inches (9.5 x 14 cm). Shown full size. Uncolored.

This miniature celestial map was published in Petrus Bertius' *Tabularum Geographicum Contractarum* printed in Amsterdam. The terrestrial maps in this volume are reduced form the Mercator/Hondius atlas. They should be described as maps by Hondius, after Mercator, from Bertius. The copper plates were obtained by Blaeu, who reissued them in 1637, and later in 1639, as illustrations in the second

editon of William Camden's *Britannia*. In this work, the maps do not have text on the verso.

Despite the small size (the photograph above shows the map in full size), there is sufficient detail to produce a useful map of the constellations.

Warner, p121.

1620 – *Magini*

N° 44   **Magini, Antonio.** Venice: 1620. *Helvetia.* Copperplate engraved map. 5 x 6.3 inches (12.5 x 17.5 cm). Uncolored.

Another late-16th-century/early-17th-century book in the older tradition is the work by Magini, *Geographiae Universae Tum Veteris. Tum Novae, Absolutissimum...* Published in Venice, it appeared in editions of 1596; 1597/98; 1608; 1617; and 1620. Smallish copperplate maps characterize the Magini atlas. In keeping with the earlier concept of a map book, they contain the maps of Ptolemy and also the modern maps. The maps are relatively well engraved and, despite the large number of editions originally published, are not overly common. Some editions (such as the 1597) have the maps as full-page illustrations, other editions (such as the 1620, source of the map shown here) have a larger page size and the map shares page space with letterpress text.

It is a well-known little atlas and when it appears on the market it is generally at an affordable price. The maps were engraved by Porro and are sometimes identified as "Porro" maps. They are very similar in appearance to those in the Porcacchi atlas.

162

c. 1623 — *Mercator's Miniature*

N° 45   **Mercator/Hondius.** Amsterdam: c. 1623. *India Orientalis.* Copperplate engraved miniature map from the *Atlas Minor*: 5.75 x 7.25 inches (14.5 x 18.5 cm). Full original color.

This atlas, the Mercator/Hondius *Atlas Minor*, was of an intermediate size. Distinctly smaller than the large folio atlases, it was, nonetheless, not quite a "pocket" atlas. Indeed, King does not consider these to be miniature maps because of their size.

The maps were well-engraved and the atlas is not uncommon on today's market. However, virtually all of the atlases were issued uncolored. Those exemplars with original color are scarce.

The atlas contained several maps related to America, including a version of the Mercator/Hondius *Virginia*.

163

1641 – *Hondius' America*

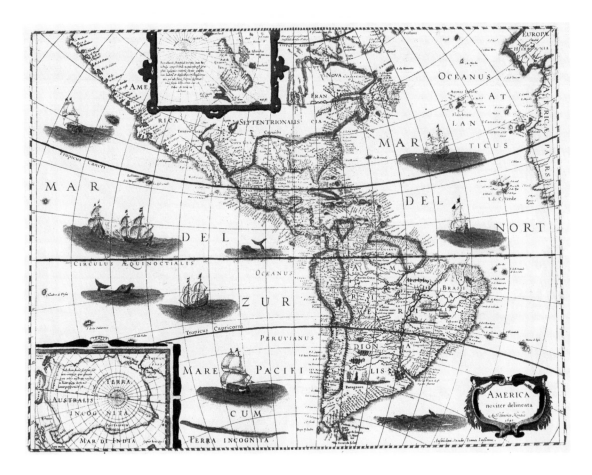

N° 49   **Hondius, Jodocus.** Amsterdam: 1641. *America noviter delineata*. Copperplate engraved map. 14.75 x 19.75 inches (37.5 x 50 cm). Full original color.

This map has a most curious history. As published originally by Jodocus Hondius in 1618, the map was surrounded on all four sides with decorative borders containing views and costumed figures. The borders were removed in approximately 1630 by Henricus Hondius to permit the image to fit into his atlas. Later the imprint was changed several times, and Jansson's name was added below the cartouche. Thus, according to Burden, there are five known states to this map. The one we illustrate is State 4, the first to show Jansson's name.

California is a peninsula and remains so on this map despite the growing trend to depict it as an island. The oceans are rich with both ships and sea creatures. The inset in the lower left shows the legendary antarctic continent and, in the upper left, an inset shows the northern regions, including a little Frisland (see Map 19).

Burden 192; Goss, *America* 27.

168

c. 1647 – *Blaeu's Terra Firma*

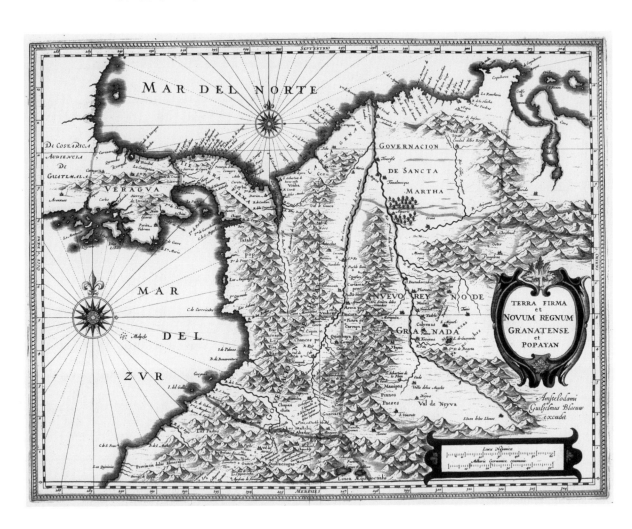

Nº 50 **Blaeu, G.** Amsterdam: c. 1647. *Terra Firma et Novum Regnum Granatense et Popayan.* Copperplate engraving. 14.5 x 19 inches (37.3 x 48.7 cm). Full original color.

A wonderfully detailed map of the Central America/ South America junction. Extensive interior detail, including city and town locations. One of the more splendid maps of the region, done with typical Blaeu brilliance.

There are two compass roses; one in each ocean. An attractive mileage scale is in the lower right.

This map often is found with a darkened centerfold. Koeman K1 Bl 34(85).

c. 1647 — *Blaeu's North Pole*

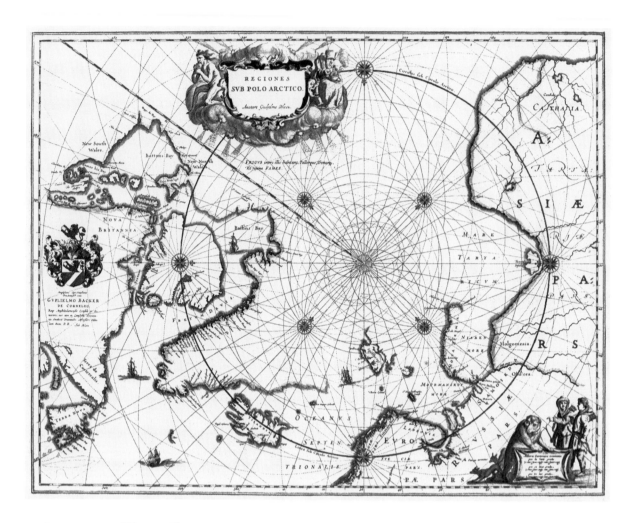

Nº 51

Blaeu, G. Amsterdam: c. 1647. *Regiones Sub Polo Arctico.* Copperplate engraving. 16 x 20.75 inches (41 x 52.5 cm). Full original color.

A particularly attractive early North Polar map. The entire Arctic Circle is shown, but offset to the right to permit showing more of North America. Hudson Bay (Buttons Bay) and James Bay are seen and represent the most westerly water known. The west coast of Greenland is shown and the northwest reaches of Baffins Bay, connecting to Hudson Bay are indicated. The true nature of Greenland, Spitzbergen and Nova Zemla has not yet been elucidated and their coastlines are incomplete. A very handsome and businesslike map with several attractive cartouches. The first state does not have the decorative arms at left center. Nordenskiold 20.

170

1650 – *Fuller's Holy Land*

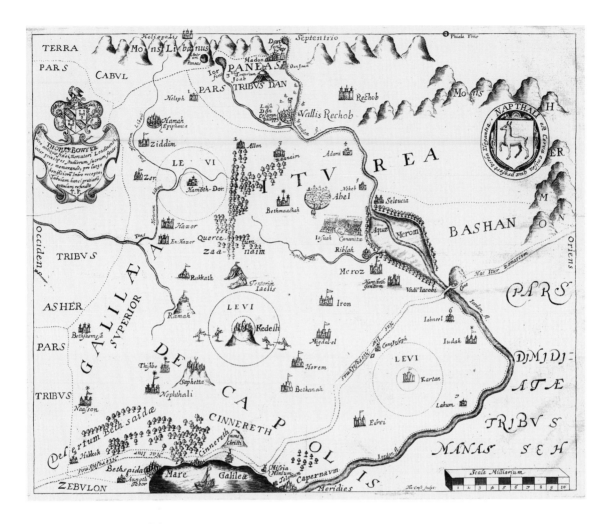

Nº 52

Fuller, Thomas. London: 1650 and later. *Napthali.* Copperplate engraved map. 11 x 13.5 inches (28 x 34.5 cm). Uncolored.

In his book, *A Pisgah-Sight of Palestine,* Fuller included a series of eighteen maps, including detailed ones of different parts of the Holy Land. The maps are engraved in a curious sparse style and are quite attractive. They were usually uncolored. The paper on which they are printed is thin and has an unfortunate tendency to age-darken.

The map shown here depicts the territory in northern Galilee allocated to the tribe of Napthali, and indicates the Tribe's emblem: a prancing deer.

Laor 283.

1654 – *Sanson's Philippines*

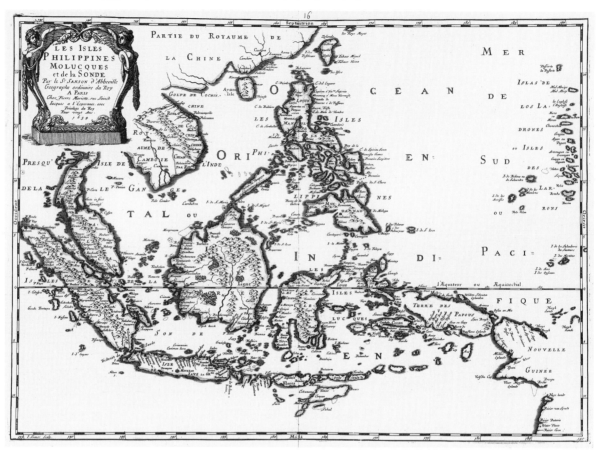

Nº 53   **Sanson, Nicolas.** Paris: 1654. *Les Isles Philippines Molucques et de la Sonde...* Copperplate engraving. 15.75 x 22 inches (40 x 56 cm). Original outline color.

Sanson is considered to be the "father of French cartography." His maps are noted for good detail and a very clear, highly focussed presentation. This map of the Philippines is one of the most detailed that appeared to date. New Guinea's shape and extent were unclear at the time and are so noted on this map.

Courtesy: Martayan Lan, New York

Nᵒ 54   **Cellarius, Andreas.** Amsterdam: 1660. *Hemisphærium Stellatum Australe, Antiquum.* Copperplate engraving. 17.25 x 20.3 inches (44 x 52 cm). Full original color, highlighted with gold.

It is widely considered that the mapping of the heavens achieved its pinnacle with the work of Andreas Cellarius. His charts of the solar system are paradigms of lucidity: they depict the various models of celestial and planetary motion with exquisite clarity, yet retain an extraordinary visual impact. The engravings of the celestial constellations are magnificent, and never equalled. Viewed from many perspectives, the allegorical figures retain all their splendor. There are two plates that show a revised "Christian" set of constellations, designed to replace the heathen ones of antiquity.

The plates were reprinted in the early 18th century by Valck and Schenck, with the addition of their names.

c. 1665 – *Grimaldi's Moon*

No 55   **Grimaldi, F.M.** c. 1665. *Figura Pro Nomenclatura, et Libratione Lunari.* Copperplate engraved moon map. Diameter 11.5 inches (29.5 cm). Uncolored.

Published in Riccioli's *Almagestum Novum*, this is an early map to show Riccioli's nomenclature. Riccioli, who first called the dark regions "maria," used the names of scholars of antiquity for north hemisphere features and those of Renaissance scholars in the south.

This map was the work of Grimaldi.

1676 – *Speed's China*

N° 56 **Speed, John.** London: 1676. *the Kingdome Of China...* Copperplate engraving. 15.5 x 20 inches (39.5 x 51 cm). Full original color.

This is the first English map with China and Japan as its dominant theme. Made by John Speed, perhaps England's most famous early cartographer, the map is highly decorative. Korea is shown as an island.

Early European concepts of Japanese and Chinese people fill the side panels and scenes and views at the top. I note that Macao, the great trading center, is represented.

Speed's maps are sought after and collected avidly. His maps of America and the detail maps of parts of the east coast of North America are prized by American collectors.

c. 1676 — *Ogilby's Road Map*

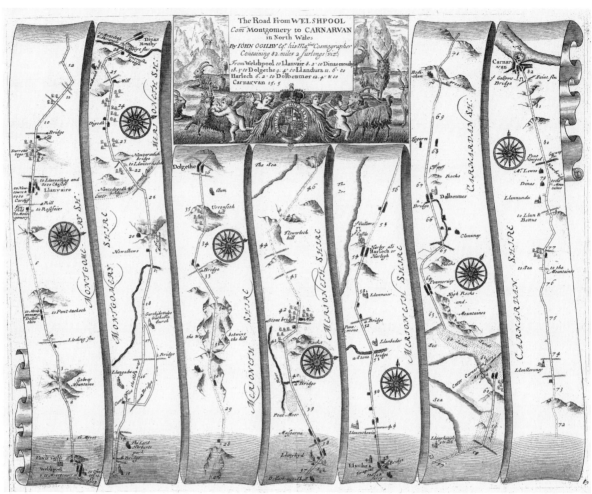

Nº 57

Ogilby, John. London: c. 1676. *The Road from Welshpool Com Montgomery to Carnarvan in North Wales...* Copperplate engraved map. 13 x 16.25 inches (33 x 41.5 cm). Later full hand color.

A splendid early road map. The map is one of a series illustrating and describing individual roads connecting major travel points. Mileage along the route is noted and unusual terrain indicated. Inns, castles and towns are marked for the convenience of travelers. Divided into individual strips, each strip has its own compass rose for orientation. In this map, the 82-mile route had two permanent gallows, one shown in the enlargement at right.

Ogilby's was the first national road atlas.

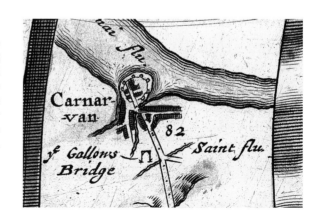

1680s — *Mallet's Miniature Maps*

№ 58   **Mallet, Allain Manesson.** c. 1680s. Left: *Terre de Iesso;* Right: *Continent Septentrional.* Copperplate engraved maps from a German edition. 5.75 x 4 inches (14.5 x 10.5 cm). Later hand color.

Allain Mallet, a French engineer, published a huge collection of miniature maps, views and portraits. Included in the collection, *Description de l'univers*, were several celestial maps and plates illustrating the known planets as well as the constellations. The portraits purport to depict various native peoples, but most of them are hardly recognizable. Many of the city views are bird's-eye views with very small cities in the distance. They reveal very little about each city and many appear to be generic views.

The maps are quite curious, and although there is very little detail, they capture a peculiar flavor: many have distorted perspective, and apparently Mallet could not decide if he was producing a map with a normal plan or a view of the land with a curious perspective.

The first edition was in French but editions were also published in German. Several states exist.

King 128.

1687 – *Brunacci's Celestial Map*

N° 59   **Brunacci, Francesco, and G.G. deRossi.** Rome: 1687. *Planisfero Del Globo Celeste.* Copperplate engraved celestial map. 17 x 22.5 inches (42 x 55.8 cm). Full color.

A brilliant and attractive celestial map. The two hemispheres show the classical constellations, predating those from the Age of Reason. The telescopic view of Saturn in the upper left shows the rings and five satellites; Jupiter shows belts and four satellites, whereas Mars shows some surface detail and Venus some of the "dusky markings" that were to confound planetary observers for hundreds of years. Sunspots are shown and the moon map shows the more prominent telescopic features.

The colors on this exemplar are beautiful; the banners are red, the engraved text at the bottom has red, green, and yellow wash, and the allegorical constellation figures are multicolored. The sun in the upper center is a brilliant yellow/gold. The map is a scarce one and much sought after because of its beauty.

Warner p44.

c. 1690 – *de Wit's World*

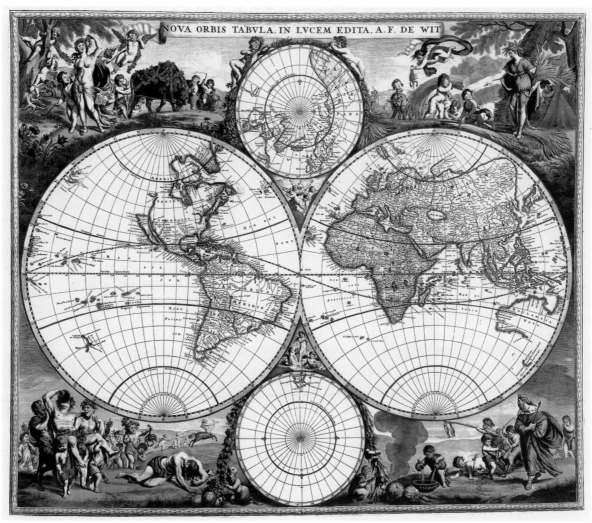

N° 60   **de Wit, Frederick.** Amsterdam: c. 1690. *Nova Orbis Tabula In Lucem Edits, A. F. De Wit.* Copperplate engraved map. 18.75 x 22 inches (47.8 x 56 cm). Later original color.

This double-hemisphere world map is considered to be among the most attractive of its time. The map shown here is State 1. A later, second state has an added frame and is more elaborate, with additional cherubs.

California is shown as an island and the Great Lakes appear as a large single body of water.

The figures in the corners depict seasons and elements. There is a spectacular debauch scene in the lower left.

Large, well-engraved world maps such as this one are avidly collected and it is increasingly difficult to find them in fine condition with original color.

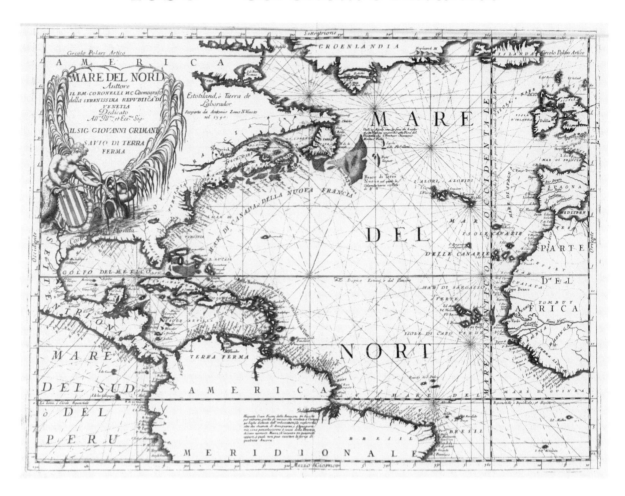

N° 61   **Coronelli, Vincenzo.** Venice: 1690. *Mare Del Nord.* Copperplate engraved map. 17.75 x 23.5 inches (45 x 60 cm). Uncolored as issued.

A splendid map. The old prime meridian is prominent in the east, passing through the Canaries and part of Iceland. The Atlantic Islands, including the Azores and Bermuda, figure prominently. The American coast is wonderful, showing those areas most well explored. The St. Lawrence ends abruptly and no Great Lakes are shown. The cartography of Florida is of the period. A great many place names are noted along the coast and some suggestions of geopolitical spheres are indicated inland. A very striking and dramatic map that, in my opinion, looks best in original uncolored condition.

1690 – *Coronelli's Pacific*

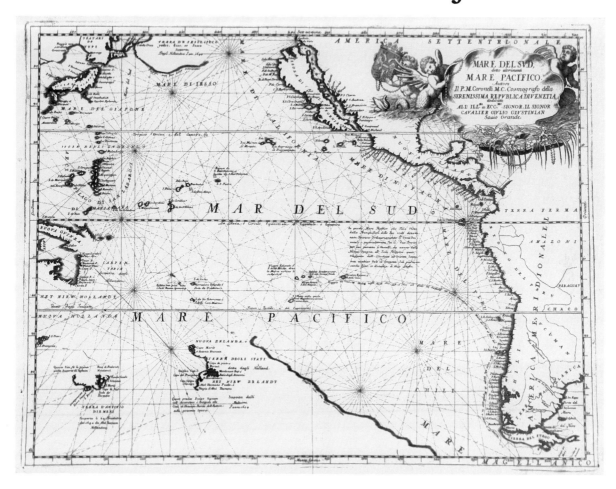

N° 62   **Coronelli, V.** Venice: 1690. *Mare Del Sud, dettro altrimenn Mare Pacifico...* Copperplate engraved map. 17.75 x 23.75 inches (45 x 60.4 cm). Uncolored as issued.

An extraordinarily bold, powerful map of the Pacific that shows the West Coast of the Americas including California as an island, a bit of Asia and a dynamic, emerging South Pacific . Tasman's recent (1644) discoveries in New Zealand are shown; the tracks of Le Maire's voyage of 1615-1617 are also indicated. Japan is charted and is situated some 50 degrees west of the California coast. A truly wonderful map with a decorative cartouche in the form of a shell, the lower half of which is filled with pearls, coral, and seaweed, all icons of the new worlds of the Pacific. The arms of Venice are prominent. The shell is supported by nereids, half winged cherub and half fish.

The map appeared first in Coronelli's *Atlas Veneto* of 1690.

McLaughlin 104; Wagner 436; Tooley, *Australia* 350.

c. 1690 – *Visscher's Northeast America*

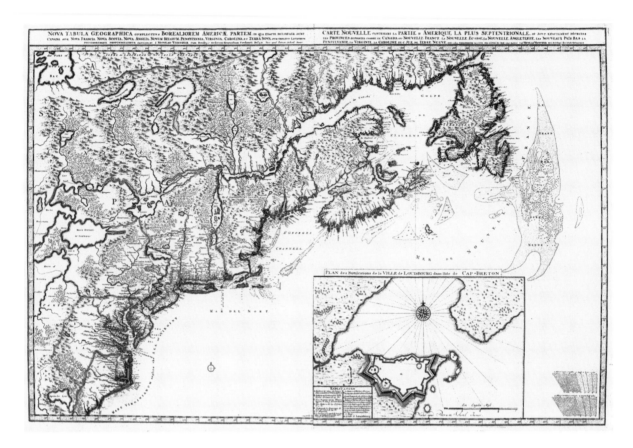

Nº 63

Visscher, Nicolaus. Amsterdam: c. 1690. *Nova Tabula Geographica Complectens Borealiorem Americæ Partem; in qua exacte delineatæ sunt Canada sive Nova Francia, Nova Scotia, Nova Anglia, Novum Belgium, Pennsylvania, Virginia, Carolina, et Terr Nova, cum Omnibus Littorum...* Copperplate engraved map in two parts, joined. 23 x 18.5 inches (56 x 46.5 cm) each. Original full hand color.

An unusual map of northeast America, published in two parts: one map shows part of Canada to Hudson Bay, west to Lake Huron, and the other the eastern tip of Canada. A line down the side of each map indicated where they could be joined; the one we show here was joined some time ago. The colors are beautiful. The sizes, shapes and positioning of the lakes are still imprecise in the 17th-century mind, and the boundaries of states are surprising, for example Pennsylvania engulfs Lake Ontario, and Florida reaches up to Lake Erie. Names of indigenous tribes occupy the unknown interior, while the settled coast has European names, soundings, shoals, etc. Tree and mountain symbols are used to indicate the topology. Altogether, a fascinating and visually bold, attractive map. When these maps are found in unjoined condition, I recommend that they stay that way. They can be displayed in a double-window mat or as two separately framed maps.

c. 1696 — *Danckert's Caribbean*

Nº 64 — Danckerts, Cornelis.

Danckerts, Cornelis. Amsterdam: c.1696. *Insulæ Americanæ, Nempe: Cuba, Hispaniola, Jamaica, Pto Rico, Lucania, Antillæ vulgo Caribæ, Barlo-Et Sotto-Vento. Etc.* Copperplate engraved map. 18.5 x 22.5 inches (47 x 57.5 cm). Original outline color; cartouche fully colored.

This is a beautiful map; the original colors have oxidized and mellowed to produce a most attractive tone. The map is characterized by bold engraving and thick outline color.

Central America sweeps diagonally across the lower left with the Yucatan Peninsula reaching toward Cuba and Florida. The coastline along the Gulf of Mexico is highly inaccurate. The Florida peninsula exhibits one of the various shapes it has assumed on maps over the centuries.

Two winged cherubs support the cartouche in the upper right. The key to the map's symbols is in the upper left, outside the neatline and left of the text. The distance scale is at the top right, also above the neatline.

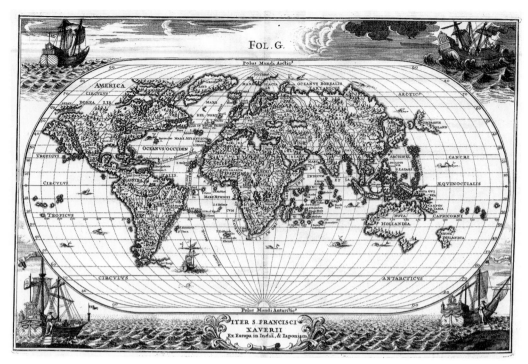

Nº 66   *Iter S. Francisci Xaverii Ex Europa in India, & Japoniam.* [Fol. G. appears at the top of the map.] 9 x 13.75 inches (23 x 35 cm). Uncolored.

Nº 67   *Representatio Geographica Itineris Maritimi Navis Victoriæ En Qua Ex Personis CCXXXVII.....*
8.75 x 13.75 inches (22.5 x 35 cm). Uncolored.

(For descriptive text see previous page.)

1704 – *Wells' English Colonies Map*

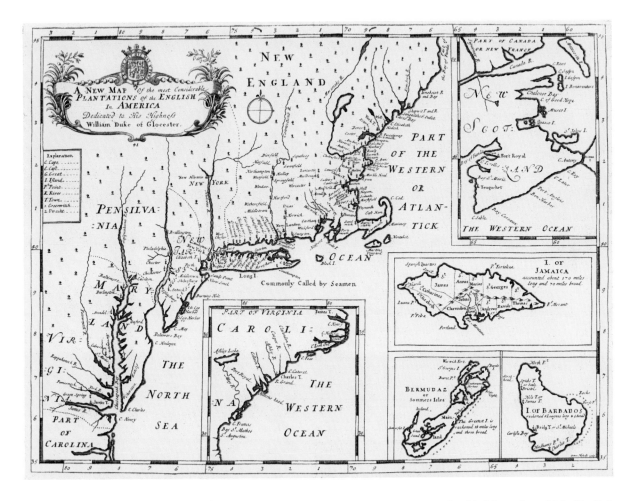

No 68   **Wells, Edward.** Oxford: 1704. *A New Map Of the most Considerable Plantations Of the English in America...* Copperplate engraved map. 14 x 18.75 inches (35.5 x 47.8 cm). Uncolored.

This map is from the 1704 edition of Wells' atlas. Many of the maps in it are of curious design and some exist in miniature version. The atlas was designed by Wells, an Oxford geographer, for instructional purposes. Wells was tutor to young William, heir to the throne. Unfortunately for Wells, William died and along with him died Wells' dream of being a friend to the King.

Highly stylized, this map shows very little substantive interior detail. The ocean is filled with small insets of other English possessions. We note Jamaica, Bermuda, and the Barbados.

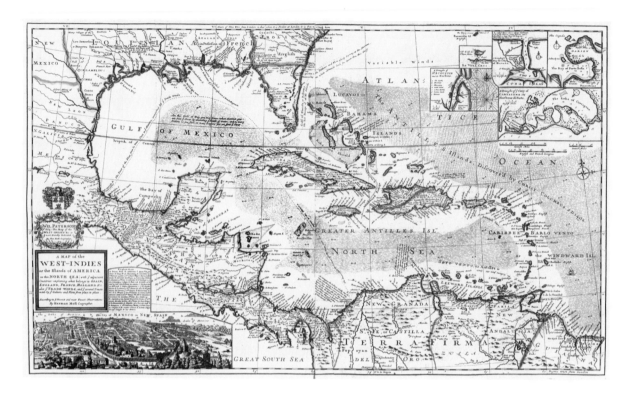

N° 69　　**Moll, H.** London: 1715. *A New Map of the West-Indies or the Islands of America...* Copperplate engraved map. Approximately 23.25 x 39.75 inches (8.4 x 101 cm). Original outline color.

A truly spectacular Caribbean map.

The scale is most impressive and this map embodies the usual Moll clarity of line and thought. The East Coast is shown from Charlestown (sic) down. Florida is a stubby peninsula and the Gulf Coast is a bit confused with the Mississippi River located far to the West, adjacent the Rio Grande. Moll includes extensive text detailing the route and timetables of the galleons bringing goods to Spain. There is wonderful coastal detail and the scene in the lower left is a city view of Mexico City.

These large Moll maps are sometimes found in poor condition because they were folded twice to fit into the atlas. The outer fold is often browned and torn, sometimes beyond reasonable repair.

Herman Moll is widely collected and his many smaller maps cost a good deal less but also display his sometimes caustic wit in the engraved text.

There were several later editions of this map.

c. 1720 – *Homann's Constantinople*

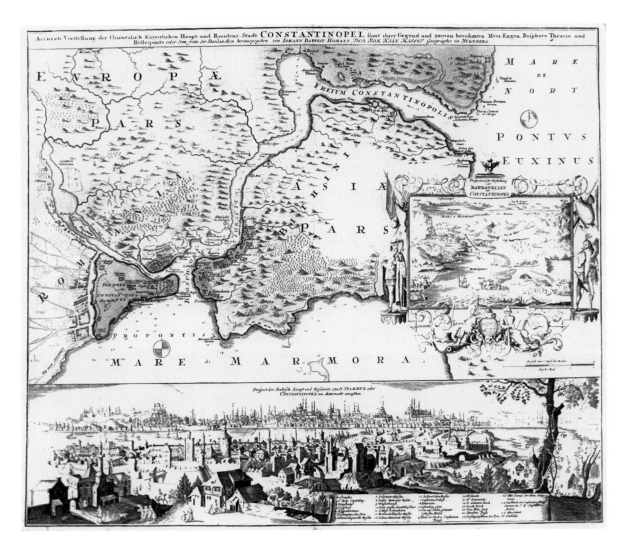

Nº 70   **Homann, Johann Baptist.** Nuremberg:1720. *Accurate Vorstellung der Orientalisch Kayserlichen Haupt- und Residenz-Stadt Constantinopel samt ihrer Gegend und zweyen berühmten Meer-Engen, Bosphoro Thracio und Hellesponto, oder dem freto der Dardanellen.* Copperplate engraved city view and map. Approximately 19 x 22.5 inches (48.5 x 57.5 cm). Original color to map; view uncolored.

This is a copperplate engraving with full original hand color to the map, and the city view is uncolored. A small scale plan of Constantinople appears in its setting on the Bosphorus. In the lower right is a cartouche with a delightful view of the Dardanelles seen from above the city. Below the map is a wonderful view of the city itself, taken from the north, probably from the heights of Pera, with a key to numbered buildings, such as the Seraglio, St. Sophia's, various mosques, etc. (13.7 x 57.5 cm). The Galata Tower appears, and we see the entrance to the Golden Horn before any bridges. The typical Homann coloring is bold and attractive. This map often has centerfold darkening.

189

1730 – *Covens and Mortier's Egypt*

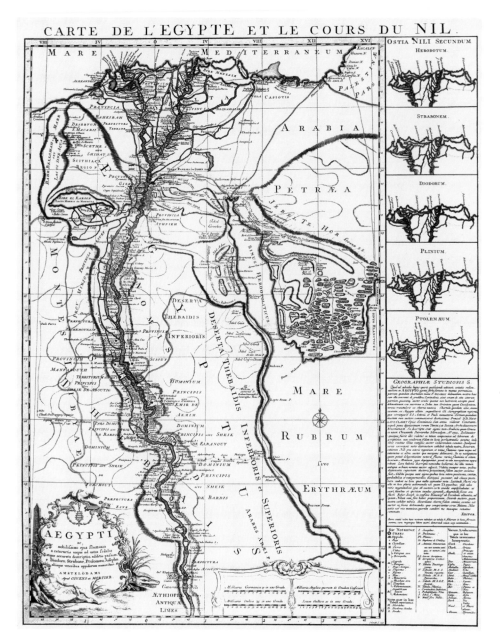

Nº 71

Covens, Jean and Corneille Mortier.
Amsterdam: 1730. *Aegypti ac nobilissimi ejus fluminis... (Carte De L'Egypte Et Le Cours Du Nil).* Copperplate engraved double-hemisphere world map. 24 x 19 inches (61 x 48.5 cm). Original outline color.

A unique Nile map. Shown in extraordinary detail, the river and its immediate surrounds are meticulously laid out. There is a tremendous amount of detail on either side of the Nile and insets at the right show the delta from Herodotus to Ptolemy.

The Nile and its source fascinated Europe from the time of antiquity and it appears as a major focus on early Africa maps. This is a much more sophisticated representation that begs the issue of the origin of the river.

c. 1730 – *Homann's Nova Hispania*

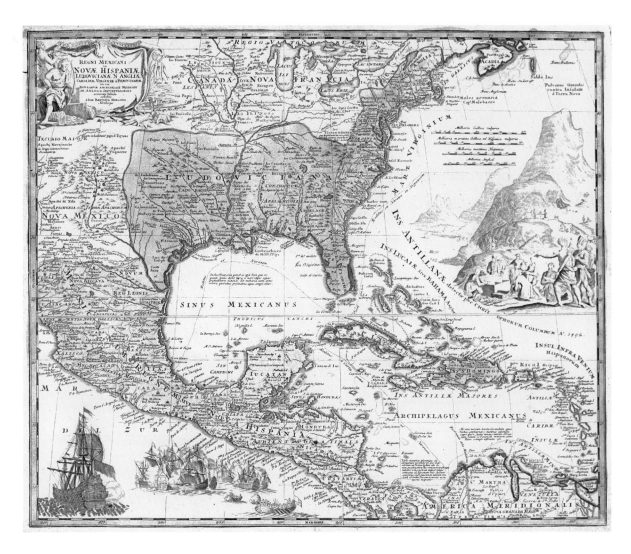

№ 72 **Homann, Johann Baptist.** Nuremberg: c. 1720. *Regni Mexicani Novæ Hispaniæ, Ludovicianæ, N. Angliæ, Carolinæ, Virginæ, et Pensylvaniæ....* Copperplate engraved map. 18.75 x 22.6 inches (47.6 x 56.6 cm). Full original color.

A dramatic map showing French influence in Louisiana in the early 18th century after the Treaty of Utrecht in 1713. The Southern boundary of Carolina intrudes into Spanish regions of Florida. The map has an unusually large amount of detail, place names and clearly delineated geopolitical boundaries.

The large uncolored vignette in the Atlantic Ocean represents the wealth of the New World. There are

splendidly engraved sailing ships in the Pacific near the Central America coast.

In this beautiful map, Homann has added much new information to de L'Isle's cornerstone map of the Southeast U.S. and Caribbean.

c. 1730 – *Homann's New England*

N° 73   **Homann, Johann Baptist.** Nuremberg: c. 1730. *Nova Anglia Septentrionali Americæ implantata Anglorumque coloniis...* Copperplate engraved map. 19 x 22.25 inches (48.5 x 57.8 cm). Full original hand color.

The cartouche in the lower right is fully colored. It is unusual to find a Homann map with the cartouche colored, but this one appears to have early color. There are traces of gum arabic on the cartouche, suggesting, perhaps, that it is early color.

Cape Cod is shown as an island. New York is so named. There is lots of nice coastal detail; inland detail is sparse. Lake Champlain is grossly misplaced. The two "Jarseys" are indicated and there is a very large lake at the head of the Delaware River.

The cartouche is wonderful. A noble native stands at the left. At the right, against a sailing ship backdrop, is a European holding a beaver skin in his right hand. With his left hand he points to a pile of trading goods at his feet, evidently trying to effect a trade with the native. As noted by Goss, this was the standard German map of the Northeast during the eighteenth century.

Goss, *America* 50. Portinaro & Knirsch 116.

N° 74   **Chatelaine, H.A.** Paris: 1719 or later. *Carte De La Nouvelle France, ou se voit le cours des Grandes Rivieres de S. Laurens & de Mississipi Aujourd'hui S. Louis, Aux Environs des quelles se trouvent les Etats, Pais, Nations, Peuples &c...* Copperplate engraved map. 16.5 x 19 inches (41.5 x 48.2 cm). Uncolored.

This map is from the *Atlas Historique* issued between 1705 and 1739. A superb America map based on the de L'Isle model. There are three insets, all in nicely engraved borders. At the lower right: a bird's-eye view of Quebec and its environs, and a view of the city from the St. Laurens; at top left: on a scale larger than that of the principal map, the Gulf coast of Louisane. Fort du Detroit is shown. Although Lake Huron is seen in an unduly horizontal alignment, the definition of the Lakes generally reflects the good sources of information available to the French. This map is wonderfully enriched with depictions of animals, canoes, sailing ships and a wealth of detail in place names, location of portages, descriptions of marshes, routes of travelers, etc. Of particular current interest is the very good detail concerning Indians and their villages. Tribal locations are given and there are several vignettes of native American activities and villages. A fascinating map.

c. 1744 — *Seutter's Small America*

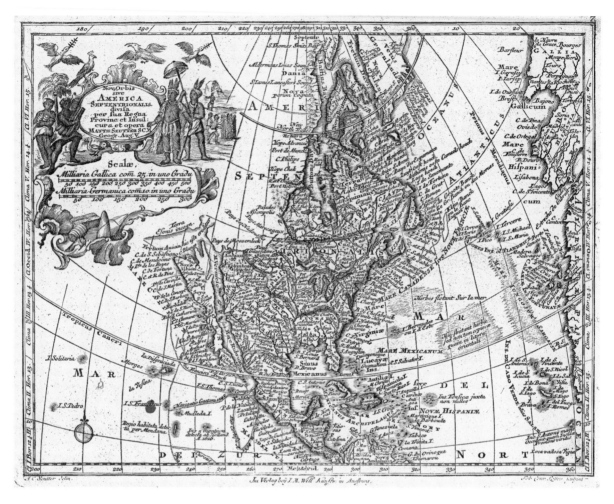

N.º 75   **Seutter, Georg Mattheus.** Augsburg: [1744]. *Nova Orbis sive America Septentrionalis, divisa per sua Regna Provinc: et Insul: cura et opera Matth: Seutter...* Copperplate engraved map. 7.5 x 10 inches (19.5 x 25.5 cm). Original hand color.

An unusual map from Seutter's scarce little atlas, *Atlas Minor*. The map was engraved by Lotter and often is attributed erroneously to a miniature Lotter atlas. The map is a reduction of part of Seutter's large America map. It has an unusual projection that gives a curious north-south elongation. California is a very thin, almost articulated, island.

This map exists in three states. The map shown here is the first state.

McLaughlin 221.

1756 – *Tirion's Japan*

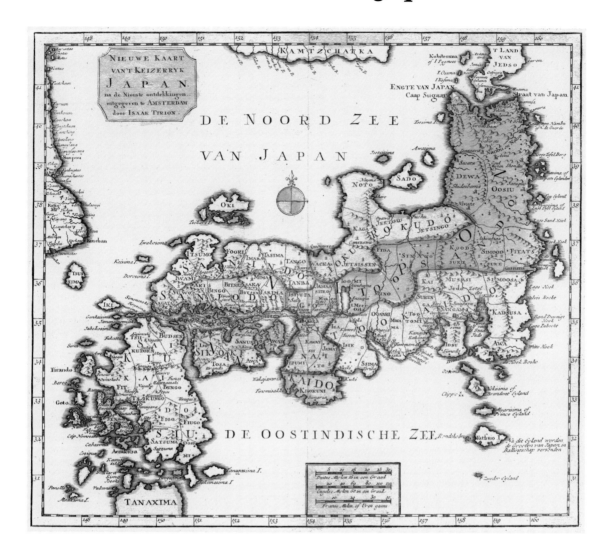

N⁰ 76 **Tirion, Isaac .** Amsterdam: c. 1740 [1756]. *Nieuwe Kaart van 't Keizerryk Japan.* Copperplate engraved map. 10.75 x 12.25 inches (27.5 x 31 cm). Full original color.

This map of Japan appeared in Isaac Tirion's *Nieuwe en Beknopte Hand-Atlas*, published in Amsterdam beginning in the mid-1740s. Some of the maps bear dates as early as the 1730s, but complete atlases seem to have been produced later. Because of the insular nature of Japan and its peoples during the Tokugawa Shogunate, Western knowledge of the islands was sparse. Foreigners were not allowed to live on the Japanese mainland; they were executed if they entered. Accordingly, maps of Japan lagged behind much of the rest of the world's cartography in accuracy. This is Tirion's second Japan map; it is based on his earlier one with modifications.

Walter 79.

c. 1756 – *Gentleman's Magazine Map*

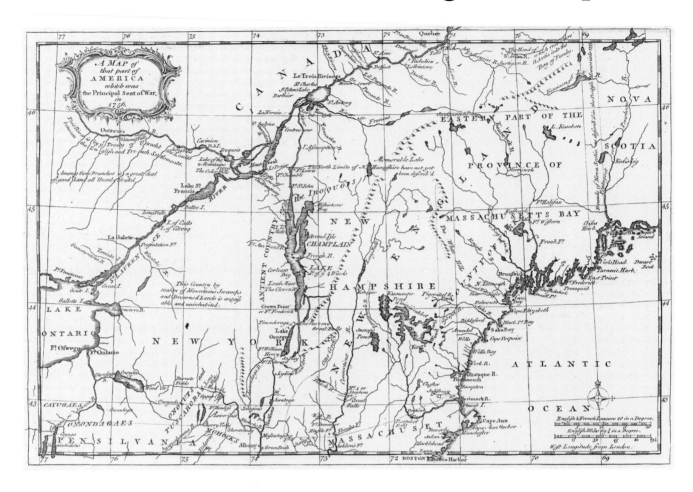

N° 77   **Gentleman's Magazine.** London: 1756. *A Map of that part of America which was the Principal Seat of War in 1756.* Copperplate engraved map. 8.5 x 13 inches (21.5 x 33 cm). Uncolored.

The *Gentleman's Magazine* was a British periodical that contained news, commentary, reviews, and especially reports of activities in far reaches of the Empire. The latter were often accompanied by well-executed, engraved maps. These maps were done for the purpose of either disseminating current information or illustrating an accompanying article. Hence, they were very clear and informative, unencumbered by the then-prevalent decoration. Some of the *Gentleman's Magazine* maps of the American colonies are the only ones available illustrating specific events, and are, accordingly, much sought after.

The larger *Gentleman's Magazine* maps are often folded and frequently have very narrow margins.

1758 — *Bellin's Northwest America*

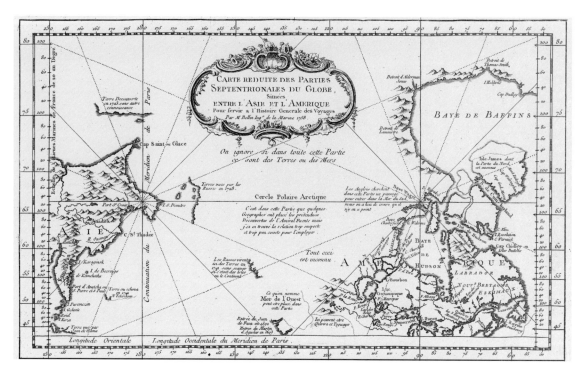

N° 78   **Bellin, J.N.** Paris: 1758. *Carte reduite des Parties Septentrionales du Globe, Situés entre l'Asie et l'Amerique...* Copperplate engraved map. 8.5 x 13.5 inches (21.5 x 34.2 cm). Uncolored.

This map is a nice example of a small, uncolored, well-engraved inexpensive map. Bellin's smaller maps, although they show detail quite well, are not collected widely enough to have caused a big price increase. These maps cover all parts of the world.

The great void in cartographic knowledge of the northern terrain between the eastern reaches of North America and the coast of Asia is demonstrated well in this map. It is a big, blank space.

1764 – *Owen/Bowen Roadmap*

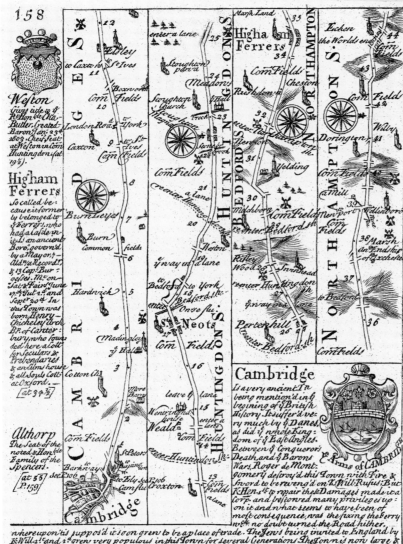

Courtesy: Lee Jackson, London

Nº 79

Bowen, Emanuel.
London: 1764. *The Road from Cambridge to Coventry.* Copperplate engraved map. 7 x 4.6 inches (17.8 x 11.6 cm). Uncolored.

This little strip road map is from a later edition of *Britannia Depicta...* with text by Owen and maps by Bowen. The first edition appeared in 1720 and is based on the maps in *Ogilby improved*. These small maps are charming and widely collected. They often appear on the market with later color. Relatively inexpensive, there is a small price premium for those showing roads from major cities such as London. The entire image, including the text, is engraved on copper.

c. 1770 – *Bonne's World*

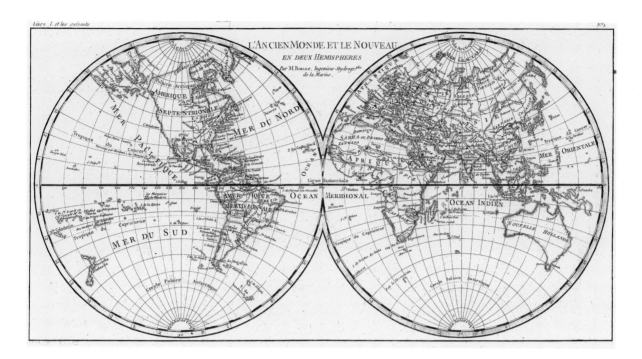

N° 80   **Bonne, Rigobert.** Paris: c. 1770. *L'Ancien Monde Et Le Nouveau En Deux Hemispheres.*
Copperplate engraved double-hemisphere world map. Approximately 8.75 x 16.25 inches
(21.5 x 41.2 cm). Each hemisphere is 8.25 inches (21 cm) diameter. Later full color.

    This is a small, rather attractive and decorative double-hemisphere world map.

    North America has a truncated, uncertain Northwest Coast, with very little interior detail in the western half of the continent. New Zealand is complete, New Guinea is not. The map predates the discovery of the Sandwich Islands (Hawaii).

    Printed on a heavy laid paper, the map is not uncommon. Since it is priced quite modestly, it represents an opportunity to acquire an 18th-century world map for a modest sum. Color, when present, is generally later.

1778 – *Antonio Zatta's Title Page*

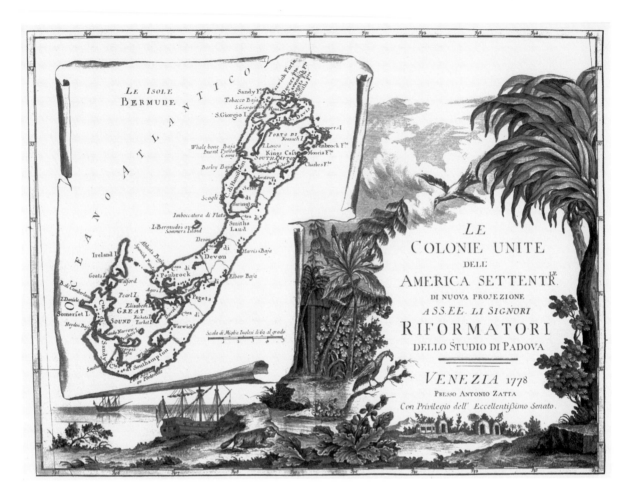

N° 81   **Zatta, Antonio.** Venice: 1778. *Le Isole Bermude.* Copperplate engraved map on title sheet *Le Colonie Unite Dell America Settentr….* Overall 12.5 x 16.5 inches (31.7 x 42.5 cm). Original color.

Antonio Zatta's atlas, *Atlante Novissimo*, was a major Italian 18th-century atlas. The illustration above shows the title page from the North American section of this atlas. The atlas contained maps of all the parts of the United States, or the United Colonies. The maps were mostly based on earlier English sources such as the John Mitchell map. Because they are derivative, Zatta maps were not, until relatively recently, collected extensively and consequently remain, more or less, underpriced. They were, however, influential in disseminating information about the lands of the new United States, and had an impact on the thinking

of the time. They seem to be collected more widely now, and their price is rising.

The Bermuda map on this title-page was probably copied from the 1763 map appearing in the *Gentleman's Magazine.*

Zatta maps generally have outline hand color. In the map above, Bermuda is colored in outline and the elaborate scene to the right is fully colored.

Unfortunately, many Zatta maps show darkening along the centerfold because of the glue used to affix the binder's guard.

200

1779 – *Roux's Miniature Harbor Charts*

N° 82   **Roux, Joseph.** Marseilles: 1779. *Barcelone.* Copperplate engraved chart. 5 x 7.6 inches (12.7 x 19.4 cm). Uncolored.

Roux issued a set of small charts of Mediterranean harbors and ports.

The maps are surprisingly detailed and give a clear view of the harbors. Prominent land features, including forts, buildings and terrain, are shown also. Compass directions are shown and each chart has enough soundings so that the suitability of the harbor for any given vessel can be determined.

c. 1780 – *Sauthier/Homann NY & NJ*

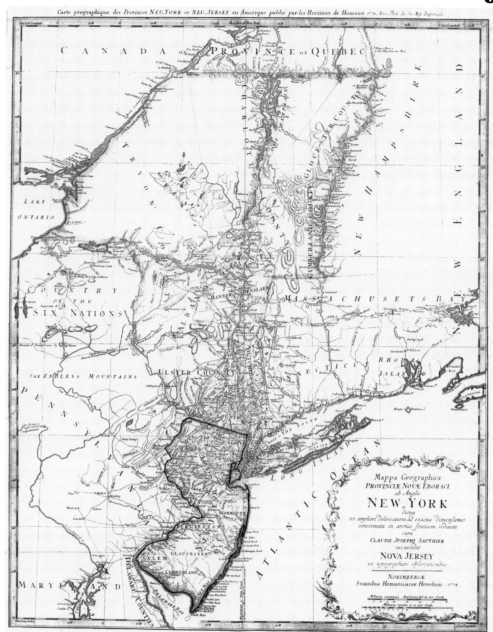

Carte geographique des Provinces NEU.YORK et NEU.JERSEY en Amerique publie par les Heritiers de Homann 1778. Avec Priv. de Sa Maj. Imperiale

N° 83　　**Homann's Heirs.** Nuremberg: 1778. Copperplate engraved map. *Mappa Geographica Provinciæ Novæ Eboraci ab anglis New-York.....* 27.75 x 22.25 inches (70.5 x 56.5 cm). Original outline color.

This detailed map is a reduction of the landmark Sauthier map, and reflects the region at the time of the American Revolution. It shows roads, towns and even some individual dwellings. The line dividing East and West Jersey is shown and there is good detail in the entire New York City area. To the west are "endless mountains" and the "Country of the Six Nations," noting Onondaga as the Meeting Town of the Six Nations. Lake Champlain is well charted. This map shows pre-Vermont New York with its eastern boundary along the Connecticut River. Settlements are shown along the Connecticut River. A fascinating map, printed on two sheets, joined. A similar map was published by Lotter.

202

1781 — *des Barre's Sea Chart*

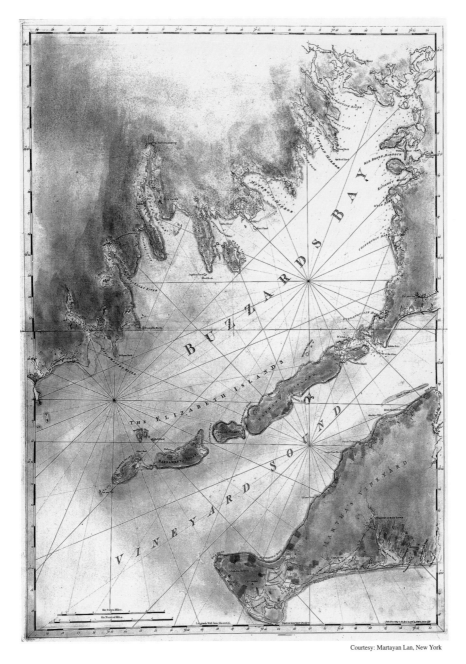

N° 84 **Des Barres, J.F.W.** London: Nov. 1, 1781. *[Large-scale chart of the region of Martha's Vinyard, Elizabeth Islands, Wood's Hole.]* Copperplate engraved chart. 41.25 x 35.75 inches (105 x 91 cm). Original full color.

This magnificent chart is from the *Atlantic Neptune*, which represents one of the great triumphs of sea chart printing. The early color made these charts particularly attractive. There is a surprising amount of interior detail on the land, and property lines and buildings are shown.

1784 – *Güssefeld's American Colonies*

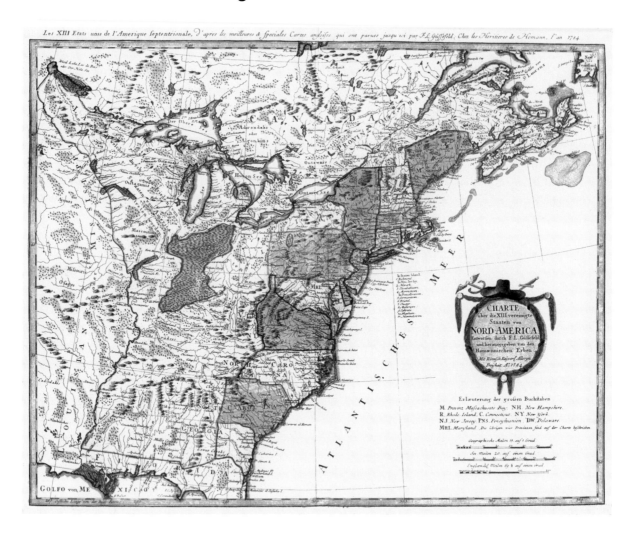

N⁰ 85   **Güssefeld, F.L.** Homann's Heirs, Nuremberg: 1784. *Charte über die XIII vereinigte Staaten von Nord-America,...* Copperplate engraved map. 17.5 x 22.5 inches (44.5 x 57 cm). Full original color.

Maps showing the United States during its very early years are relatively uncommon; this is a nice example. With only the original states colored, the map gives a good impression of the relatively small size of the fledgling country with respect to the rest of the land westward. The borders of the states are clearly indicated by the colors, but Güssefeld indicates the westward claims of many of them by dotted lines extending to the Mississippi. There is a wealth of detail, including Indian tribes, and western outposts. We note that a major part of present-day New York is still "Iriquois Nation." Maine is shown as part of Massachusetts, and Vermont is part of New Hampshire. The Western Territories are colored a pale yellow and show much detail.

The exemplar used for the photograph above had expert, minor repairs and has some wrinkling along the centerfold, but lacks the centerfold darkening that is common on this map.

1786 – *Bevis' Star Charts*

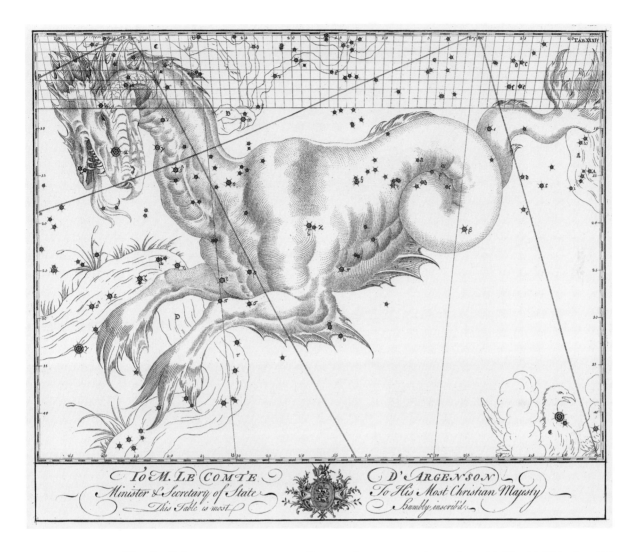

N° 86   **Bevis, John.** London: c. 1786. Copperplate engraved star chart. 12.25 x 14.75 inches (31 x 37 cm). Uncolored.

In the 1740s John Bevis, an English physician, began work on a large-scale star atlas. It was based on the Bayer atlas (see Map 40) and Bevis added stars from various later catalogues, such as those by Hevelius, Halley and Flamsteed. The maps are particularly attractive and well engraved. They are quite scarce since Bevis died before the atlas was published. After his death the copper plates disappeared and his heirs entered into a protracted legal dispute. Only a few sets of charts were assembled and bound into the *Atlas Celeste or the Celestial Atlas...* Accordingly, complete atlases are only rarely found. The plates were issued uncolored (we have seen one atlas with seemingly early color throughout), but many have been colored later for decorative effect.

Warner p22.

1794 – *Herschel's Saturn*

Philos. Trans. MDCCXCIV. *Tab.* VI. *p.* 32.

N° 87   **Herschel, William.** London: 1794. [The Planet Saturn.] Mezzotint. 7.3 x 5.9 inches (19 x 15 cm). Uncolored.

Sir William Herschel, one of the great figures in the history of astronomy, discovered the planet Uranus. He was a diligent observer who applied early statistical techniques to the study of the stars, but also observed the planets.

This is a spectacular mezzotint plate of Saturn. Herschel, in the accompanying text, describes the belts seen on the surface and also notes the transit of one of Saturn's satellites which he readily sees, in addition to actually resolving its shadow, which he can measure. Telescopes had been improved sufficiently, by the 18th century, to permit such detailed mapping of planet surfaces. In particular, the moon was a fruitful subject, although Herschel had little interest in it.

This plate appeared in the *Philosophical Transactions* 1794.

1797 – *Sotzmann's Pennsylvania*

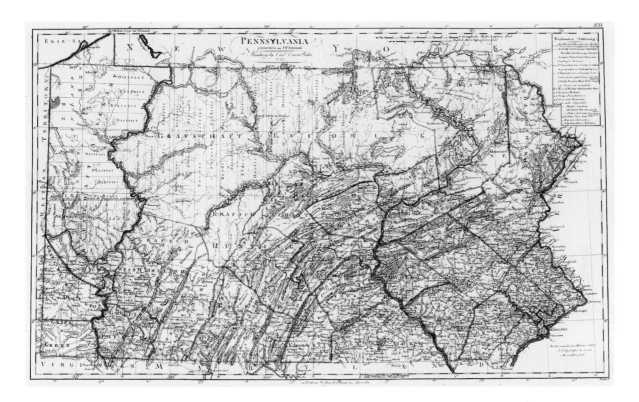

No 88 **Sotzmann, Daniel Friedrich.** Hamburg: 1797. *Pennsylvania*. Copperplate engraved map. 16 x 27.25 inches (40.8 x 69.2 cm). Original full hand color.

In the late 18th century Christoph Ebeling began the monumental project of mapping the new United States. Sotzmann, the Berlin cartographer, was to produce the copperplate maps, one of each of the new states and territories. Maps of New Hampshire, Vermont and Connecticut were produced with the 1796 date; Massachusetts was undated but likely 1796. Pennsylvania, shown here, is dated 1797. Only ten states were completed.

The project was never finished and the Sotzmann maps of the states are *among the scarcest of all state maps. There are very few known complete sets of the ten published maps and individual ones are of the greatest scarcity.*

The maps are very well engraved and are quite beautiful.

THE CENTURY OF TRANSITION

The 19th century opened with mapmakers making maps in the traditional way, using engraved copper plates. But a great revolution in printing had occurred, in 1799, with Senefelder's invention of lithography. Soon, most maps were being produced by this process.

Although lithography was cheaper and faster than copperplate engraving, it produced what many consider to be an æsthetically inferior image. The engraved line has thickness, a dimensionality that the lithographed line does not have and consequently it results in a distinctly flatter image. We cannot but mourn the passing of the much more attractive maps done by copper engraving, while we understand the implications of this technological advance in the history of mapmaking.

For many years, among map collectors, nineteenth-century maps were considered the "penny-stock" maps. Many of the maps in this period were eschewed by "serious" collectors, so they remained undervalued. However, in recent years nineteenth-century maps have come into their own. Now, the 19th century is perhaps the fastest-growing area of map collecting. Many factors seem to have contributed to this development.

For Americans, the 19th century is the century of the American map. American mapmakers came into their own and the century is replete with names such as Bradley, Mitchell, Johnson, and Colton, among others. Their atlases, once plentiful on the market, have become relatively scarce and the maps derived from them have steadily appreciated in price. They seem to be destined to continue upward. Many younger American collectors, unfamiliar with Latin, or even modern European languages, do not relate particularly well to earlier maps in these languages and prefer the (to them) readable American maps of the 19th century.

Certainly, too, the history of North America is sufficiently old to have a cachet and a dollop of romance, yet it is recent enough so that many can relate to it with greater ease than they can to earlier maps. For those to whom the cartography of some of the early maps is too esoteric, the cartography of the 19th century is understandable, yet "different"

enough to seem quaint. In my own business I have noticed, over the years, an increasing preference for 19th-century American maps of, for example, the East coast, over their earlier counterparts, such as the Blaeu or Jansson maps of the 17th century.

A very decent collection of 19th-century maps showing the evolution of one's home county or state can be assembled for relatively little money. The transition from territory to state and the subsequent progressive subdivision into counties and townships are all mapped. Indeed, a chronological sequence of Kansas or Nebraska maps shows the inexorable spread of counties rolling westward like a creeping artillery barrage. Such assemblages of American history can be built for less than the cost of a single Jansson Virginia, for example. The beginning collector should be aware, however, that there are exceedingly scarce and important 19th-century maps that are as difficult to find and as expensive to collect as any of the earlier trophy maps, and some of them will sell for six figures.

The large-scale folding map, sometimes called a pocket map, became popular in the 19th century. American production of folding maps was quite brisk. Large wall, or roller, maps were produced inexpensively and many have become very scarce and sought after.

Technology made the atlas cheap in the 19th century. Steam presses and lithography finally brought the price of printed images to where common people could afford them. This happened with a vengeance in the United States, where as the country enlarged and changed, the demand for up-to-date maps was enormous. Maps from some of the more common American atlases, notably the Mitchell, Johnson and Colton atlases, are plentiful on the market. These too, were long viewed with scorn by sophisticated collectors and dealers. Indeed, a criterion of a "serious" map dealer used by many collectors was that they did not have these maps in stock. My, but how the world changes!

The situation with maps seems to parallel that of prints. When I read some of the earlier literature on print collecting, I note the only slightly veiled scorn that was heaped on those artists whose works were plentiful. Those were the days when the best prints of, for example, Eisan, Utamaro, or Hiroshige were affordable in excellent condition. Of course it was possible to then denigrate Toyokuni III and

Yoshitoshi. More recent collectors have to be revisionists by necessity. In Western prints, even the Sadlers are collected now, despite what was said about them fifty years ago. The map world has changed far more rapidly, probably because of the smaller total number of maps in the pool. What took over fifty years in the print world has taken but a decade in the map world.

Maps such as the southwest map from the Johnson atlas, which shows California, New Mexico and Arizona, Colorado and Nevada, were updated almost continually and many states exist, each freezing a frame of the geographical film that was playing so quickly in those days. This particular map is selling for five times its price of but a few years ago. Fashion dictates much of this situation. For example, a decade ago one could barely sell a map of Ohio, or for that matter, of many of the Midwestern states. Today, while the trade isn't brisk, these maps are collected and cost has begun to rise. In general, those maps of the more settled areas of the eastern half of the country are less in demand than those that document the dramatic geopolitical changes of the West. However, there are many subspecialities in map collecting; railroads and canals are but two, and I don't doubt that in time these maps will be prized as highly as any of the maps from the earlier centuries.

The century began with a struggle between the more ornate and decorated maps that traced their lineage to the Golden Age of mapmaking, and those maps that were harbingers of the machine age. The desire to show decoration manifested itself in the works of Levasseur, Tallis, and the ornate decorative borders of Mitchell and Colton. Vignettes showing scenes as varied as building façades or burning prairies were common. While the stylistic changes were developing, the technological ones were happening also, and copperplate engraving almost completely gave way to lithography, cerography and "modern" reproductive processes, an evolution occurring still.

1801 — *Wallis' Linen-backed London*

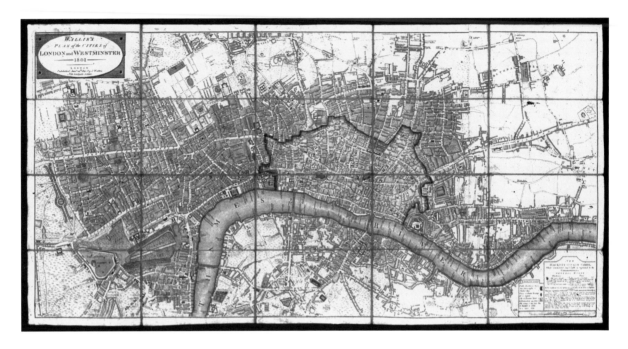

Nº 89
Wallis, James. London: 1801. *Wallis' Plan of the Cities of London and Westminster 1801.* Folding linen-backed city plan. 16.5 x 33.75 inches (42 x 86 cm). Original outline hand color.

Maps such as these were used extensively by travelers. The inset in the lower right gives hackney coach fares. Wallis published several such maps and they are chronicled in Howgego.

The map is accompanied by its original slipcase. Although these maps are robust, their usefulness was short-lived as the city and coach fares changed. Large numbers were probably discarded and they have become somewhat scarce. Just a few years ago these maps were not widely collected; now they are.

This map is a variant of Howgego, *Printed Maps of London,* 214(4); it predates the addition of the Isle of Dogs sheet.

1801 — *Bellin's Northern Cuba Chart*

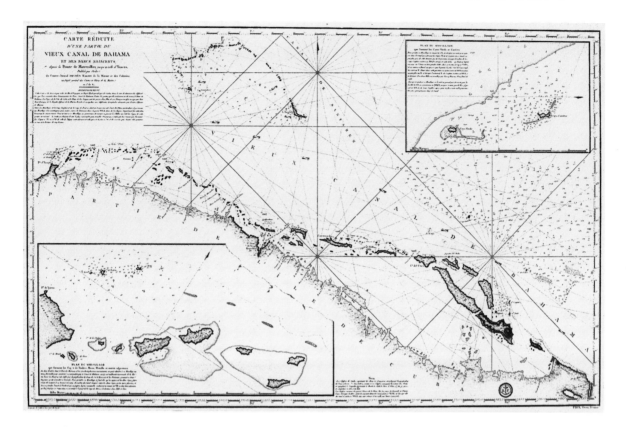

N° 90   **Bellin, J.N. (Dépôt général de la Marine).** Paris: 1801. *Carte Réduite De Une Parte Du Vieuz Canal De Bahama Et Des Bancs Adjacents depuis la Pointe de Maternillos jusqu'à celle d'Ycacos.* Copperplate engraved map. 23.75 x 35.75 inches (57.5 x 91 cm). Uncolored.

In addition to making small maps and plans, Bellin was involved with the production of large maps and sea charts. These are big, beautifully functional charts that cover much of the world.

Illustrated here is a large, dramatic sea chart showing part of the northern coast of Cuba and the nearby islands.

The coastal detail is superb and the islands are noted in the detail necessary for navigational use. The chart is thick with soundings and we note two insets, each with more detail. Rhumb lines crisscross the chart.

1801 – *Schroeter's Moon*

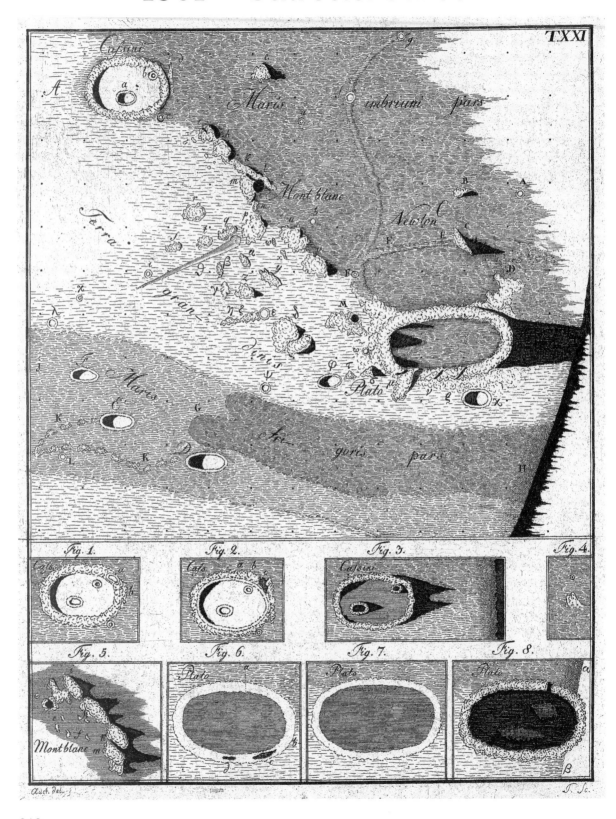

1801 – *Pichon's Paris*

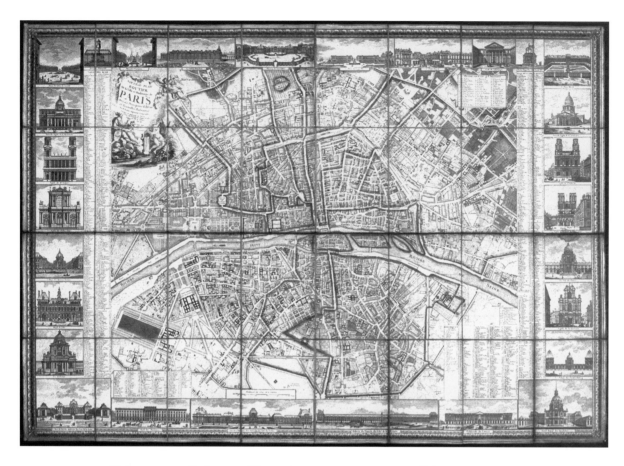

Nº 91   **Schroeter, Johann H.** (left) Paris: 1801. *Table XXI [The area of the Moon around Plato]*. Copperplate engraved lunar detail. 8.5 x 6.6 inches (21.6 x 16.8 cm). Uncolored.

Nº 92   **Pichon, M.** Esnault, Paris: An. 10: 1802. *Nouveau Plan Routier De la Ville et Faubourgs de Paris...* Folding linen-backed city plan. 40 x 56 inches (101 x 142 cm). Original hand outline color.

Schroeter's careful observations of lunar detail are preserved in a skillfully rendered series of forty-two copperplates that accompany his lengthy text, *Selenotopographische Fragmente zur Genauern Kenntniss Der Mondfläche...* published in Göttingen in 1791. A second volume appeared in 1802. His studies lasted from 1781-1797, and opened a new era of lunar mapping by concentrating on specific detail under various angles of illumination, a procedure that revealed much new detail. This lunar atlas, while not complete (Schroeter's complete observations were never published) is an important one in the chronology of lunar studies.

Schroeter's observatory was destroyed by Napoleon's troops in 1813. His published lunar work is scarce and is often available only in poor condition.

This is a splendid folding city plan of Paris published early in the 19th century. Executed by Pichon in 1801 (year 9 of the Republic), it was published a year later. Every block and street is shown and identified. The map was issued originally in a slipcase, missing in the present exemplar.

The map is particularly attractive with its surround of 26 views and facades of important buildings.

Pedley illustrates a fully colored version of this map, dated 1789, in *Bel et Utile*.

1811 – *Cary's Northwest Territory*

N° 93   **Cary, John.** London: 1811. *A New Map of Part of the United States of North America, Exhibiting The Western Territory...* Copperplate engraved map. 18 x 20 inches (46 x 51.2 cm). Full original hand color.

A very nice English-language map of this area which corresponds to a relatively narrow geographical window, capturing the ephemeral Western Territory before it was divided up into states. The map has an extraordinary wealth of information concerning travel – portages; size of vessels appropriate to different rivers (for example, there is detail about the Wabash); length of rivers; and brief early chronology of settlements, such as for Kentucky (where Col. D. Boon is noted). In addition we note Indian boundaries according to treaties, settlements, routes, and a good Great Lakes representation. This map is often found with a discolored centerfold.

1811 – *Cary's Northeast U.S.*

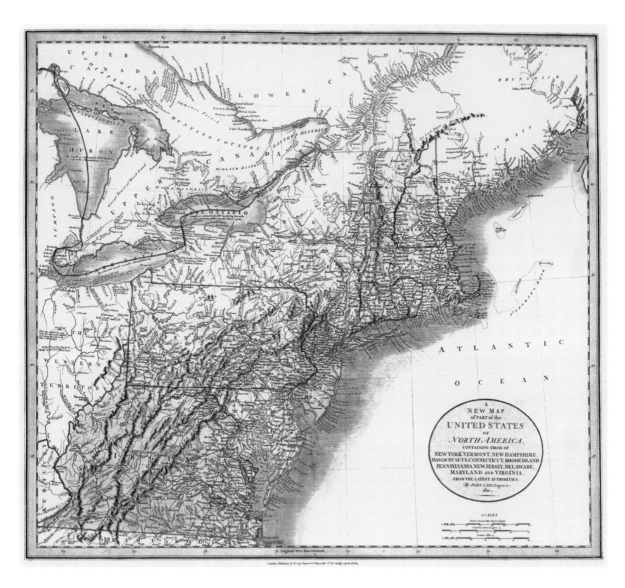

N̊ 94   **Cary, John.** London: 1811. *A New Map of Part of the United States of North America, containing Those Of New York, Vermont, New Hampshire, Massachusetts, Connecticut, Rhode Island, Pennsylvania, New Jersey, Deleware, Maryland and Virginia.* Copperplate engraved map. 18 x 20.4 inches (45.8 x 51.5 cm). Original full hand color.

A copperplate engraved map in exceptionally nice condition. The color is superb and there is no darkening along the centerfold, as is commonly found in these maps. From Cary's *Universal Atlas.* This map includes part of the Great Lakes and a bit of the Western Territories. Maine blends into Canada in a diplomatic ambiguity. This is a wonderful very early 19th-century British map made just before the War of 1812.

1814 – *Italy*

Nº 95

Mawman, Joseph. London: 1814. *Italy From the Original Map By Rizzi-Zannoni.* Copperplate engraved folding linen-backed map. 16 x 40 inches (40.5 x 100 cm). Original outline color.

This is a very detailed map of Italy designed to be used for a classical tour through that country conducted by the Rev. J.C. Eustace. The engraving is superb and the mountains have a three-dimensional appearance. A remarkable effort was expended to produce this map for what must have been a relatively small number of tourists.

c. 1815 – *German Folding Postal Map*

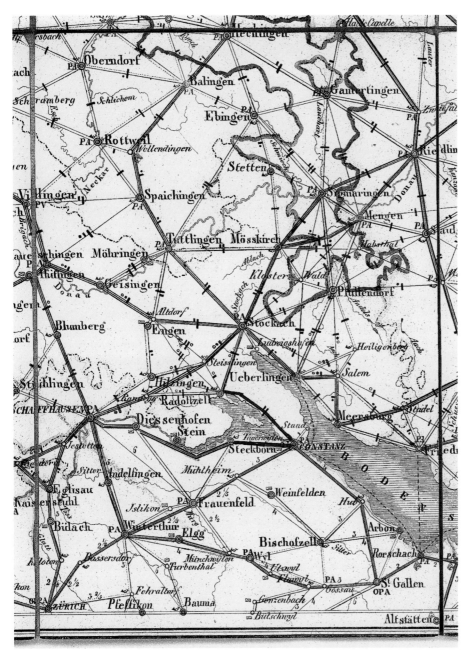

Nº 96

(German postal map). c. 1815. *Postkarte des Grossherzogthums Baden und des Königreichs Würtemberg Nebst Theilen der angrenzenden Länder.* Folding lithographed linen-backed map. 17.5 x 17 inches (44.5 x 43.2 cm). Original hand outline color.

This is a very detailed German folding map showing cities and towns, postal routes, post offices and connecting roads. Towns are keyed by size and their postal facilities. Roads are shown and mail express routes *(eilwagen kursen)* are noted. Water routes, such as on the Boden See, are shown connecting land points.

This is a pre-railroad map, and it demonstrates impressively the thoroughness of the postal system and the number of express routes, even through rural areas.

1817 – *Roman Europe*

N° 97

Thomson, John.
Edinburgh: 1817.
Orbis Romani Pars Occidentalis. Copperplate engraved map. 11.5 x 8.3 inches (30 x 22.5 cm). Original hand outline color.

This map illustrates the boundaries of the western part of the Roman Empire. As such, it is a retrospective map and does not purport to show "modern" or emerging cartography. Maps such as these, while interesting to many, are not widely collected and generally sell at a substantially lower price than their counterparts that show the same lands at the time the maps were made. There are many examples of classical atlases and maps that show the lands known to the ancients. Even the splendid examples by Ortelius seem to have a smaller market. This, of course, means that some very interesting and old maps can be bought for modest sums.

This map is from a little-known atlas by Thomson, and contained maps based on those by Wyld.

218

1822 – *Carey & Lea's Arkansas*

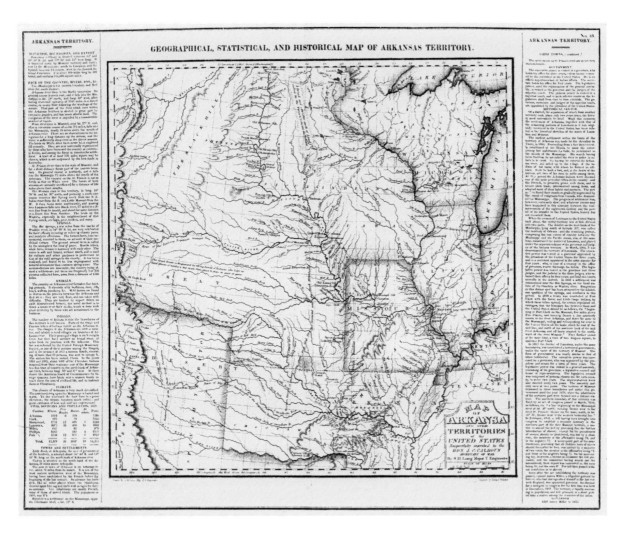

N° 98   **Carey, H.C. & Lea, I.** Philadelphia:1822 [1823]. *Map of Arkansa and other Territories of the United States Respectfully inscribed to the Hon. J.C. Calhoun Secretary Of War By S.H. Long Major T. Engineers.* Copperplate engraved map, surrounded on two sides by letterpress text. Map is 14.5 x 14.75 inches (37 x 37.5 cm). Overall image (map and text) 16.5 x 20.5 inches (42 x 52 cm). Original hand color.

Considered to be the first Arkansas map, the region covered extends well into the Northwest Territory. Space does not permit even a brief listing of the myriad data found on this map. However, we do note the Proposed National Road slicing westward just below the tip of Lake Michigan; the Dubuques Lead Mines; and the fact that The Great Desert is frequented by roving bands of Indians... Also shown are the several roads and trails leading to the southwest. Often ignored in descriptions of this map are the faint dotted lines that show the limits of various types of geological formations, such as limestone, coal and sandstone. As such it is an early geologic map as well. The letterpress text is fascinating and adds immeasurably to the interest.

Carey & Lea maps are often found with centerfold splitting and severe centerfold discoloration.

1822 – *Carey & Lea's Cuba*

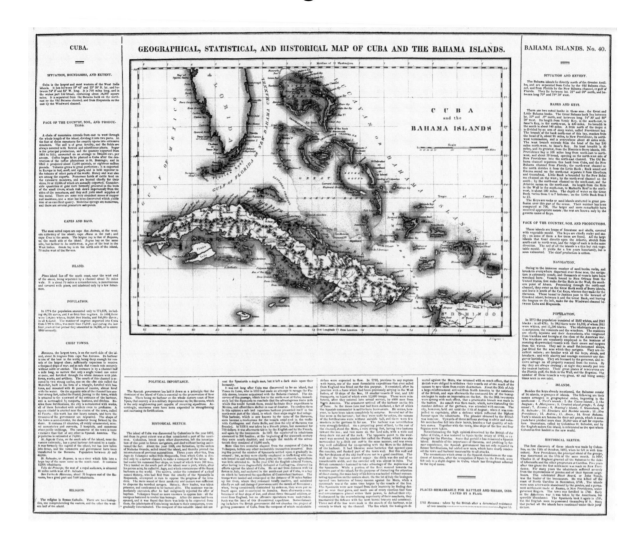

N° 99 **Carey & Lea.** Philadelphia: 1822. *Geographical, Statistical, and Historical Map Of Cuba And The Bahama Islands.* Copperplate engraved map. Surrounded on three sides by letterpress text. Map is 11.5 x 9.6 inches (23.5 x 35 cm) overall image (map and text)16.5 x 20.5 inches (42 x 52 cm). Original hand color.

The maps from the Carey & Lea atlas are characterized by extensive letterpress containing information relevant to the map. The map shown here is of Cuba and the surrounding islands. The juxtaposition of text and map results in a very decorative image.

The map shown here is from the first edition of the atlas.

1822 – *Jamieson's Celestial Maps*

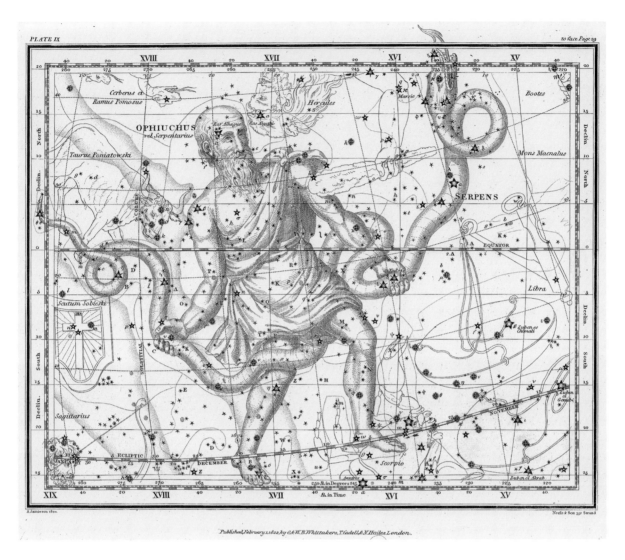

N° 100   **Jamieson, Alexander.** London: 1822. *(Celestial Map.)* Copperplate engraved map. 7 x 9 inches (17.8 x 23 cm). Uncolored.

Jamieson's celestial atlas, published in London by G.& W.B. Whittaker, contained thirty separate star maps and an extensive text and catalogue of interesting objects.

The allegorical constellation figures are still prominent and many of the stars are shown with their Bayer (see Map N° 40) numbers.

Jamieson's atlas was issued in both colored and uncolored versions. Many of the maps appearing separately have been colored by hand recently.

c. 1833 – *S.D.U.K.*

N° 101   **Society for the Diffusion of Useful Knowledge (S.D.U.K.).** London: c. 1833. *China and the Birman Empire With Parts Of Cochin-China And Siam.* Copperplate engraved map. 12.25 x 14.5 inches (31 x 37 cm). Original hand outline color.

The S.D.U.K. atlas is one of the great 19th-century atlases, produced entirely by copperplate engraving. The maps are exceptionally clear and unencumbered by unnecessary decoration. They present a very "modern" appearance. Although I have seen a few intact atlases that had full color which, to my eye, appeared to be of the period, most of the atlases had sparse outline color only. Some were produced uncolored.

222

c. 1833 — *S.D.U.K.*

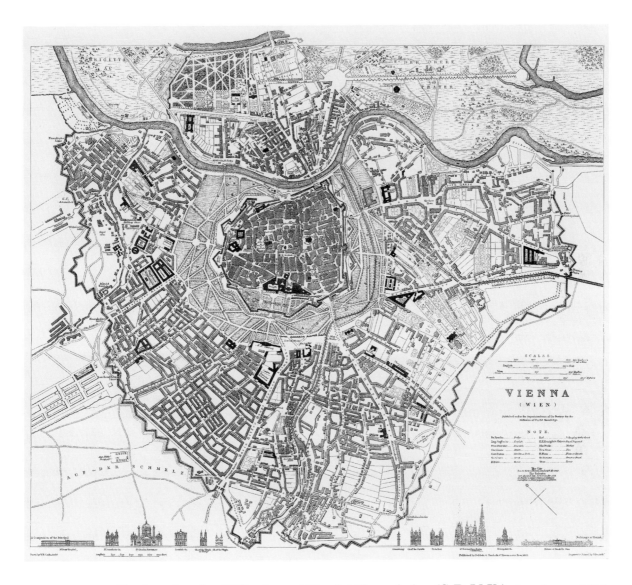

Nº 102   **Society for the Diffusion of Useful Knowledge (S.D.U.K.).** London: c. 1833. *Vienna (Wien).* Copperplate engraved city plan. 12.5 x 14.75 inches (32 x 37.5 cm). Original hand color.

The S.D.U.K. atlas contained over forty individual plans of some of the world's great cities. These generally had more color than the maps. In these plans rivers were generally blue and the cities' open spaces and parks were colored green. Many plans, such as that of Vienna illustrated here, had vignettes of important building facades; some had view details.

These plans are still rather common and inexpensive, yet they are among the nicer 19th-century city plans. Most are single-sheet, but some, such as Venice, Paris and London, are double-sheet plans.

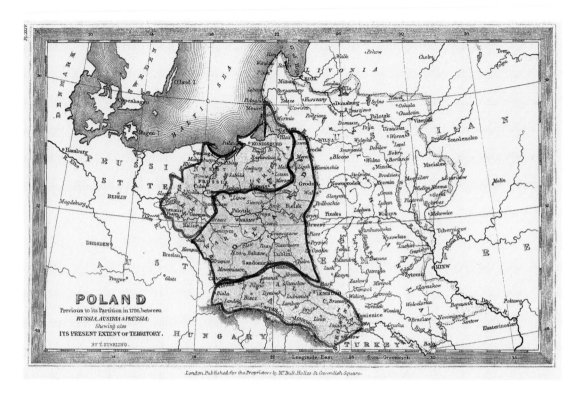

POLAND
Previous to its Partition in 1795, between
RUSSIA, AUSTRIA & PRUSSIA;
Shewing also
ITS PRESENT EXTENT OF TERRITORY.
BY T. STARLING.

London, Published for the Proprietors by M^r. Bull, Holles St. Cavendish Square.

Nº 103   **Starling, Thomas.** London: 1833. *Poland.* Copperplate engraved map. 3.5 x 5.5 inches (8.8 x 14.3 cm). From the *Geographical Annual or Family Cabinet Atlas.* Original color.

Also known as *Royal Cabinet Atlas*, this volume contains a series of very attractive little maps, each hand colored. The atlas contains a pair of small "Mountains of the World" and "Rivers of the World" plates, also fully hand colored.

Prices for the European maps, such as the Poland illustrated above, are much lower than prices for the New World maps.

Nº 104   **Mitchell, S. Augustus.** (Facing page.) Philadelphia: 1834. *The Tourist's Pocket Map of the State of Ohio Exhibiting Its Internal Improvements Roads Distances &c. By J.H. Young.* Copperplate engraved folding pocket map. 15 x 12.5 inches (38.5 x 32 cm). Original hand color.

This is a typical American 19th-century "pocket map." It folds upon itself and has a self-cover with a decorative embossed design, and the 1830 Ohio census on the inside. The map has colored counties and there is a profile of the Ohio and Erie Canal along the bottom. I chose this as a pocket map example because it has some of the typical defects these maps exhibit. There is minor staining along some folds and minor breaks in the paper where the folds intersect. It is unusual to find these maps in fine condition.

1834 – *Mitchell Pocket Map*

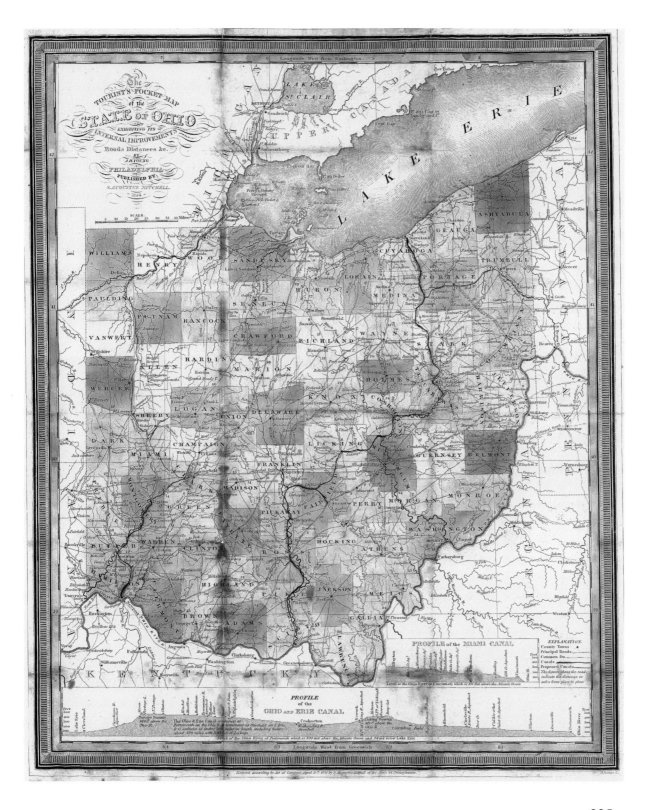

1835 – *Burritt's Star Charts*

N° 105   **Burritt, Elijah H.** New York: 1835-56. *Northern Circumpolar Map.* Copperplate engraved map. 12.5 inches diameter (32 cm). Original hand color.

This is a map from a very attractive American star atlas. The early editions all have hand color; by 1856 they were issued with an unattractive brown and pink color. One of the last star atlases to show allegorical constellations, the atlas was very popular, containing seven colored charts. The maps seem to be based on those of Pardies published a hundred years earlier.

226

1838 – *Tallis' Street Maps*

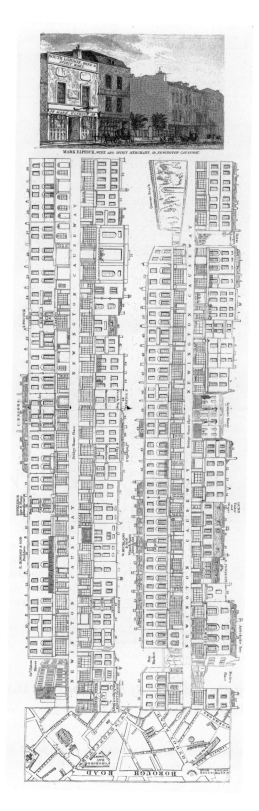

MARK ELFRICK, WINE AND SPIRIT MERCHANT, 59, NEWINGTON CAUSEWAY.

Nº 106

Tallis, John B. London:1838(9). *Newington Causeway.* Copperplate engraved street map. 16.3 x 4.6 inches (41.5 x 11.8 cm). Uncolored.

   Here is a street map from a set of London maps separately issued by Tallis. There is only one known complete set (located in the Guildhall) of all the sheets published, although individual exemplars are not uncommon. In these maps, each individual building is drawn and one could walk down the street identifying structures as one passed them. Each strip has one or more vignettes and some of the buildings carry advertising.

   The inset below shows a portion of the map at 75% of the full size. It provides a good idea of the great detail that was included in these strip maps.

227

1866 – *U.S. Land Office Report*

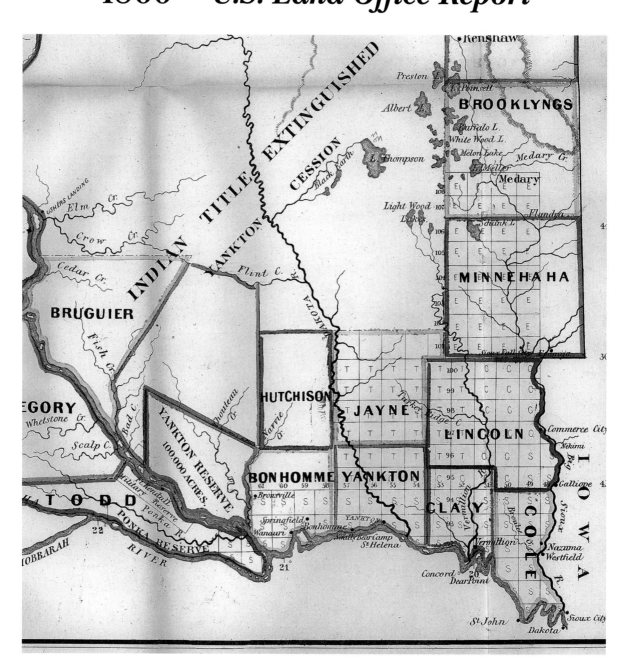

N° 113   **United States Land Office. Annual Report 1866.** *Dakota Territory.* Lithographed map. 23.5 x 20.25 inches (59.7 x 51.5 cm). Original outline color.

The Government issued maps of the ever-enlarging and subdivided United States in the 19th century. Many were singularly unattractive. However, this report, issued in 1866, contained 23 large maps of the states and territories that are unusually attractive. Printed on slightly better paper than the earlier Reports, the 1866 maps are quite desirable. One particularly scarce state of these maps was issued linen-backed. In this detail from the Dakota Territory map, we see counties and townships in the southeast and the large area that had, until recently, belonged to the Indians.

1860s – *Johnson's Southwest*

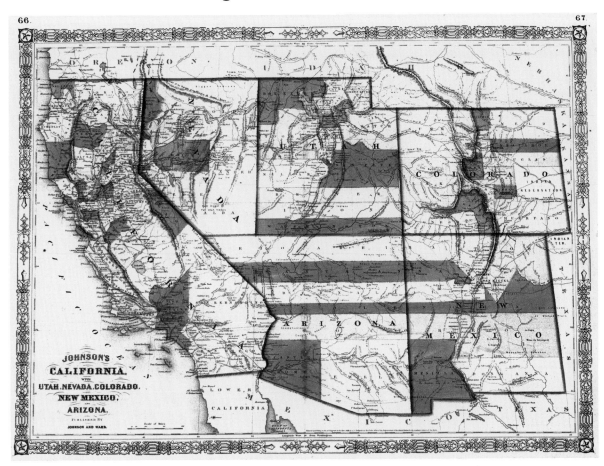

N° 114   **Johnson and Ward.** New York: [1860s.] *Johnson's California, Utah, Nevada, Colorado, New Mexico and Arizona.* Lithographed map. Approximately 11 x 17 inches (28 x 43 cm). Original full hand color.

This is a very interesting map of the American Southwest. It occupies a temporal crossroads, after California statehood, and before adjacent territories became states and before the railroad touched the Pacific.

There are many different states of this map, and they show dramatic and rapid changes in counties. The map is of historical value; it shows Pony Express routes, the locations of gold and silver mines, Indian tribes, wagon trails, and much more relating to the geography and history of the region. Large areas are devoid of detail indicating that exploration was not yet over in this region.

A large double-page map, the colors are typical Johnson pink, yellow, and green, making it an attractive map as well.

235

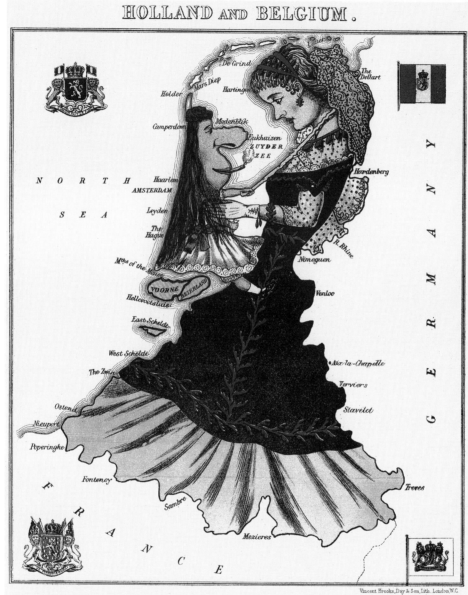

HOLLAND AND BELGIUM.

Dame Holland, trick'd out in her gala clothes,
And Master Belgium, with a punchy nose;

Seem on the map to represent a land,
By patriot worth, and perfect art made **grand**.

Vincent Brooks, Day & Son, Lith. London W.C.

N° 115

Aleph (pseudonym for William Harvey).
London: 1869.
Geographical Fun or Humorous Outlines of Various Countries.
Lithographed map. 9.5 x 8 inches (24 x 20.2 cm). Full printed color.

Published in 1869 by Hodder and Stoughton, these maps have become quite scarce. The author, the London physician William Harvey(!) writing under the pseudonym "Aleph," put forth a series of twelve maps showing various countries as curious people in the great tradition of English caricaturists. The maps themselves are anonymous, and Harvey states in the introduction that they are to be used to teach geography. It is assumed that they were not intended as political satires.

Printed by Vincent Brooks, Day & Son, the maps are an example of Victorian chromolithography. The set was reprinted in 1990 and collectors should be aware that reproductions exist.

1869 — *Teaching Geography*

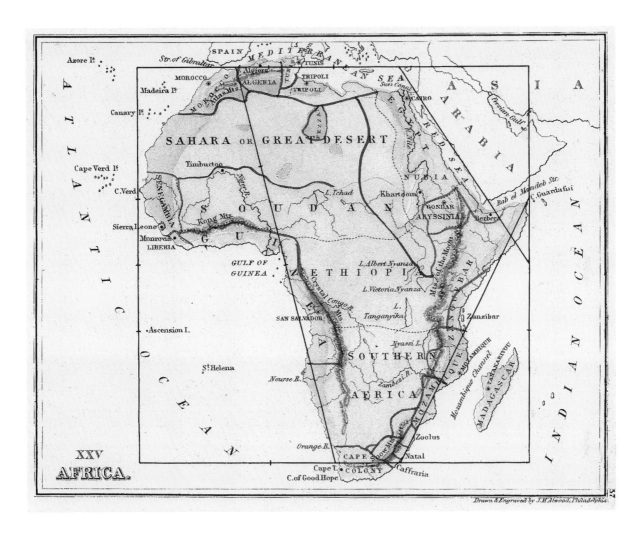

N⁰ 116 **Keam, Peter and John Mickleborough.** Philadelphia: 1869. *Africa.* Lithographed map. 5.25 x 6.75 inches (13.5 x 17.2 cm). Printed color.

Keam and Mickleborough were two teachers in the Cincinnati school system. They wrote a little book, *A Hand Book of Map Drawing* (J.H. Butler, Philadelphia), that was used to teach students to draw maps. The book had several maps such as the one shown above. In each case, the maps were described by a series of geometric figures. Opposing pages contained just the geometric figures and students were to use them as guides for drawing the map.

A number of such teaching methods were developed in both the 18th century and the 19th century and the maps provide a significant insight into the importance of geography in the schools of the era.

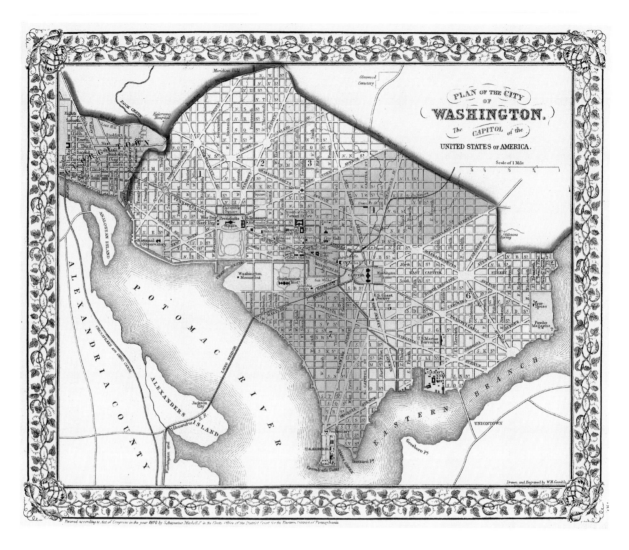

Nº 117   **Mitchell, S. Augustus.** Philadelphia: 1870. *Plan of the City of Washington, The Capitol of the United States of America.* Lithographed city plan. 11 x 13.5 inches (27.8 x 34 cm). Full printed color.

Mitchell's atlases were among the most popular of the 19th-century American atlases. His *New American Atlas* and *New General Atlas* were issued between 1846 and 1893. In addition to maps of states, territories and foreign countries, they included city plans of New York, Boston, Baltimore, Philadelphia and Washington, D.C.

The atlases were all issued with colored maps. The color varied from pale and drab to brilliant and vibrant. The city plan shown here has a "grape leaf" border typical of Mitchell maps.

Once very common and inexpensive, these atlases have been collected avidly in the past decade or so and are now commanding substantial prices; the individual maps have followed and they are no longer as cheap as they were.

c. 1874 – *A Spanish Thematic Map*

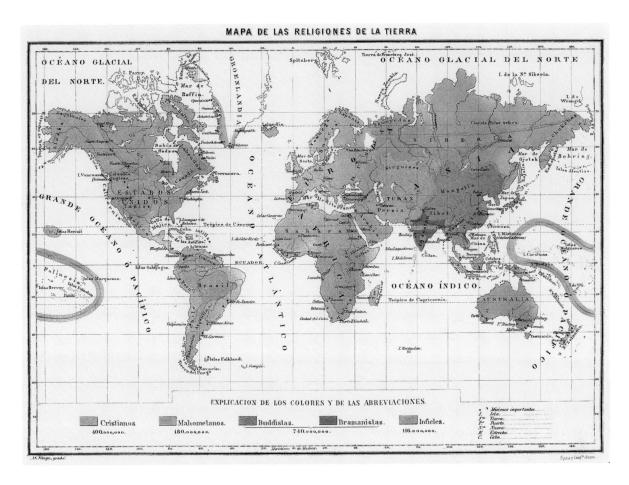

MAPA DE LAS RELIGIONES DE LA TIERRA

Nº 118 **(Anon).** (Madrid?): c. 1874. *Mapa De Las Religiones De La Tierra*. Lithographed map. 8.25 x 11.75 inches (21 x 30 cm). Full printed color.

Very typical of the vast number of loose maps that can be found in many dealer's drawers, this one shows the global distribution of different religions.

A thematic map is a map that portrays a particular subject, such as rainfall, animal life, or, as in this case, religion. Thematic maps can form the nucleus of a most interesting collection, and are often relatively inexpensive. The 19th century was replete with them, but as in the present case, it is often very difficult to track down their origin. A dealer would find it uneconomical to try to determine the source of this map; a collector could find it fascinating.

239

1875 – *Linen-backed Railroad Map*

Nº 119   **Stanford, Edward.** London: 1875. *Stanford's Smaller Railway Map of the United States Distinguishing The Unsettled Territories; The Railways; The Cities & Towns according to Population; also the State Capitals & County Towns.* Linen-backed folding map in self-cover. Approximately 15.5 x 27.5 inches (39.5 x 70 cm). Full hand color.

There was much European interest in the emerging rail lines in the United States after the Civil War. This English map shows the railroads in dark red against a pale color wash delineating the different states and territories. Similar maps were issued at frequent intervals as the expanding system required.

Railroad maps are a collecting subspecialty and many collectors concentrate on this topic. Maps such as the one shown here are somewhat scarce and priced accordingly, but many local maps, derived from atlases and also showing railroads, are still in plentiful supply and quite inexpensive.

1878 – *Schmidt's Moon*

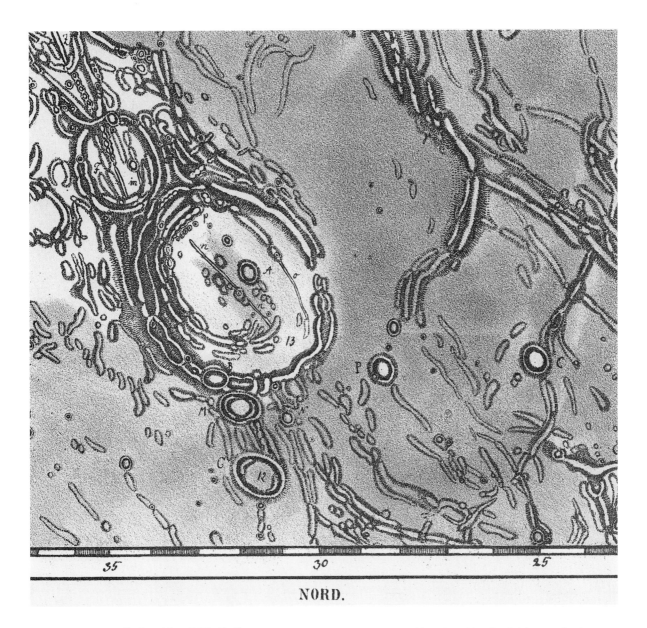

NORD.

N° 120   **Schmidt, J.F. Julius.** Berlin: 1878. *Charte der Gebirge Des Mondes*. Lithographed map. Detail, full size. Approximately 16 x 16 inches (40.7 x 40.7 cm). Uncolored.

Shown full size, this is one of the great 19th-century moon maps. Schmidt, director of Athens Observatory, had previously published a finely-engraved edition of Lohrmann's earlier moon map. He then completed this monumental work, based on observations made during 1840-70, published in only a small number in Berlin. The atlas consisted of a portfolio of loose sheets, a text volume and abbreviated text. British selenography was ascendant during this period and Schmidt's work was largely ignored by English-speaking lunar observers, most of whom were amateurs.

North is at the bottom, since the image is shown as it is seen in an astronomical (inverting) telescope.

1896 – *Monteith's Africa*

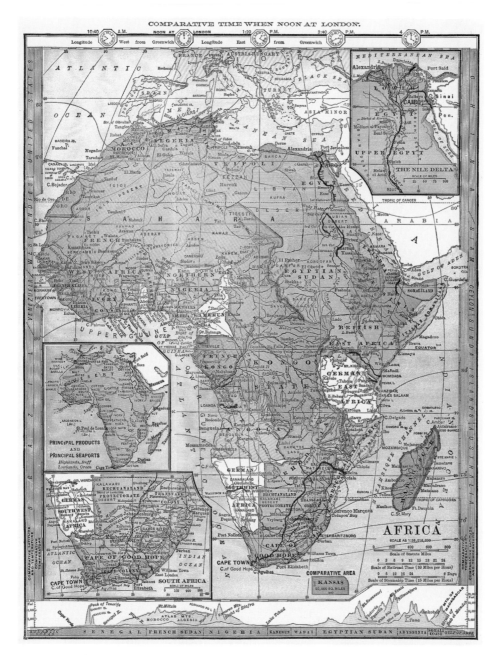

Nº 121

Monteith, James. New York: 1896. *Africa.* Lithographed map. Approximately 11 x 18.5 inches (27.7 x 21.5 cm). Full printed color.

The continent of Africa was, during the late 19th century, largely divided up among European colonial powers. These geopolitical changes, as well as the rapid mapping and European exploration of the interior, makes this an exciting century for Africa map collectors.

Most 19th-century maps of Africa are, to put it bluntly, cheap. The continent is undercollected and the maps plentiful. The map shown here is from *Barnes's Complete Geography* and, unlike most school geography maps of the period, has brilliant and attractive printed color. It should be noted that maps from this period were generally printed on inferior paper and should be deacidified to avoid continuing deterioration.

1912 — *Motoring tour map*

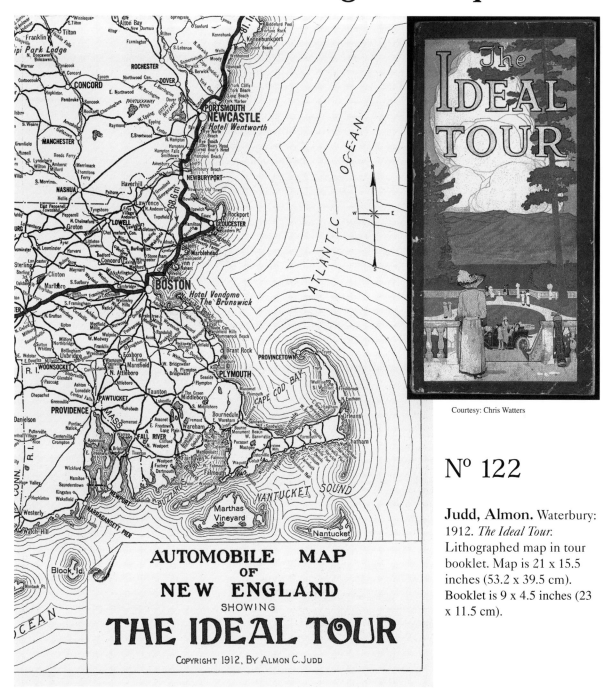

Courtesy: Chris Watters

Nº 122

Judd, Almon. Waterbury: 1912. *The Ideal Tour.* Lithographed map in tour booklet. Map is 21 x 15.5 inches (53.2 x 39.5 cm). Booklet is 9 x 4.5 inches (23 x 11.5 cm).

With the advent of serious motoring, a large number of private individuals, associations, and tourist organizations issued maps, many (such as this one) sponsored by hotels along the route. These ephemeral publications provide an insight into the activities and perceptions of the period, and also provide a record of roads that were suitable for motoring in the post-bicycle and pre-interstate highway period. The booklets or covers were often decorated nicely as we see in the inset above.

Maps such as these appeared in the United States and in Europe as well. They are still quite affordable but are not as common as just a few years ago. A few map dealers are beginning to stock this type of map; they can also be found at ephemera fairs.

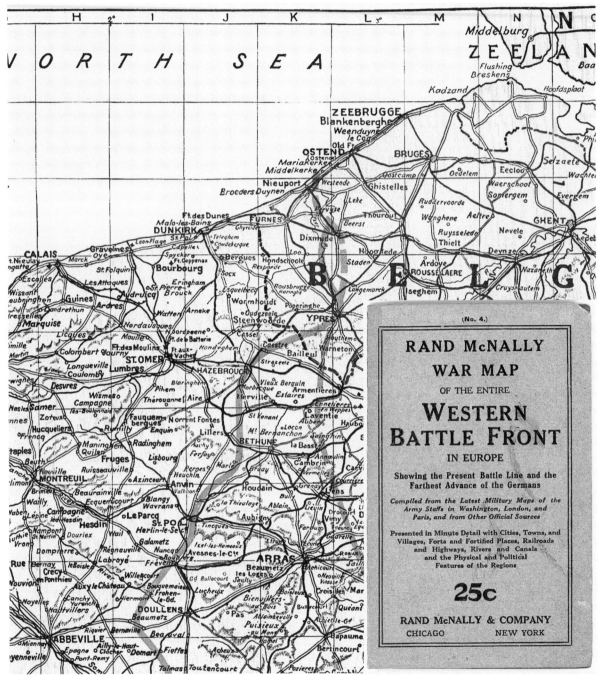

Courtesy: Chris Watters

Nº 123

Rand McNally. Chicago: nd. *War Map Of The Battleground Of Liberty.* Lithographed color map in self-cover (inset). 23 x 26 inches (58.5 x 66 cm).

During the Great War, map companies published lithographed maps showing the current trench lines in Europe. These very detailed maps are an indication of how stagnant the conflict became. After the war, the companies published maps for tours of the battlefields.

1918 – *Doubleday Burma*

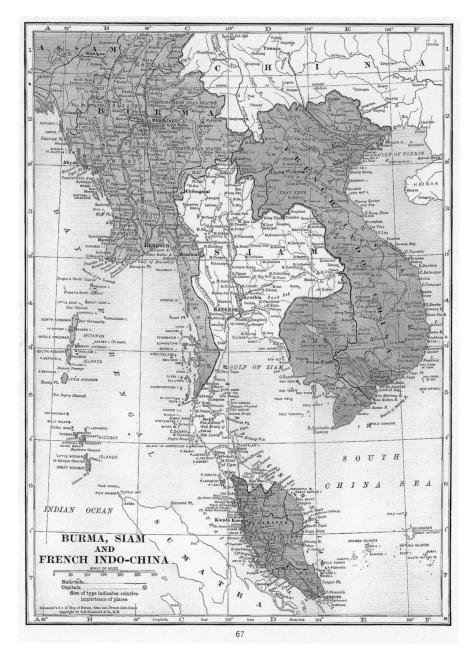

Nº 124

**Doubleday, Page &
Co.** Garden City:
1918.
*Burma, Siam And
French Indo-China.*
Lithographed map.11
x 8 inches (28 x 20.2
cm). Printed color.

Although published by Doubleday *(Geographical Manual and New Atlas)*, the map is a Hammond map. The full printed color is pale, and typical of 20th-century atlas maps. In an age of global war, when geopolitical boundaries changed rapidly, these maps can provide fertile ground for collecting. The shrinking European empires were chronicled on maps such as these, and fascinating sequences can be assembled inexpensively.

Printing technology had advanced substantially by the early 20th century and atlases became relatively cheap. Thus mass-produced maps from this period are plentiful and relatively inexpensive. The collector is cautioned that the paper is not of very high quality and should be routinely deacidified. In many cases, it is already browned.

c. 1920 — *Two 20th-Century Postcards*

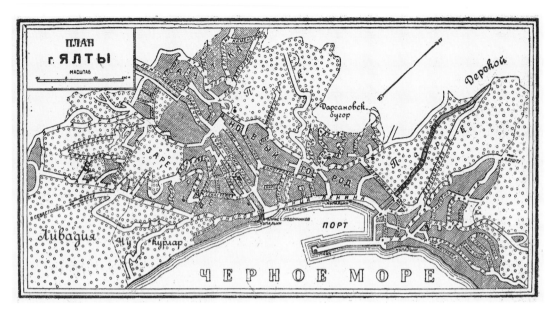

Nº 125, 126 — Two printed postcards. *Yalta* (top) and *Alicante* (bottom). Shown full size.

Shown full size, these two postcards are representative of the map-postals commonly found at postcard and ephemera fairs. The top one, uncolored, Russian, undated, but definitely post-1917 and pre-1940, shows Yalta on the Black Sea. The lower card, with original full printed color, printed in Barcelona, shows Alicante. Both are lithographed images. The variety of such postcards is staggering and they are quite cheap.

246

1920s-1940s — *American Road Maps*

Courtesy G. Robinson

Nº 127, 128

Pair of American Highway Maps. Pre-interstate highway maps, c.1940 (above); 1920s (next page). Folded lithographed maps; various sizes.

These rather typical American road maps were produced in very large numbers by oil companies (as above) and tourist service companies (next page) for free distribution. All the major oil companies produced similar maps and gasoline stations would have racks of them available for patrons to select.

Although many were discarded or "used up," relatively large numbers survive in good condition, and these maps are among the least expensive maps that have a wide collecting base today.

The cover illustration on the Sinclair map above is typical of the attractive lithographed designs that appeared on highway maps. Today, the colors and design evoke nostalgia and they are collected for these images as well as for the maps within.

Road maps were printed on cheap paper and folded. The paper should be deacidified and the maps handled carefully to prevent separation along the folds.

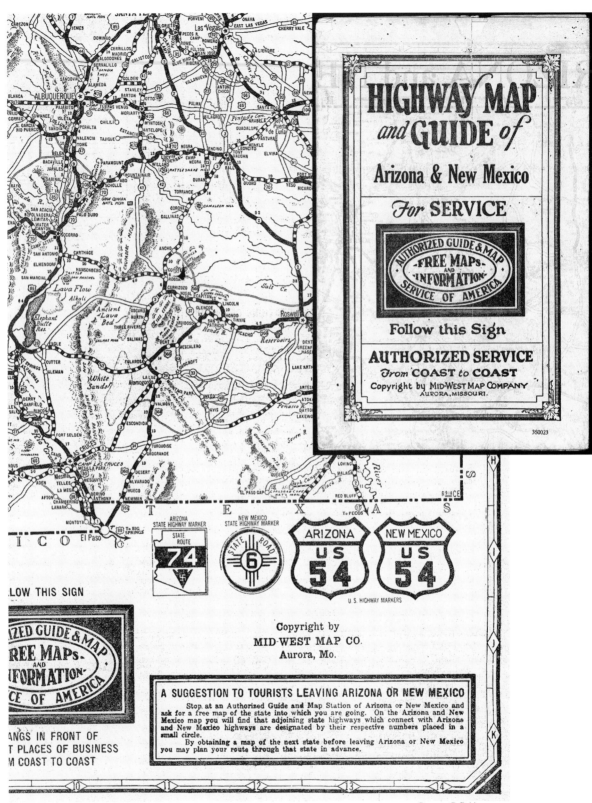

Courtesy G. Robinson

(For text, see previous page.)

248

1937 – *Paris Metro*

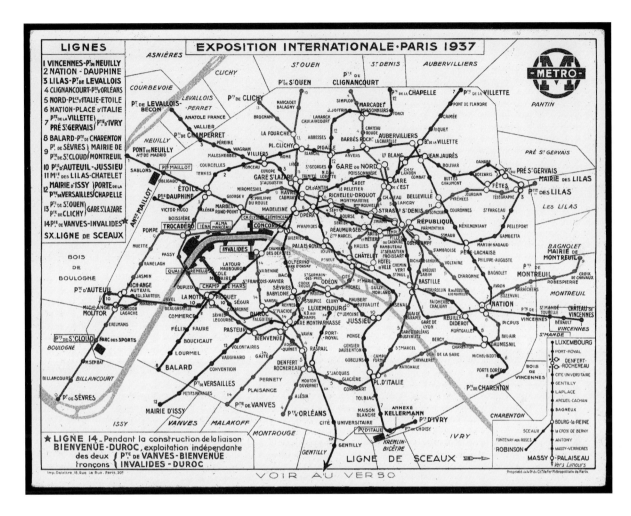

Nº 129 **Ch. de Fer Métropolitain de Paris.** Delattre, Paris: 1937. *Exposition Internationale • Paris 1937.* 5.6 x 7.5 inches (14.2 x 19 cm). Printed in color.

Ephemeral maps such as this one are generally not to be found in inventories of larger map dealers, but rather are to be discovered by burrowing through stacks of ephemera found on display at some book fairs and, certainly, at ephemera fairs.

In addition to being an attractive little artifact, maps such as these can be used to help date larger more detailed city plans that may have been issued without dates. Publishers sometimes printed city plans and guides without publication dates so that they had a longer sales life – without a printed date the buyer would not know that the map might be outdated.

1945 – *Military Map on Cloth*

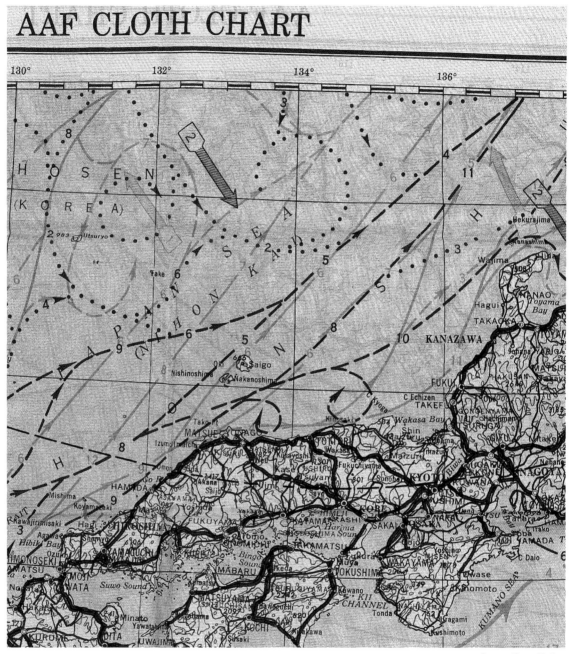

AAF CLOTH CHART

Courtesy: George Ritzlin

Nº 130   **WW II Cloth Chart.** Army Map Service, March 1945. *Japan and South China Sea* on recto, *East China Sea* on verso. 18.5 x 32.5 inches (47 x 82.5 cm). Printed in color, both sides, on woven rayon (?). (Photographed against black background to accentuate weave.)

    Military maps printed on silk are known from the early 18th century.

    Maps such as the one above were issued to fliers for use if forced down. They are lightweight, more durable than paper, and fold compactly. The illustration shows full-size detail. Similar maps, some on silk, were in use in both theaters of war and also during the Vietnam conflict. Many of these are fairly common and can be acquired inexpensively.

Part III

Appendices

Appendix A

The Makers of Maps

Who made all these maps? Who were the cartographers, publishers, engravers and sellers? This is a far more complex question than we might first imagine. Many of the cartographers were also engravers and publishers and ran their own map businesses. Sometimes some of these functions were farmed out. There was also a high degree of intermarriage among map families. We cannot begin to explore this fascinating topic here, other than to indicate that it exists. But we do present a much attenuated list of names and dates and briefly abstracted contributions. *This list is not complete.* It will, however, be adequate for many purposes and I have tried to include many of the more commonly encountered names.

Aa, Pieter van der. *Dutch map publisher, worked in Leiden. 1659-1733.*
Atlas nouveau et curieux... 2 volumes. Leyden. 1714,1728.
La galerie agréable du monde... 66 volumes, from 6-48 maps per volume. Leyden. 1729.
Nouvel Atlas... 1714.
Représentation où l'on voit un grand nombre des isles, cotes, rivières et ports de mer... Leyden. c. 1730.
Cartes des itineraires et voïages modernes... Leyden. 1707.

Adrichom, Christian (Adriche, Adrichomius). *Dutch theologian and surveyor. 1533-1585.*
Theatrum Terrae Sanctae. This is one of the earliest atlases of the Holy Land, containing 12 maps. Copied extensively by later cartographers.

Ainslie, John. *British surveyor and map engraver. 1745-1828.*
Several Scottish maps.
A New Atlas... A. Constable and Company, Edinburgh. 1814.

Åkerman, Andreas. *Swedish engraver of maps, celestial and terrestrial globemaker. 1723?-1778.*
Atlas Juvenilis... Uppsala. 1789?
Atlas Hydrografica... 1768.

Allard (Allardt), Carel. *Dutch map publisher. 1648-1709.*
Atlas Minor. Amsterdam. 1696.
Atlas Major. 3 volumes, c. 1705.
Nova Tabula India Orientalis. Amsterdam. c. 1680.

Andriveau-Goujon. *French cartography and publishing company; produced maps and globes. 19th century.*
Atlas Classique et Universal de Géographie... Paris. 1843-1844.
Atlas Élémentaire Simplifié de Géographie... (with E. Soulier) Paris. 1838.
Atlas de Choix... Paris. 1841-1862.

Apianus, Petrus (Apian, Peter). *Astronomer Royal and cartographer, produced maps, celestial and terrestrial globes, as well as astronomical charts. 1495-1552.*
World map (woodcut) in: *Joannis Camertis Minoritani...* by C. S. Solinus. Joannes Singrenius, Vienna. 1520. This is the second map to bear the name "America," the earliest being by Waldseemüller. Apianus also wrote a little cosmography (*Cosmographia Petri Apiani...*) that appeared in many editions, the first in 1522. From 1544 on it included the famous cordiform world map based on that of Gemma Frisius.

Arrowsmith, Aaron, Jr. *British cartographer, son of Aaron, Sr. "Hydrographer to his Majesty." 19th century.*
Orbis terrarum veteribus noti descriptio... London. 1828, 1830. (Published for use at Eaton College)
Arrowsmith's comparative atlas... (with S. Arrowsmith) London. 1829, 1830.
Outlines of the world... S. Arrowsmith, London. 1828.

Arrowsmith, Aaron. *British publisher and cartographer. 1750-1823.*
Various maps of American counties and localities.
A map exhibiting all the discoveries in the interior parts of North America. Faden, London. 1802.
A new map of Mexico... London. 1810.
Chart of the West Indies... London. 1803.
Map of the world... London. 1794, 1799.
A New General Atlas... A. Constable and Company. Edinburgh, London. 1817.
A New and Elegant General Atlas... (with S. Lewis). J. Conrad and Co. Philadelphia. 1804; Thomas and Andrews, Boston. 1805, 1812, 1819.
Nouvel Atlas Universel-portatif de Géographie Ancienne et Moderne. H Langlois, Paris. 1811. Maps engraved by Ambroise Tardieu, Bondeau, Semen. Maps by Arrowsmith, J.B. Poirson, Bonne, Lorin, d'Anville. Arrowsmith published a great number of large-scale maps, including those of Africa, the U.S.A. (1796 and thereafter); Persia, 1813; Pacific Ocean, 1798; Panama Harbour, 1806; etc.

Arrowsmith, John. *British cartographer and publisher. Nephew of Aaron, Sr. 1790-1873.*
The London Atlas. London. 1834 and other editions.
The London Atlas of Universal Geography. London. 1840 and other editions.
Flinder's and Light's Maritime Portion of S. Australia. London. 1839.
Light's District of Adelaide. London. 1839.

Arrowsmith, Samuel. *British cartographer, son of Aaron, Sr. active c. 1840.*
The Bible Atlas... London. 1835.

Baldwin and Cradock. *Publishers, London, 19th century.*
Published *Wyld's Atlas* and some editions of the S.D.U.K. atlas, from 1831 on.

Barbié du Bocage, Jean Denis. *French cartographer and globemaker. 1760-1825.*
Recüeil de cartes géographiques... Sanson and Company, Paris. 1791.

Barents, William (Barentszoon; Barendsz; Bernard, Guillaume). *Dutch cartographer and navigator. 1550-1597.*
Nieuwe Beschryvinge Ende Caertboek vande Midlandtsche Zee... C. Clae, Amsterdam. 1595.
Description de la Mer Méditerranée... C. Nicolas, Amsterdam. 1607.
Sea-charts and maps in Waghenaer's *Spieghel der Zeevaert.*

Behaim, Martin. *German geographer and terrestrial globemaker. 1459-1506.*
Behaim made the oldest surviving modern terrestrial globe.

Bellin, Jacques Nicholas (Belin). *French cartographer and publisher. 1703-1772.*
Le Petit Atlas Maritime. Paris. 1764.
Atlas Maritime. 1751.
Neptune Français. 1753.
Carte de Cours de Fleuve de Saint Laurent depuis Quebec. Paris. 1761.
Partie Occidentale de la Nouvelle France, ou du Canada. Paris. 1745.
Partie Orientale de la Nouvelle France, ou du Canada. Paris. 1745.
Carte Réduit de l'océan Oriental... Paris. 1740.
Recüeil des Villes, Ports d'Angleterre (with J. Rocque). Desnos, Paris. 1766.

Bertelli: Donato, Fernando, and others. *Venetian publishers. 16th century.*
The Bertellis produced copies of Gastaldi's oval world map, 1562, 1565, 1568; sold composite atlases and worked with Camocio, Forlani and Zenoi. The Bertellis sometimes signed with their initials, e.g. "F.B.", "D.B."

Bien, Julius. *Publisher. 1829-1909.*

Bertius, Petrus (Berts, Pierre; Bertii, Petri). *Belgian publisher and cartographer; cosmographer to Louis XIII. 1565-1629.*
Tabularum Geographicam. 1600,1602,1606; J. Hondius, Amsterdam. 1616.
Variae Orbis Universi... (maps engraved by Melchoir Tavernier). Paris. 1628?

Black, Adam and Charles. *British publishers, A. Black, 1784-1874.*
General Atlas. Edinburgh and London. 1840, subsequent editions.
Atlas of Australia. 1853.

Blackie, Wallis Graham. *British cartographer. Died 1906.*
The Imperial Atlas of Modern Geography. London. 1860.

Blackwood, William and Sons. *British publishers. Active 1839-1853.*
Atlas (engraved by Lizars). 1838, later editions.

Blaeu, Willem (Janssonius, Jans Zoon, Guglielmo). *Founder of the Blaeu family firm in Amsterdam. 1571-1638.* He studied under Tycho Brahe and became a globe and instrument maker, mapmaker and publisher. He issued his first atlas in 1630 using plates he had acquired from Hondius.
Holland. 1604.
Spain. 1605.
World. 1605.
Continents. 1608.
Licht der Zeevaert. 1608.

Blaeu, Joan and Cornelius. *Dutch cartographers and publishers. Sons of Willem. 1596-1673; 1610-1648.* The Blaeu firm was burned out in the Great Fire of 1672.
Atlas Major. 9 or 11 vols 1662.
Town Plans. 1649-1653.
Also published wall maps, now rare.

Blome, Richard. *British cartographer and topographer. Died 1705.*
Britannia. T. Roycroft. 1673.
Speed's Maps Epitomised. 1681, 1685.
Cosmography and Geography. T. Roycroft, London. 1682, 1693.

Blunt, E. and G. W. *American cartographers and publishers. 1802-1878.*
Blunt's Charts of the North and South Atlantic Oceans... New York. 1830. Some charts were based on those of des Barres and others.

Bonne, Rigobert. *French engineer, and globemaker. 1727-1795.*
Atlas de Toutes les Parties Connues du Globe Terrestre... J.L. Pellet, Geneva. 1780.
Atlas Portatif. Paris. c.1785.

Atlas Encyclopedique (with N. Desmarest). Paris. 1787-1788.
Atlas Maritime. Lattre, Paris. 1762.

Bordone, Benedetto. *Italian cartographer, worked in Venice. 1460-1531.*
Published "Isolario," 1528, 1532, 1534, 1537, 1547.

Bowen, Emanuel. *British printseller, publisher, engraver; engraver of maps to George III and Louis XV. Died 1767.*
Brittania Depicta or Ogilby Improved (with J. Owen). 1720 and subsequent editions.
Complete System of Geography. 1744-1747.
Complete Atlas or Distinct View of the Known World. 1752.
Maps for Harris's Voyages. 1744.
The Royal English Atlas. 1762 and subsequent editions.
An Acurate Map of the Island of Barbadoes. London. 1747.
The Large English Atlas (with T. Kitchen). Bowles, London. c.1760, 1767.

Bradford, Thomas Gamaliel. *American cartographer. 19th century.*
Atlas Designed to Illustrate the Abridgement of Universal Geography... W.D. Ticknor, Boston; F. Hunt and Co., New York. 1835.
A Comprehensive Atlas... W. D. Ticknor, Boston; Wiley and Long, New York. 1835.
An Illustrated Atlas... of the United States. E.S. Grant and Co., Philadelphia. c. 1838.
A Universal, Illustrated Atlas... (with S.G. Goodrich). C. D. Strong, Boston. 1842.

Breese, Samuel. *American cartographer. 19th century.*
The Cerographic Atlas of the United States. S.E. Morse and Co., New York. 1842.
Morse's North American Atlas. Harper and Bros., New York. 1842-45.

Brion de la Tour, Louis. *French cartographer. 18th century.*
Atlas Général... Paris. 1766, 1767.
Coup d'oeil Général sur la France... (based on Desnos). Grangé, Guillyon and others, Paris. 1765.

Briot, B. J. *French map publisher. Active c. 1660.*
Atlas de France... Paris. 1820-28.
Atlas Universel... Paris. 1816...

Brué, Adrien Hubert. *French publisher. 1786-1832.*

Buache, Philippe. *French cartographer, celestial and terrestrial globe maker. 1700-1773.*
Atlas géographique de parties du monde. 1769-1799.
Atlas géographique et universelle. 1762.
Cartes et tables de la géographique physique. 1754.
Considérations géographiques... Ballard, Paris. 1753-1754.

Bünting, Heinrich. *German cartographer. 1545-1606.*
Itinerarium Sacrae Scripturae. 1581 and later. This is the atlas that contains the famous Clover-leaf World map, Europe as a virgin, and Asia as Pegasus.

Burdett, P. P. *British cartographer. Acitve 1777-1794.*

Burr, David H. *American cartographer. 1803-1875.*
The American atlas... J. Arrowsmith, London. 1839.
An atlas of the state of New York... New York. 1829, 1838.
A new universal atlas... D.S. Stone, New York. 1835?

Bussemacher, Johann. *German publisher and printer. Active in Cologne. fl. 1580-1613.*
Noted for publishing Quad's Atlas, 1592, 1600 and later.

Butler, Samuel Bishop of Lichfield and Coventry. *British cartographer and antiquary. 1774-1839.*
An atlas of ancient geography... Carey and Son, Philadelphia. 1831; Carey, Lea and Blanchard, Philadelphia. 1834; Blanchard and Lea, Philadelphia. 1855.
An atlas of modern geography... Longman, Brown, Green and Longmans, London. 1844, etc.

Camden, William. *British antiquary and historian. 1551-1623.*
Brittania. 1607; editions into the 1700s.

254

Carey, Henry Charles; and Lea, Isaac. *American map publishers. 1793-1879; 1792-1886 respectively.*
A complete historical, chronological and geographical American atlas... 1822. Philadelphia. 1823, 1827.
Family cabinet atlas. Philadelphia. 1832, 1834.

Carey, Mathew. *American cartographer. 1760-1839.*
Carey's American atlas... Philadelphia. 1785, 1809.
Carey's American pocket atlas... Philadelphia. 1796, 1801, 1805, 1813, 1814.
Carey's general atlas... Philadelphia. 1796, 1802, 1814, 1817, 1818.
A scripture atlas... Philadelphia. 1817.
A general atlas for the present war... Philadelphia. 28 January 1794.

Carleton, Osgood. *Surveyor and publisher. Active in Boston. late 18th-early 19th century.*
Noted American mapmaker, produced large-scale maps of the U.S. and the Northeast.

Carver, Captain W. *British surveyor and cartographer. 18th century.*
American atlas. 1776.
A new map of the province of Quebec... Sayer and Bennett, London. 1776.

Cary, George; and John II. *British publishers, cartographers, celestial and terrestrial globe makers. died 1859; c. 1791-1852 respectively. .*

Cary, John. *British engraver, publisher, celestial and terrestrial globe maker. c.1754-1835.*
Issued many atlases. Engraved maps for Camden's "Brittania," 1789.

Cellarius, Christophorus (Keller). *German cartographer. 1638-1707.*
Harmonia Macrocosmica. 1660 (later reprinted by Valk & Schenk).

Chatelaine, H.A. *French cartographer and publisher. 1884-1743.*
Atlas Historique 7 volumes. 1705-1720 and later.

Cluverius, Philip (Cluver, Clüver). *German geographer and mapmaker. 1580-1623.*
Introductionis in Universam Geographicam. Jansson-Waesberg, Amsterdam. 1661, 1676, 1682. Joannem Pauli, Amsterdam. 1729; G. Muller, Brunswick. 1641; Elzevir Press, Amsterdam. 1659; J. Wolters, Amsterdam. 1697; I. Buonem, Wolfenbüttel. c. 1667; *Germaniae Antiqae UrbiTres.* Ex officina Elziviriana, Leyden. 1631.

Coignet, Michel. *French mathematician, cartographer. 1549-1623.* Published and edited editions of Ortelius' *Epitome* from 1601.

Collins, Captain Greenville. *British surveyor, cartographer, and Hydrographer to the King. Active 1669-1696.*
Great Britain's Coasting Pilot. London. 1693-1792.

Colton, G.W. *American cartographer, publisher. 19th century.*
An influential American publisher of atlases, pocket maps and wall maps.
Numerous maps and atlases.

Cook, Captain James. *British navigator and cartographer. 1728-1779.*
The North American Pilot (with M. Lane). Sayer and Bennett, London. 1779.
A General Chart of the Island of Newfoundland (with M. Lane). London. 1775.
Chart of the N.W. Coast of America and N. E. coast of Asia. Faden, London. 1784.
Le Pilot de Terre-neuve... (with M. Lane). Paris. 1784.
Chart of the Southern Hemisphere... Strathan and Cadell, London. 1777.

Coronelli, P. Vincenzo Maria. *Italian theologian, cartographer, geographer, celestial and terrestrial globemaker. 1650-1718.*
Atlante Veneto. Venice. 1690, 1691, 1695-1697.
World map. 1695.
Corso Geografico Universale. Venice. 1692.
Gli Argonauti. Venice. 1693.
Isolario dell'Atlante Veneto. Venice. 1696-1697.

Coulier, P. J. *French cartographer. 19th century.*
Atlas Général... Paris. 1850.

Covens, Jean. *Dutch publisher. Peak activity c. 1740.*
L'Amerique Septentrionale. Amsterdam. 1757.
Atlas Nouveau... Amsterdam. 1683-1761.
Mappe Monde ou Globe Terrestre... Amsterdam. 1703?
Nieuwe Atlas... Amsterdam. 1730-1739, 1740-1817.
Nouvel Atlas... 1735.
Veteris Orbis Tabulae Geographique. Amsterdam. 1714
Covens worked with Corneille Mortier as "Covens and Mortier," published maps by a number of cartographers and engravers including J. Condet, Cassini, Delisle, van der Aa, Allard, Sanson, H. Jaillot, Visscher, etc.

Cowperthwait, DeSilver, & Butler. *American publishers active in Philadelphia. mid-19th century.* Replaced the firm of Thomas Cowperthwaite & Co. This company published many editions of atlases by Mitchell.

Cram, George. *American publishers. 1841-1928.*
Unrivalled Family Atlas. 1883.
Unrivalled Atlas. 1887.

Cramer, Zadok. *1773-1813*
Navigator. 1806; through 1818.

Cruchley, George Frederick. *British printer, mapseller, engraver, celestial and terrestrial globe maker. Active 1822-1875.*
Published the 1862 edition of J. Cary's *New and Correct English Atlas.*

Cummings, Jacob Abbott. *American geographer. 1772-1820.*
Cummings Ancient and Modern Geography. Boston. 1813 and later editions up to c. 1821.
School Atlas to Cummings Ancient and Modern Geography... Cummings and Hilliard, Boston. 1818.

Dalrymple, Alexander. *British cartographer and publisher. 1737-1808.* Created charts and plans of various parts of the world; specialized in the East Indies.

Dancker[t]s, Cornelis Hendrik, Justus (Dankers, Danquerts, Danckertz). *Family of Dutch publishers, globe-makers. Late 16th to early 17th century.*

d'Anville, Jean Baptiste Bourguignon. *French cartographer. 1697-1782.*
Nouvel Atlas de la Chine, de la Tartarie Chinoise, et du Thibet. H. Scheurleer, le Haye. 1737.
Atlas Général. 1740 and subsequent editions.
Géographie Ancienne Abrégée. 1769-1820.
Atlas Antiquus Danvillanus Minor... Schneideri-Weigeliana. 1798-1799.
A Complete Body of Ancient Geography. Sayer, London. 1771; Sayer and Bennett, London. 1775 and later editions.
A New Map of North America. R. Sayer, London. 1763.
The Western Coast of Africa... Sayer, London. 1789.
North America... London. 1775.
Carte des Isles de l'Amerique... Homann, Nuremberg. 1740.
Atlas and Geography of the Ancients... T. Chaplin for J. Davis, London. 1815.
Atlas to the Ancient Geography... R. M'Dermut and D.D. Arden, New York. 1814.

Danz, C. F. *German cartographer. 19th century.*
Acht tafeln zur physisch-medicinischen topographie des kreises schmalkalden... (with C.F. Fuchs). N.G. Elwert, Marburg. 1848.

d'Après de Mannevillette, Jean Baptiste Nicolas Denis (Manneville). *French cartographer. 1707-1780.*
The Oriental Pilot. Laurie and Whittle, London. 1797.
Instructions sur la Navigation des Indes Orientales et de la Chine... Dezauche, Paris. 1775.
The East-India Pilot... Laurie and Whittle, London. 1795.
Le Neptune Oriental... J.Robustel, Paris. 1745 and subsequent editions.
Supplément au Neptune Oriental... Demonville, Paris; Malassis, Brest. 1781.

Darby, John. *British cartographer and publisher. Active 1677.*
Darby worked on John Seller's *English Pilot* and *Atlas Maritimus.*

Darton, William (William and Sons). *British geographers and publishers. Active 1810-1837.*
Darton's New Miniature Atlas. London. 1820, 1825.
Union Atlas. London. 1812.

de Bry, Theodore. *Engraver, publisher, worked in Frankfurt on Main. c. 1527-1598.*
His *Collection of Voyages, Grands et Petits Voyages* is well known for early detailed maps of Americas as well as unusual copperplate engravings of views and native Americans.

de Champlain, Samuel. *French cartographer. 1567-1635.*
Les Voyages de le Seur de Champlain... Paris. 1613.

de Fer, Antoine. *French mapseller. Active 1640s.*

de Fer, Nicolas. *French geographer, publisher and engraver. 1646-1720.*
La France Triomphante Sousla Régne Louis le Grand. 1693.
Atlas Curieux. 1700-1705.
Cartes et Descriptions Générales et Particuliéres au Sujet de la Succession de la Corunne d'Espagne. 1701-1702.
Petit et Nouveau Atlas. 1705.
Les Postes de France et d'Italie. 1700, 1728, 1760.
Introduction à la Géographie... Denet, Paris. 1717-1740.
Atlas ou Recüeil de Cartes Géographiques... Paris. 1709-1728, 1746-1753.
Atlas Royal... Paris. 1699-1702.

de Herrera y Tordesillas, Antoine. *Spanish cartographer. 1559-1625.* Responsible for the first known printed map showing California as an island (1622).
Description des Indes occidentales. Madrid. 1601; Emmanuel Colin, Amsterdam; Michel Soly, Paris. 1622; N.R. Franco, Madrid. 1726.
Historia general de los hecho de los castellanos en las islas i tierra firme del mar oceano... N.R. Franco, Madrid. 1726.
Nievve werelt anders ghenaempt West-Indien. M. Colijn, Amsterdam. 1622.
Novus orbis, sive descriptio Indiae occidentalis... M. Colinium, Amsterdam. 1622.
Novi orbis pars duodecima... Theodore de Bry, Frankfurt. 1624.

de Jode, Gerard (de Iudaeis; de Iudoeis, Gerardi; Judaeus). *Dutch cartographer and publisher. 1509-1591.*
Speculum orbis terrarum. Antwerp. 1578; (with Cornelis de Jode) 1593-1613. The 1593 Cornelis de Jode reissue appeared as *Speculum orbis terrae.*

de Jonghe, Clemendt. *Dutch cartographer. 17th century.*
Europae nova discriptio. Amsterdam. 1661.

de la Feuille, Daniel. *Cartographer, worked in Amsterdam. 1640-1709.*
Atlas portatif... Amsterdam. 1706-1708.

de la Feuille, Jacobo. *Cartographer. 1668-1719.*

de la Feuille, Paul. *Cartographer. 18th century.*

de la Fossé, J. B. *Cartographer. 18th century.*

de la Harpe, Jean Francois. *French cartographer. 1739-1803.*
Abrégé de l'histoire générale des voyages... E. Ledoux, Paris. 1780, 1820.

de la Tour. *French? cartographer. 18th century.*
Plan of New Orleans. Jefferys, London. 1759.

de Laet, Joannes. *Dutch geographer. 1593-1649.*
Nieuvve Wereldt... I. Elzever, Leyden. 1625, 1631, etc.
Beschrijvinghe van West-Indien... Elzivir, Leyden.
Historie ofter Iaerlijck verhael van de verrichtinghen der geoctroyeerde West-Indische compagnie. Elzevir, Leyden. 1644.

de las Cases, Emmanuel (Marie Joseph Auguste Emmanuel Dieu-Donné, Comte de las Cases; A. Le Sage). *French cartographer. 1766-1842.*

De Lat, Jan. *Dutch cartographer. 18th century.*
Atlas portatif trés exact... J. Keyser, Deventer. 1747.

de Medina, Pedro. *Spanish navigator. Active 1540.*
Arte de navegar... (with a map of the New World). 1545.

de Rapin-Thoyras, Paul. *British historian. 1661-1725.*
Atlas to accompany Rapin's History of England... J. Harrison, London. 1784-1789. This atlas contains many views of cities and fortifications.

de Ségur, Louis Philippe, Comte. *French cartographer. 1753-1830.*
Atlas pour l'histoire universelle... A. Eymery, Paris. 1822; Furne et Cie, Paris. 1840.

De Silver. (see Cowperthwaite).

de Wit, Frederick (Witt). *Dutch cartographer and publisher.*
Zee karten. Amsterdam. *1675.*
Nieut kaert-boeck. Amsterdam. c. 1720.
Ducatum LIvoniae et Curlandia. Amsterdam. c. 1690.
Regni Norvegia. Covens and Mortier, Amsterdam. c. 1744.
Regni Suecial... Covens and Mortier, Amsterdam. c. 1730.
Atlas & zee-karten. Amsterdam. c. 1660.
Urbis Moskvae. Amsterdam. c. 1660.
Atlas. Amsterdam. 1670 and later.
Atlas major... Amsterdam. After 1688.
Atlas minor... Amsterdam (sold by Christopher Browne). 1708.

Delamarche, Alexandre. *French cartographer, made celestial, terrestrial globes and armillary spheres. 1815-1884.*
Atlas de la géographiques ancienne, du moyen age, et moderne... A. Grosselin et Cie, Paris. 1850.

Delamarche, Charles François. *French cartographer, made celestial and terrestrial globes and armillary spheres. 1740-1817.*
Institutions géographiques, ou description générale du globe terrestre... Bellin, Paris. 1795.

Delamarche, Félix. *French publishers. 19th century.*
Atlas de la géographie ancienne du moyen age, et moderne... Paris. 1820.

Delisle, Guillaume (de L'Isle; del Isle). *French cartographer, geographer, and globe maker. 1675-1726.*
Atlas de géographie. Paris. 1700-1712.
Atlas géographique des quatre parties du monde (with P. Buache). Dezauche, Paris. 1769-1799,1789, 1831.
Atlas géographique et universel... Paris. 1700-1762, 1781-1784, 1789-1790.
Atlas nouveau... Covens and Mortier, Amsterdam. 1730, 1733.

Delisle, Joseph Nicolas. *French cartographer, astronomer. 1688-1768.*
Atlas russicus... St. Petersburg. 1745.

des Barres, J. F. W. *British publisher. 18th century.*
The Atlantic neptune. London. 1779
A chart of the harbour of Rhode Island... London. 1776. Des Barres' charts of the eastern American seacoast are among the most prized.

Desnos, L. C. *Danish cartographer, chartmaker, and globe maker. Active mid-18th century.*
Almanach géographique... Paris. 1770.
Atlas chorographique... Despilly, Paris; Savoye. 1763-1766.
Atlas général, contenant le detail des quatre parties du monde... Paris. 1767-1769, 1790-1792.

Disturnell, John. *American (New York) publisher. 1801-1877.*
30 Miles Round NY. 1839.

Dix, Thomas. *British surveyor and geographer. Active 1799-1821.*
A complete atlas of the English counties... (with William Darton). London. 1822.

Doncker, Hendrick. *Dutch cartographer. 17th century.*
De zee-atlas ofte water-woereld... Amsterdam. 1600-1661, 1665, 1666.
The lightning columne, or sea mirrour... (with T. Jacobsz and H. Goos). C. Loots-Man, Amsterdam. 1689-1692.

Doolittle, Amos. *American engraver. 1754-1832.*
Responsible for many early American maps including ones in Carey's *American Atlas,* 1795 and the *American Pocket Atlas* 1813.

Doppelmayr, Johann (Gabriel; Doppelmaier). *German physicist, cartographer, chartmaker, and globe maker. 1677-1750.*

Drury, Andrew. *British cartographer and publisher. 18th century.*
A new and universal atlas...engraved by Mr. Kitchin and others. A. Drury and R. Sayer, London. 1761; A. Drury, R. Sayer, and C. Bowles, London. 1763.

du Caille, Louis Alexandre. *French cartographer. 18th century.*

du Pinet, Antoine Sieur de Noroy. *French cartographer. 1510?-1566?*
Plantz...de plusieurs villes... I. d'Ogerolles, Lyons. 1564.

du Val, Pierre (du Val d'Abbéville). *French cartographer and geographer. 1619-1683.*
Cartes de géographie... A. de Fer, Paris. 1662; 1688-1689.
Cartes et tables de géographie... Paris. 1667.
La géographie françoise. Paris. 1677.
La monde ou la géographie universelle... N. Pepingue, Paris. 1670; du Val, Paris. 1682.

Dufour, Auguste Henri. *French cartographer. 1798-1865.*
Atlas de géographie numismatique... Crozet, Paris. 1838.
Le globe; atlas classique... J. Renouard et Cie, Paris. 1835.

Dunn, Samuel. *British cartographer and astronomer. Died 1794.*
A map of the British Empire in North America. London. 1774
A new atlas of the mundane system. R. Sayer, London. 1774; Laurie and Whittle, London. 1786-1789, 1796, 1800.

Elwe, Jan Barend. *Amsterdam publisher. Late 18th century.*
Reis Atlas...Duitschland. 1791.
Atlas. 1792.

Emory, William Hemsley. *American topographical engineer, Lt. Col. 1813-1875.*
USA — Mexico Boundary. 1857.

Evans, Lewis. *British surveyor. c. 1700-1756.*
Maps for Jefferys's *American Atlas.* 1776.
Map of the middle British colonies in America. Philadelphia. 1755.
A new and general map of... the United States of America. Laurie and Whittle, London. 1794.

Faden, William. *British cartographer, engraver, publisher. 18th-19th century.*
A map of a part of Yucatan... within the bay of Honduras... London. 1787.
Plan of Quebec. London. 1776
The North American atlas. London. 1777.
Le petit neptune français. London. 1793.
Map of the Ottoman dominions in Asia... London. 1822.
The course of the Delaware river... London. 1778.
The United States of North America. London. 1785.
Atlas minimus universalis... London. 1798.
Atlas of battles of the American Revolution... Bartlett and Welford, London. 1845?

General atlas... London. 1797, 1799, 1750-1836. Succeeded Thomas Jefferys as Faden and Jefferys.

Field, Barnum. *American cartographer. 19th century.*
Atlas designed to accompany the American school geography... W. Hyde and Co., Boston. 1832.

Findlay, Alexander George. *British cartographer and engraver. 1812-1875.*
General chart of the Mediterranean sea. Laurie, London. 1839.
A modern atlas... W. Tegg and Co., London. 1850.

Finé, Orance (Finaeus, Orontius; Finnaeus). *French astronomer, mathematician, cartographer. 1484-1555.*
Single-heart shaped woodcut world map. 1519.
Double-heart shaped in *Novus orbis regionum...* by S. Grynaeus and J. Huttich. Paris. 1531.

Finley, Anthony. *American publisher, cartographer. 19th century.*
Atlas classica... Philadelphia. 1829.
A new American atlas... Philadelphia. 1826.
A new general atlas... Philadelphia. 1824, 1829, 1830, 1831, 1833.

Frémont, John Charles. *American explorer. 1813-1890.*
Missouri to Rockies. 1842, 1843.
Gold region W. Kansas. 1860.

Fry, Joshua; and Jefferson, Peter. *British cartographers. 18th century.*
A map of the inhabited part of Virginia... Sayer and Jefferys, London. 1754? This is a very important and sought after map. A reduced version was included in deVaugondy's atlas.

Fullarton, Archibald. *British printer, publisher. 19th century.*

Furlong, Lawrence. *American cartographer. 18th-19th century.*
The American coast pilot... E. M. Blunt, Newburyport. 1809.

Gastaldi, Giacomo (Gastaldo, Jacobo; Castaldo). *Italian cartographer and cosmographer. c. 1500-c. 1565.*
World map. Venice. 1546. Gastaldi's world map was subsequently copied by de Jode, Forlani, F. Bertelli, Camocio, Duchetti, Valgrisi and Ortelius. His (1848) atlas has become quite scarce; it contains several New World maps that are much sought after.

Gendron, Pedro. *Spanish cartographer. 18th century.*
Atlas ô compendio geographico del globo terrestre... Barthelemi, Madrid; Bonardel, Cadiz. 1756-1758.

Geographisches Institut Weimar. *Austrian celestial and terrestrial globemakers. Nineteenth century.*

Gibson, John. *British engraver. Active 1750-1787.*
Atlas minimus....revised, corrected and improved by Eman. Bowen, geographer to his majesty. J. Newberry, London. 1758; T. Carnan and F. Newberry, London. 1774-1779, 1792; M. Carey, Philadelphia. 1798.

Goodrich, Charles Augustus. *American geographer. 1790-1862.*
Atlas accompanying Rev. C. A. Goodrich's outlines of modern geography... S. G. Goodrich, Boston; M'Carty and Davis, Philadelphia. 1826.

Goodrich, Samuel Griswold. *American cartographer and publisher. 1793-1860.*
Atlas designed to illustrate the Malte-Brun school geography. H. and F.J. Huntington, Hartford. 1830, 1838.
A general atlas of the world... C. D. Strong, Boston. 1841.

Goos, Abraham. *Dutch engraver; worked on celestial and terrestrial globes. Active c. 1640.*
Africae described... Published by John Speed, editions 1626-76.

Goos, Pieter. *Dutch cartographer. Active 1654-1666.*
Pas-kaarte van de zuyd-west-kust van Africa. Amsterdam. 1669.
Pascaerte van nova Hispania Chili... Amsterdam. 1666.
Paskaarte van het zuy de lijckste...van Rio de la Plata... Amsterdam. 1666.
Paskaarte vertonende alle de zekusten van Europa... Amsterdam. c. 1665.

Pascaertevan Groen-Landt... Amsterdam. 1669.
Noordoost cust van Asia van Japan tot Nova Zemla. Amsterdam. 1669.
Paskaerte zynde t'oosterdeel van Oost Indien... Amsterdam. 1669.
Paskaart van Brazil... Amsterdam. 1669.
De zee atlas, ofte water-weereld... Amsterdam. 1666 and other editions.
L'atlas de la mer... Amsterdam. 1670.
Le grand et nouveau miroir ou flambeau... Amsterdam. 1671.
The lighting column or sea mirrour. Amsterdam. 1660-1661.
De nieuwe groote zee-spiegel... Amsterdam. 1676.
Pascaart van de noort zee. Amsterdam. 1669.
Pas caerte van Nieu Nederlandt... Amsterdam. 1669.
Paskaerte van de zuyt en noort... in Nieu Nederlandt. Amsterdam. 1669.
Pascaert vande Caribes eylanden. Amsterdam. 1669.
Pascaerte van West Indien... Amsterdam. 1669.
Nieuwe werelt kaert. Amsterdam. c. 1669.

Gordon, Thomas F. *American mapmaker. 1787-1860.*
Gazetteer of the State of New York... Philadelphia. 1836.

Götz, Andreas (Goetzio, Andrea). *German cartographer. 1698-1780.*
Brevis introductio as geographiam antiquam... B. J. C. Weigel, Nuremberg. 1729.

Gray, O.W. & Son. *American engineers, collaborated with Walling to produce maps and atlases.* Latter half 19th century.

Greenleaf, Jeremiah. *American cartographer. 1791-1864.*
A new universal atlas... G. R. French, Brattleboro, Vermont. 1842.

Greenleaf, Moses. *American cartographer. 1777-1834.*
Atlas accompnying Greenleaf's Map and statistical survey of Maine... Shirley and Hyde, Portland, Maine. 1829.

Greenwood, Christopher and John. *British cartographers. 1786-1855; active 1821-1840, respectively.*
Atlas of the counties of England... J. and C. Walker, 1834.

Guicciardini, Lodovico (Luigi). *Italian cosmographer, historian and cartographer. 1521-1589.*
Descrittione de... tutti i paesi bassi... 1567, editions to 1660. Contains many attractive city views.

Güssefeld, F. L. *German cartographer. Active 1797.*
Charte von Nord America. Homann's Heirs, Nuremberg. 1797.
Charte über die XIII vereinigte staaten von Nort-America. Heritières de Homann, Nuremberg. 1784.

Guthrie, William. *American geographer. 1708-1770.*
Atlas universel pour la géoraphie de Guthrie. H. Langlois, Paris. 1802.
General atlas for Guthrie's Geography... B. Warner, Philadelphia. 1820.
The atlas to Guthrie's System of Geography... C. Dilly and G. G. and J. Robinson, London. 1795.
The general atlas for Carey's edition of Guthrie's Geography improved... Mathew Carey, Philadelphia. 1795.

Hall, Sidney. *British engraver, cartographer, chartmaker. Active 1818-1860.*
A travelling county atlas... Chapman and Hall, London. 1842.
Black's general atlas... A. and C. Black, Edinburgh. 1840, 1841.
A new general atlas... Longmann, Rees, Orme, Brown and Green, London. 1830; Longman, Brown, Green and Longmans, London. 1857.

Halley, Edmund (Halleius, Edmundus). *British astronomer, globemaker. 1656-1742.*
A new and correct chart...in the western and southern oceans... London, 1701.
Novum & accuratissimen totius terrarum orbis tabula... R. and I. Ottens, Amsterdam. 1740.

Hart, Joseph C. *American cartographer. Died 1855.*
A modern atlas of fourteen maps... R. Lockwood, New York; J. Grigg and A. Finley, Philadelphia. 1828, 1830; and other editions.

Hawkesworth, John. *British. 1715?-1773.*
An account of the voyages undertaken...by commodore Byron, captain Wallis, Captain Carteret and Captain Cook...atlas. W. Strahan and T. Cadell, London. 1773.

Hennepin, R. P. Ludovico. *German cartographer. 18th century.*
Amplissimae regionis Mississippi... Homann, Nuremberg, c. 1720.

Hinton, John. *British bookseller and publisher. Died 1781.*
Hinton sold many of Emanuel Bowen's maps.

Hoefnagel, Joris (Hufnagel, Georg). *Belgian cartographer. 1542-1600.* Engraved many of Braun & Hogenberg's views.
Brightstowe. Braun and Hogenberg, Cologne. c. 1575.
Canterbury. Braun and Hogenberg, Cologne. c. 1575.

Hogg, Alexander. *London publisher. Late 18th century.*
Published several large-scale plans of London.

Holland, Major Samuel. *British cartographer, surveyor. 18th century.*
A map of the island of St. John. London. 1775.
A chorographical map of the country between Albany, Oswego, Fort Fontenae, and les trois rivieres. London. 1775.
A new and accurate chart of the North American coast... Laurie and Whittle, London. 1808.
Chart for the navigation between Halifax and Philadelphia. Laurie and Whittle, London. 1798.

Hollar, Wenceslaus. *London etcher and engraver, born in Prague. 1607-1677.*

Homann, Johann Baptist (Homanno, Bpt.). *German cartographer, publisher, celestial and terrestrial globe maker. 1663-1724.*
Africa. Hasio. c. 1737.
Totius Africae nova repraesentatio. c. 1720.
Totius Americae... Nuremberg. c. 1720.
Agri Parisiensis. Nuremberg.. c. 1730.
Virginia, Maryland, et Carolina... Nuremberg. c. 1759.
Planiglobii terrestris. Nuremberg. c. 1702.
Atlas Germaniae specialis... Nuremberg. 1753.
Atlas novus terrarum... Nuremberg. 1702-1750.
Grosser atlas. Nuremberg. 1716, 1737-1770.
Kleiner atlas scholasticus... Johann Hubern, Nuremberg and Hamburg. 1732.
Neuer atlas... Nuremberg. 1710-1731; 1712-1730.

Homann's Heirs (Héritières de Homann, Homannianos Heredes). *German publishers. 18th and 19th century.*
Atlas compendiarus... 1752-1790.
Atlas geographicus... Nuremberg. 1759-1784.
Bequemer hand-atlas... Nuremberg. 1754.
Homannischer atlas... Nuremberg. 1747-1757.
Kleiner atlas... Nuremberg. 1803.
Major atlas scholasticus... Nuremberg. 1752-1773.
Schul-atlas... Nuremberg. 1743, 1745-1746.
Städt-atlas... Nuremberg. 1762.

Hondius, Henricus (Hondt; de Hondt). *Dutch cartographer, chartmaker, globe maker. 1587-1638.*
Africae nova tabula. J. Jansson, Amsterdam. c. 1650.
India quae orientalis dicitur... J. Jansson, Amsterdam. c. 1652.
Nova totius terrarum orbis geographica... Amsterdam. 1630.
Nouveau théâtre du monde... J. Jansson, Amsterdam. 1639-1640.

Hondius, Jodocus (Hondt; de Hondt; Jodok). *Dutch cartographer, engraver, chartmaker, globe maker. 1563-1611.*
Africae, nova tabula. J. Jansson, Amsterdam. 1632.
Nieuwe caerte van...Guiana. Amsterdam. c. 1605.
Andaluziae nova descripttio. Antwerp. c. 1607
Nova totius Europae descripttio (with Petrus Kaerius). Amsterdam. 1613.
Nova et accurata Italiae hodiernae descriptio. Abraham Elzivir, Batavorum. 1627.

Honter, Johann (Honterus, Johannes; Grass, J.). *German cartographer. Active c. 1540.*

258

Rudimentorium cosmographicorum. Kronstadt. 1542; Froschouer, Zürich. 1549, 1564, 1570. Noted for his cordiform world map.

Hornius, Georg (Horn). *Dutch cartographer. 1620-1670.*
Accuratissima orbis antiqui delineato. J. Jansson, Amsterdam. 1654.
Accuratissima orbis delineato. J. Jansson, Amsterdam. 1660.
A compleat body of ancient geography... T. Osborne, London; The Hague. 1741.

Horsburgh, James. *British cartographer, publisher. 1762-1836.*
Atlas of the East-Indies and China Sea. London. 1806-1821.
Steel and Horsburgh's New and complete East-India pilot (with Penelope Steel). Steel and Goddard, London. 1817.

Houzé, Antoine Philippe. *French cartographer. 19th century.*
Atlas universel historique et géographique... P. Duménil, Paris. 1848, 1849.

Hoxton, Walter. *British cartographer. 18th century.*
...Mapp of the bay of Chesepeack... W. Betts and E. Baldwin, London. 1735.

Hulsius, Lieven. *Dutch cartographer. Active 1597.*

Humboldt, Friedrich Heinrich Alexander von, Baron. *German explorer, inventor of the isothermal line. 1769-1859.*
Atlas du Nouveau Continent. Paris. 1814-34.

Hutchins, Thomas. *Geographer to the United States (first and only!). 1730-1789.*

Jaillot, Charles Hubert Alexius (Alexis Hubert; Hubert). *French engraver, cartographer, globe maker. 1640-1712.*
Europe. Covens and Mortier, Amsterdam. c.1730.
Le Canada. Paris. 1696.
Atlas... Paris. 1696.
Atlas nouveau... Paris. 1689.
Atlas françois... Paris. 1695-1697; 1695-1701.

Jamieson, Alexander. *British astronomer. Active 1822.*
Celestial Atlas. London. Various editions.

Jansson, Jan (John; Ieansson, Iean; Janssonius, Joannes; Janssonium, Joannem; Janssen). *Dutch publisher and printer. 1596-1664.*
America nova delineata. Amsterdam. 1652.
Nova et accurata poli arctici... Amsterdam. c.1652.
Insularum Indiae orientalis nova descriptio. Amsterdam. c. 1652.
Mar di India. Amsterdam. c.1650.
Accuratissima Brasiliae tabula. Amsterdam. c.1650.
Essexiae descriptio. Amsterdam. 1652.
Vectis Insula. Amsterdam. 1646.
Insula Zeilan. Amsterdam. 1652.
Chili. Amsterdam. c.1652.
China. Amsterdam. 1652.
Guina sive Amazonium regio. Amsterdam. 1650.
Guinea. Amsterdam. c. 1650.
Nova Hispania et Nova Galicia. Amsterdam. 1657.
Nova Belgica et Anglia Nova. Amsterdam. 1652.
Belgii novi... Amsterdam. 1657.
Virginiae partis australis... Amsterdam. 1649.
Le nouvel atlas ou théâtre du monde... Amsterdam. 1647, 1639-1640.
Nuevo atlas... Amsterdam. 1653.
Nouveau théâtre du monde ou Nouvel atlas. Amsterdam. 1647-1657.
Novus atlas... Amsterdam. 1657-1658.
Atlantis majoris... Amsterdam. 1650, 1657.
Atlas minor... Amsterdam. 1651.
La guerre d'Italie... Amsterdam. 1702.
Joannis Janssonii atlas contractus... Amsterdam. 1666.
Die neuwen atlantis... Amsterdam. 1639, 1644; French, 1639, German, 1640.
Nieuwen atlas... 3 volumes: 1642-1644; 5 volumes: 1652-1653; 6 volumes: 1657-1658.
Tooneel der vermaarste koop-steden en handelplaatsen van de geheelde wereld... Amsterdam. 1682.

Jansson's name is sometimes coupled with Waesburg: Jansson-Waesburg. The Jansson lineage is complex and will not be listed here. Jansson, along with H. Hondius published the Hondius atlases. After H. Hondius died, he continued to enlarge and publish them.

Jefferys, Thomas (Jefferies). *British engraver and publisher. 1695?-1771.*
American atlas. London. 1775, 1776, 1778.
A plan of the town of Northhampton... London. 1746.
An exact chart of the river St. Laurence... London. 1775.
A new map of Nova Scotia... London. 1775.
A description of the Spanish islands and settlements...West Indies. London. 1762.
The West India atlas. Sayer and Bennett, London. 1775; Julien, Paris. 1777.
Mappa ou carta geographica... Faden, London. 1790.
A map of South Carolina... London. 1757.
The west coast of Florida and Louisiana. R. Sayer, London. 1775.
A map of the most inhabited parts of New England... London. 1774.
A complete pilot for the West Indies... Laurie and Whittle, London. 1794-1795.
A description of the maritime parts of France... Faden and Jeffrys, London. 1774.
The West-India islands... Laurie and Whittle, London. 1775.
A general topography of North America and the West Indies... Sayer and Jefferies, London. 1762, 1768.

Johnson, A.J. *American publisher. Mid-19th century.*
*Johnson's New Illustrated (steel plate) Family Atlas...*1861 to 1880s under various titles.

Johnston, Alexander Keith. *British publisher. 1804-1871.*
Stanford's library map of Australia. London. 1879.

Julien, Roch-Joseph. *French publisher. 18th century.*
Atlas géographique et militaire de la France... Paris. 1751.
Le théâtre du monde. Paris. 1768.

Keere (Kaerius). *Dutch cartographer, bookseller and engraver in Amsterdam; worked in London. Hondius' brother-in-law. 1571-1646.*

King, Clarence. *American. Active 1860s-1870s.*

Kircher, Athanasius. *German astronomer; Jesuit. 1602-1680.*
Mundus Subterraneus. Amsterdam. 1665 and later. This had the first map to show ocean currents; first to plot location of volcanoes.

Kitchin, Thomas (Kitchen). *British engraver and publisher. 1718-1784.*
A plan of the navigable canals... Liverpool. c.1770.
A general atlas... R. Sayer, London. 1768-88; c. 1783; Sayer and Bennett, London. 1787. Laurie and Whittle, London. 1797.
The small English atlas... (with T. Jefferys). London. 1749, 1751, 1775, 1785, 1787.
The large English atlas... R. Wilkinson, London. 1762.
The royal English atlas... (with Emmanuel Bowen). R. Wilkinson, London. 1762, 1778.
England illustrated. London. 1764.

Köhler, Johann David (Koehler; Koeleri, Io Davidis). *German cartographer. 1684-1755.*
Bequemer schul- und reisen-atlas... C. Weigelin, Nuremberg. 1734?

Kruse, Karsten Christian. *German cartographer. 1753-1827.*
Tabellen und charten zur allgemeinen geschichte der drey letzen jahrhunderte... Leipzig. 1821.

Lafreri, A. (Lafréry, Antoine). *Italian cartographer. 1512-1577.*
Geografia tavole moderne... Rome. c.1575.

Langenes, Barent. *Dutch cartographer. Active at the end of the 16th century.*
Caerte-thresoor... C. Claesz, Amsterdam. 1599.
Hand-boeck... C. Claesz. Amsterdam. 1609.

Lapié, Alexander Émile. *French cartographer. 19th century.*
The French empire. Faden, London. 1811, 1813.

Atlas universel... (with P. Lapie). Eymer, Fruger et Cie, Paris. 1829-1833.

Lapié, Pierre. *French cartographer. 1777-1851.*
Atlas classique et universel... Magimel, Picquet, Paris. 1812.

Latour, Arsène Lacarrière. *American cartographer. Died 1839.*
Atlas to the historical memoir of the war in West Florida and Louisiana... J. Conrad and Co., Philadelphia. 1816.

Lattré, Jean. *French publisher and engraver. Active 1772-1735.*
Atlas tôpographie des environs de Paris... Paris. c. 1762.

Laurie, Robert. *British publisher and engraver. c.1755-1836. Operated in partnership with James Whittle as Laurie and Whittle.*
A new and correct map of the British colonies in North America... London. 1797.
A new and general map of the southern dominions... United States of America. London. 1794
Laurie and Whittle's New improved English atlas. London. 1807.
The country trade East-India pilot... London. 1799, 1803.
A new and elegant imperial sheet atlas... London. 1796; 1798; 1800, 1808, 1813-1814.
A new juvenile atlas... for J. Melish, J. Vallance, and H.S. Tanner by G. Palmer. Philadelphia. 1814.
Laurie and Whittle's New and elegant general atlas... London. 1804.

Lavoisne, C. V. *American cartographer. 19th century.*
A new historical, chronological and geographical atlas... M. Carey and Son, Philadelphia. 1820, 1821.

Le Moyne de Morgues, Jacques. *French engraver. 16th century.*
Engraved a Florida map in de Bry's *Brevis narratio corum quae in Florida...* Frankfurt am Main. 1591.

Le Rouge, George Louis. *French cartographer and publisher. Active 1741-1779.*
Atlas Ameriquain septentrional... Paris. 1778.
Atlas nouveau portatif... Crepy, Paris. 1748, 1756-1759, 1767-1773.
Pilote Américain septentrional... Paris. 1778.
Recueil des plans de l'Amerique septentrionale... Paris. 1755.
Recueil des villes, ports d'Angleterre. Paris. 1759.
Atlas général... Paris. 1741-1762.

Lea, Philip. *British cartographer, publisher, and globe maker. Active 1666-1700.*
A new mapp of America. London. c. 1686.
A new mapp of Asia. Lea and Overton, London. c. 1686.
A new map of Carolina. London. c. 1686

Lea and Blanchard. *American publishers located in Philadelphia. Mid-19th century.*

Leigh, Samuel. *British publisher. Active 1820-1842.*

Lelewel, Joachim. *Polish cartographer. 1786-1861.*
Atlas de J. Lelewela... Vilna and Warsaw. 1818.

Levasseur, V. *French cartographer. Active c. 1847.*
Atlas national...de la France... A. Combette, Paris. 1847.

Lewis, Meriwether. *American explorer. 1774-1809.*

Linschóten, Jan Huygen. *Dutch cartographer. 1563-1610.*
Histoire de la navigation aux Indes Orientales. Cloppenburch, Amsterdam. 1619.
Navigatio ac itinerarium Johannis Hugonis Linscotani... Alberti Henrici, The Hague. 1599.

Lizars, William Home. *British engraver and publisher. 1788-1859.*
Engraved Blackwood's *Atlas,* 1830.

Lotter, Tobias Conrad. *German publisher and cartographer. 1717-1777.*
A map of the most inhabited part of New England. Augsburg. 1776.

Recens edita totius Novi Belgii... Augsburg. 1760.
Atlas géographique... Homann's Heirs, Nuremberg. 1778.
Atlas minor... Augsburg. c. 1744.
Atlas novus... Augsburg. c. 1722.

Lucas, Fielding, Jr. *American cartographer and publisher. 1781-1852.*
A general atlas... Baltimore. 1823.
A new and elegant general atlas... Baltimore. 1816?
A new and general atlas of the West India islands... Baltimore. 1824.

Luyts, Jan. *Dutch cartographer. 1655-1721.*
Introductio ad geographiam novam et veterem... F. Hana, Trajecti ad Rhenum. 1692.

Magini, Giovanni Antonio (Magin, Anthoine). *Italian geographer. 1555-1617.*
Italia. Nic. Tebaldini, Bologna. 1620, 1642. Magini's *Italia* is apparently the first map of Italy by an Italian.

Mallet, Alain Manesson. *French cartographer. 1630-1706?*
Description de l'univers... D. Thierry, Paris. 1683; other editions and dates.

Malte-Brun, Conrad. *German cartographer. Active 1775-1786.*
Atlas complet... F. Buisson, Paris. 1812; T. Lejeune, The Hague, Brussels. 1837.
A new general atlas... Grigg and Elliot, Philadelphia. 1837.
Universal Geography... A. Finley, Philadelphia. 1827-1829; J. Laval, Philadelphia. 1832.

Marshall, John. *American historian. 1755-1835.*
Life of George Washington...maps... C. P. Wayne, Philadelphia. 1807; J. Crissy, Philadelphia. 1832.

Mawman, Joseph. *English bookseller and, later, publisher. Active late 18th-early 19th century.*

Mayer, Johann Tobias, the Elder. *German cartographer and astronomer. 1732-1762.*

Mela, Pomponius. *Roman geographer. Active A.D. 50.*
Pomponii Melae de situ orbis... S. Birt, London. 1739.

Melish, John. *American cartographer. 1771-1822.*
A military and topographical atlas of the United States... G. Palmer, Philadelphia. 1813, 1815.

Mercator, Arnold. *Dutch cartographer. 1537-1587.* Son of Gerhard Mercator.

Mercator, Gerhard (Merkator). *Dutch cartographer, and globe maker. 1512-1594.*
Atlas minor... (with J. Hondius). Amsterdam. 1607, editions through 1738.
Atlas ofte afbeeldinghe vande gantsche weerldt... J. Jansson, Amsterdam. 1634.
Atlas or a geographical description... H. Hondius and John Johnson, Amsterdam. 1636.
Atlas sive cosmographicae... Dusseldorf. 1595; Hondius, Amsterdam. 1611, 1613, other editions.
Gerardi Mercatoris atlas. Hondius, Amsterdam. 1607, 1619, 1630, and later.
Gerardi Mercatoris et I. Hondii atlas... Hondius and Jansson, Amsterdam. 1633, 1683.
Historia mundi or Mercator's atlas... Michael Sparke, London. 1637.

Mercator, Gerhard the Younger. *Dutch cartographer. Active 1595.*
Irlandiae regnum... Amsterdam. 1638.
Anglia regnum... Amsterdam. 1638.

Mercator, Rumold. *Dutch cartographer. 16th century.* Son of Gerhard Mercator the Elder.

Meyer, Joseph. *German cartographer. 1796-1856.*
Neuster grosser schulatlas... Hildburghausen; New York. 1830-1838.
Neuster universal-atlas... Philadelphia. 1830-1840.

Miller, Robert. *British publisher. Active 1810-1821.*
Miller's new miniature atlas. London. 1810, 1820.

Mitchell, John (Mitchel). *British cartographer. Died 1768.*
A map of the British and French dominions in North America.... London. 1755.

Mitchell, Samuel Augustus. *American cartographer. 1792-1868.*
Mitchell's ancient atlas. T. Cowperthwait and Co., Philadelphia. 1844; E. H. Butler and Co., Philadelphia. 1859.
Mitchell's Atlas of outline maps... T. Cowperthwait and Co., Philadelphia. 1839.
Maps of New Jersey, Pennsylvania... Philadelphia. 1846.
A new universal atlas. Philadelphia. 1849, 1850.
Mitchell's school atlas... T. Cowperthwait and Co., Philadelphia. 1839.

Mogg, Edward. *British publisher and cartographer. Active 1808-1826.*

Moll, Herman. *German geographer, publisher, bookseller, and globe maker. 1688-1745.*
Map of Africa. London c. 1714.
A new and exact map... within ye limits of ye South sea... London. c. 1713.
Map of North America... London. c. 1714.
Map of Europe. London. 1708.
A new map of the north parts of America... London. 1720.
Map of South America (in collaboration with B. Lens). Herman Moll, Bowles, Overton and King, London. c. 1713.
Map of Asia. London. c. 1714.
A map of the East Indies... London. c. 1715; Dublin. c. 1720.
A new map of Great Britain. London. c. 1715.
Atlas minor. London. 1729 and later.
A system of geography. Churchill, London. 1701.
The world described... John Bowles, London. c. 1727.
A new map of Denmark and Sweden... London. c. 1715.
A new and exact map of Spain and Portugal. London. 1711.
Florida... London. 1728.
A new and exact map of the dominions of the king... on ye continent of North America. Bowles, London. c. 1713.
Virginia and Maryland. London. 1729.
A map of the West Indies... Moll and King, London. c. 1713.
A new and correct map of the world... London. c. 1728.
A new description of England and Wales... London. 1724.
Atlas manuale... A. and J. Churchill, London. 1709; I. Knapton, P. Knaplock, London. 1723.
Forty-two new maps of Asia, Africa and America... J. Nicholson, London. 1716.
A set of thirty-two new and correct maps of the principal parts of Europe... London. 1727?
A system of geography. T. Childe, London. 1701.
The world described. J. Bowles, London. 1709-1720; 1709-1736.
A new and compleat atlas... London. 1708-1720.

Montanus, Henri. *Flemish cartographer. Born 1556.*

Montanus, Petrus (Also known as Pieter Van der Berg). *Dutch cartographer and publisher working in Amsterdam. 17th century.*
Wrote text for Mercator-Hondius atlas of 1606.

Montresor, Major. *British cartographer. 18th century.*
Province de New York. Le Rouge, Paris. 1777.

Morden, Robert. *British cartographer, bookseller, publisher, and globe maker. Died 1703.*
A new map of the British empire in America. London. c. 1695.
Geography rectified. London. 1680, 1688.
A newe description of the whole worlde in two hemispheres... London. c. 1688.
To Capt. John Wood this map of the world... is humbly dedicated (with William Berry). London. c. 1688.
Fifty-six new and accurate maps of Great Britain... London. 1708.
Atlas terrestris... London. 1695.
Map of Essex (with J. Pask) London. c. 1700.

Morse, Jedidiah. *American geographer. 1761-1826.*
The American geography... J. Stockdale, London. 1794.

Modern atlas... (with S. E. Morse). J. H. A. Frost, Boston. 1822; Collin and Hannay, New York. 1828.
A new universal atlas of the world... (with S. E. Morse). Howe and Spalding, New Haven. 1822.

Morse, Sidney Edwards. *American cartographer and publisher. 1794-1871.*
The cerographic bible atlas... New York. 1844, 1845.
A new universal atlas of the world... N and S. S. Jocelyn, New Haven. 1825.
Nuevo sistema de geografía... (with J.C. Brigham). Gallaher and White, New York. 1827-1828.
The cerographic missionary atlas. New York. 1848.

Mortier, David. *Dutch cartographer and publisher. 18th century.*
Nouveau théâtre de la Grande Bretagne... London. 1715-1728.

Mortier, Pieter (Peter; Petrum). *Dutch publisher. Died 1711.*
Atlas nouveau de cartes géographiques choisies... Amsterdam. 1703.
Les forces de l'Europe, Asie, Afrique et Amérique... Amsterdam. 1702.

Moule, Thomas. *British topographer, bookseller, and publisher. 1784-1851.*
The English counties delineated... George Virtue, London. 1836, editions to 1839.

Mount, Richard (Mount, R. and Page, T.; Mount, W. and Page, T.; Mount and Davidson; Page, T. and Mount, W. and F.; Thornton and Mount). *British publishers. 18th and 19th century.*
The sea coasts of France... London. c. 1715.
A chart of the island of Hispaniola. London. c. 1737.
The island of Jamaica. London. 1737.
A new mapp of Carolina. London. c. 1737.
A chart of the Caribe islands (unsigned). London. c. 1737.
Atlas maritimus novus... London. 1702.
The publishing house was founded by Richard Mount.

Mouzon, Henry. *British cartographer. 18th century.*
An accurate map of North and South Carolina. Sayer and Bennett, London. 1775.

Moxon, James. *British engraver and cartographer. Active 1647-1696.*
A new description of Carolina. John Ogilby, London. 1671.

Moxon, Joseph. *British polymath, and globe maker. 1627-1700?*

Müller, Gerhard Friedrich. *German cartographer. 18th century.*
Nouvelle carte des decouvertes faites par des vaisseaux Russe aux côtes inconnues de l'Amerique septentrionale... St. Petersburg. 1754, 1758.

Müller, Johann Cristoph. *German cartographer. 1673-1721.*
Le royaume de Boheme... Covens and Mortier, Amsterdam. 1744.

Münster, Sebastian. *Swiss theologian and cartographer. 1489-1552.*
Cosmographiae universalis... H. Petri, Basle. 1540 and later.

Nantiat, Jasper. *British. 18th and 19th century.*
A new map of Spain and Portugal. Faden, London. 1810.

Nolin, Jean Baptiste; Sr. and Jr. *French cartographers, publishers and globe makers. 1648-1708, 1686-1762, respectively.*
Regnum Portugalliae... Homann, Nuremberg. 1736.
Atlas général à l'usage des collèges et maisons d'education... Mondhare, Paris. 1783.
Nouvelle édition de théâtre de la guerre en Italie... Paris. 1718.
Le théâtre du monde... Paris. 1700?-1744.

Nollet, Jean Antoine. *French cartographer and globe maker. 1700-1770.*

Norden, John. *British cartographer, topographer, surveyor. 1548-1626.*

Norie, John William. *British publisher and hydrographer. 1772-1843.*
The complete East India pilot. London. 1816.

Ogilby, John. *British printer, geographer. 1600-1676.*
Britannia... London. 1675.
Britannia depicta or Ogilby improv'd... T. Bowles, London. 1720, 1736, 1751; C. Bowles, London. 1764.
Translated Montanus. *Africa, Asia, America.* 1670...

Olney, Jesse. *American cartographer. 1798-1872.*
A new and improved school atlas... D. F. Robinson and Company, Hartford. 1829 and other editions.
Olney's school atlas... Pratt, Woodford and Company, New York. 1844-1847.

Ortelius, Abraham. *Dutch cartographer. 1527-1598.*
Theatrum orbis terrarum... Coppenium Diesth, Antwerp. 1570, 1571, 1573, 1574, and many other editions. Appeared in Latin, English, French, German, Spanish and Dutch.
Various supplements were issued; *Addidamentum iii. Theatrus orbis terrarum,* Antwerp. 1584; *Addidamentum iv...* Plantin, Antwerp. 1590.
Assorted epitomes and abridged editions; *Abrégé du Théatre...* I.B. Vrients, Antwerp. 1602; *An epitome of Ortelius.* John Norton, London. 1602; *Epitome theatri orbis terrarum...* Keerberg, Antwerp. 1601; Plantin, Antwerp. 1612; later editions.

Ottens, Reiner Joachim and Joshua. *Dutch cartographers and publishers. 18th century.*
Atlas sive geographia compendiosa... Amsterdam. 1756.
Atlas minora sive geographia compendiosa... Amsterdam. 1695-1756.

Overton, Henry. *Brtitish publisher and bookseller. Active 1706-1764.*
Reprinted Speed in 1743, also published Bowen's *Royal English atlas* c. 1764.

Overton, I. *British cartographer. 17th century.*
A new mapp of America. London. c. 1686.

Overton, Philip; John; Henry. *British printers and mapsellers. 18th century.*

Owen, John. *British cartographer. Collaborated with E. Bowen. 18th century.*
Britannia Depicta. London. 1720

Palairet, Jean. *French cartographer. 1697-1774.*
Carte des possessions Angloises & Françoises... de l'Amerique septentrionale (unsigned). English by Thomas Kitchin; London, Amsterdam, Berlin and The Hague. 1755.
Atlas méthodique... J. Nourse and P. Vaillant, etc., London. 1755. Maps engraved by J. Gibson and T. Kitchin.
Bowles' universal atlas... Carrington Bowles, London. 1775-80; Bowles and Carver, London. 1794-1798.

Palmer, William. *British map engraver. Active c. 1766-1800.*
Engraved maps for Sayer, and Laurie & Whittle, and Faden's *Atlas Minimus.*

Payne, John. *American cartographer. Active c. 1800.*
A new and complete system of universal geography...[atlas]. Low and Wallis, New York. 1798-1800.

Perthes, J. *German publisher active in Gotha. 19th century.*
Known for his splendidly produced atlases.

Petermann, August Heinrich. *British cartographer. 19th century.*
The atlas of physical geography... (with Thomas Milner). Ward and Lock, London. 1850.

Philipps, Sir Richard. *British cartographer. 1767-1840.*
Philipps published under the pseudonym "Rev. J. Goldsmith."
A geographical and astronomical atlas... by the Rev. J. Goldsmith. G. and W. B. Whitaker, London. 1823.

Piccolomini, Alessandro. *Italian astronomer. 1508-1578.*
De le Stelle Fisse (1570) was the first printed star atlas.

Pigafetta, Philip (Filippo). *Italian cartographer, traveler. 1533-1603.*
Translated Ortelius' text into Italian.

Pinkerton, John. *British cartographer. 1758-1826.*
Maps to accompany Pinkerton's Modern geography... Conrad and Company, Philadelphia. 1804?
A modern atlas... T. Cadell and W. Davies; Longmann, Hurst, Orme and Brown. London. 1815; T. Dobson and Son, Philadelphia. 1818.

Pitt, Moses. *British map publisher. Active 1654-1696.*
A description of places next to the north-pole... Oxford. 1650.

Plancius, Peter (Platevoet). *Flemish cartographer, celestial and terrestrial globe maker, theologian. 1552-1622.*
Mapmaker to the Dutch East India Company; prolific.

Plantin, Christopher (Plantijn). *Printer and bookseller, worked in Lyons, Paris, prior to going to Amsterdam. 1514-1589.* Plantin was a very important cartographic printer who printed for Ortelius (*Theatrum* as well as other works), and many other cartographers.

Popple, Henry. *British cartographer. Active c. 1730.*
A map of the British Empire in America. Stephen Austen, London. 1733.

Porcacchi, Tomaso (Thomaso). *Italian cartographer. 1530-1585.*
L'isole piv famose del mondo... S. Galignani and G. Porro, Venice. 1572, 1576, 1590, 1604; heirs of Simon Galignani, Venice. 1605; P. and F. Galignani, Padova. 1620 and other editions.

Porro, Girolamo. *Venetian engraver and publisher. Active c. 1560s to 1590s. Engraver for Magini, 1596...*

Potter, P. and Company. *American cartographical publisher. 19th century.*
Atlas of the world... Poughkeepsie. 1820.

Pownall, Thomas. *British geographer. 1722-1805.*
A topographical description...of North America... J. Almon, London. 1776. With a large *Map of the middle British colonies in North America.*

Preuss, Charles. *American; assisted Frémont. 1803-1854.*

Proctor, Richard Anthony. *British astronomer. 1834-1888.*
Wrote a number of popular astronomy books; star atlas.

Ptolomaeus, Claudius (Ptolemy; Tolomeo, Claudio). *Egyptian geographer and astronomer. Died A. D. 147?*
Cosmographia. Bologna. 1477.
Liber geographiae. Jacobus Pentius de Leucho, Venice. 1511.
Geographia... Nicolaus Laurentii, Florence. 1482, 1511; Schott, Strasbourg. 1513, 1535. The 1477 *Cosmographia* (incorrectly dated 1462) is the first edition with engraved maps.
The 1511 *Liber geographiae* contains the first printed map showing parts of the North American continent.
Laurentii's 1482 edition of the *Geographia* contains verse written by Francesco di Niccolo Berlinghieri (1440-1501); Schott's 1513 edition contains not only the first map of Switzerland, but the first map printed in three colors.

Purchas, Samuel. *British author. c. 1575-1626.*
Purchas his pilgrimage... London. 1625 and other editions.

Quad, Matthais (Quadum). *German cartographer and globe maker. 1557-1613.*
Geographisch handtbuch. Johann Bussemacher, Cologne. 1600.
Europae totius orbis terrarum partis praestantissimae... J. Bussemacher, Cologne. 1592, 1594.
Fasciculus geographicus... J. Bussemacher, Cologne. 1608. The eighty-six maps which appear in the *Fasciculus geographicus* are the same as those which appeared in Bussemacher's *Geographisch handtbuch* from 1600.

Ramusio, Giovanni Battista. *Venetian geographer. 1485-1557. First printed plan of Montreal (Hochelaga), woodblock western hemisphere.*

Rand McNally & Co. *American map publishers. Traces origins to 1850s. Noted for publishing guidebooks, maps and for use of wax engraving.*

Ratzer, Bernard. *American draftsman. 18th century.*
City of New York. 1776.
Province of New Jersey. 1777.

Raynal, Guillaume Thomas François. *French cartographer, Jesuit. 1713-1796.*
Atlas portatif... E. van Herrenvelt, D.J. Changuion, Amsterdam. 1773; J. L. Pellet, Geneva. 1780, 1783-1784 and other editions. Pellet's editions were renamed *Atlas de toutes les parties connues du globe terrestre...*

Rees, Abraham. *British cartographer. 1743-1825.*
The cyclopaedia; or universal dictionary of arts, sciences and literature... atlas and modern atlas. S.F. Bradford; Murray, Fairman and Company. London. 1806; Longman, Hurst, Rees, Orme and Brown; F.C. and J. Rivington. London. 1820.

Reichard, Heinrich August Ottokar. *German cartographer. 1752-1828.*
Atlas portatif et itinéraire de l'Europe. Weimar. 1818-1821.

Reid, Alexander. *British cartographer. 1802-1860.*

Reisch, Gregor. *German cartographer and mathematician. 1470-1525.*
World map. Frieburg. 1503 and later.

Renard, Louis. *French cartographer and publisher. Active 1715.*
Atlas de la navigation... Amsterdam. 1715; R. and J. Ottens, Amsterdam. 1739.
Atlas van zeevaert en koophandel door de geheele weereldt... R. and J. Ottens, Amsterdam. 1745.

Ricci, Father Matteo. *Italian missionary and cartographer. 1552-1610.*

Riccioli, Giovanni Battista. *Italian astronomer. Active 1598-1671.*

Riedinger, Johan Adam (Reidiger, Rüdinger). *Swiss cartographer and globe maker. 1680-1756.*

Rizzi-Zannoni, Giovanni Antonio. *Italian cartographer. 1736-1814.*
Nuova carta della Lombardia... Naples. 1795.
Atlante marittimo delle due Sicilie... Naples. 1796.
Atlas géographique... Lattré, Paris. 1762.
Atlas géographique et militaire ou théatre de la guerre présente en Allegmagne... Lattré, Paris. 1763.
Atlas historique de la France... Desnos, Paris. 1765.
Le petit neptune françois... Desnos, Paris. 1765.

Rocque, John. *British (born in France?) surveyor, engraver, and publisher. Died 1762.*
A general map of N. America... London. 1761.
Recüeil des villes ports d'Angleterre... (with J.N. Bellin). Desnos, Paris. 1766.
An exact survey of the cities of London and Westminster, the borough of Southwark... London. 1746.
The small British atlas... London. 1753, 1764.
A set of plans and forts on America... London. 1763.

Roggeveen, Arent. *Dutch cartographer. Died 1679.*
Het eerste deel van het brandende veen verlichtende alle de vaste kusten ende eylanden van geheel West-Indien of te rio Amasones. P. Goos, Amsterdam. 1675.
La primera parte del monte... P. Goos, Amsterdam. 1680.

Rosaccio, Giuseppe. *Italian astronomical and terrestrial cartographer. c. 1530-1620.*
Il mondo... F. Tosi, Florence. 1595.
Mondo elementare et celeste... E. Deuchino, Trevegi? 1604.
Teatro del cielo... Discepoli, Viterbo. 1615; G. Righettini, Trivegi? 1642.
Teatro del mondo... A. Pisarri, Bologna. 1688, 1724.

Roux, Joseph. *French cartographer and publisher. 18th century.*
Nouveau recuiel des plans des ports...de la mer Mediterranée... (with others). Y. Gravier, Gênes. 1779, 1838; Marseilles. 1764.
Carte de la mer Mediterranée. Marseilles. 1764.

Russell, John and J.C. *British cartographers and globe-makers. 18th century.*
An American atlas... H.D. Symonds and J. Ridgeway, London. 1795.
New general atlas... Baldwin and Cradock, London. 1810.

Russell, P. *British topographer. Active 1769.*

Ruysch, Johannes. *Dutch cartographer and astronomer. Active 1506.*

Sanson, Guillaume. *French cartographer and globe maker; son of Nicolas, Sr. Died 1703.*

Sanson, Nicolas fils. *French cartographer. 1626-1648.*

Sanson, Nicolas père. *French cartographer and geographer. 1600-1667.*
L'Afrique en plusiers cartes nouvelles... Paris. 1656.
Judea seu terra sancta... Covens and Mortier, Amsterdam. 1744.
L'Amérique en plusiers cartes... Paris. 1657.
L'Amérique en plusiers cartes nouvelles... Paris. 1662, 1667?
L'Asie en plusiers cartes nouvelles... Paris. 1652-53.
Atlas nouveau... Hubert Jaillot, Paris.1689-1690, 1692-1696, 1696.
Cartes générales de toutes les parties du monde... Pierre Mariette, Paris. 1658.
Cartes particulières de la France... Mariette, Paris. 1676.
Cartes générale de la géographie ancienne et nouvelle... (with N. Sanson *fils* and Guillaume Sanson). Mariette, Paris. 1675.
Géographie universelle... (with N. Sanson *fils* and Guillaume Sanson). Paris. 1675?
Description de tout l'univers... (with N. Sanson *fils*). F. Halma, Amsterdam. 1700.

Sanson, Pierre Moulard. *French cartographer; grandson of Nicolas Sr. 18th century.*

Santini, P. *Italian publisher and cartographer. Active c. 1776.*
Atlas portatif d'Italie... Venice. 1783
Atlas universel... Remondini, Venice. 1776-1784.

Sauthier, Claude Joseph. *British cartographer. 18th century.*
A map of the inhabited part of Canada... Faden, London. 1777.
Mappa geographica provinciae Novae Eboraci... Homann's Heirs, Nuremberg. 1778.
A topographical map of the northern part of New York island... Faden, London. 1777.
Sauthier occasionally collaborated with Bernard Ratzer; their work was signed "Sauthier & Ratzer."

Savanarola, Raffaelo. *Italian cartographer. 18th century.*
Universu terrarum orbis... 2 volumes. B. Conzatti, Patavi. 1713. Pseudonym of Lasor à Varea.

Saxton, Christopher. *British cartographer and surveyor. c. 1542- c. 1610.*
County maps of England and Wales, 1574-1579.
An atlas of the counties of England and Wales... London. 1579.
The shires of England and Wales described... "Sold by Philip Lea." London. 1693.
The traveller's guide... London. 1687.

Sayer, Robert. *British mapseller and publisher. Active 1780-1810.*
North America and the West Indies... London. 1783.
d'Anville and Robert's New map of North America... London. 1763.
Chart of Greenland. London. 1783.
The American military pocket atlas. London. 1776.
The seat of war in New England... London. 1775.
An English atlas... London. 1787.
Atlas britannique... London. 1766.
General atlas... London. 1757-1794.
Sayer was also in partnership with John Bennett, under the name "Sayer and Bennett."

Schedel, Hartmann. *German historian, geographer and publisher. 1440-1514.*
Liber cronicarum... Nuremberg. 1493. ("The Nuremberg Chronicle.")

Scheibel, Johann Ephraim. *German mathematician. 1736-1809.*

Scheiner, Christopher. *German astronomer. Born 1575.*

Schenk, Peter (Schenck). *Dutch publisher and engraver. c.1660-1719.*
Americae. By A.F. Zürner, Amsterdam. 1709.
L'Amerique Septentrionale. By G. Delisle, Amsterdam. 1708.
Petri Schenkii Hecatompolis... Amsterdam. 1702.
Atlas contractus... Amsterdam. 1705?, 1709?
Atlantis sylloge compendiosa... Amsterdam. 1709.
Atlas anglois... D. Mortier, London. 1715.

Scherer, Heinrich. *German mathematician and cartographer. 1628-1704.*
Atlas marinus... M.M. Rauchin, Monachii. 1702.
Critica quadripatria... Mathias Riedl, Monachii. 1710.

Scheuchzer, Johann Jakob. *Swiss mathematician and cartographer. 1672-1733.*

Schieble, Erhard. *French engraver. 19th century.*
Engraved maps for Dufour and Andriveau-Gouijon.

Schraembl, Franz Anton. *Austrian cartographer. 1751-1803.*
Allgemeiner grosser atlas... P.J. Schalbacher, Vienna. 1786-1800.

Schreiber, Johann Georg. *German cartographer. 1676-1745.*
Atlas selectus... Leipzig. c. 1749.

Schropp, Simon. *German map publisher. 19th century.*

Schröter, Johann Hieronymous. *German astronomer. 1745-1816.* Noted for lunar and planetary observations.

Scull, N. and Heap, G. *American cartographers; official Pennsylvania surveyors. 18th century.*
A plan... of Philadelphia. W. Faden, London. 1777.

Scull, William. *American cartographer. 18th century.*
A map of Pennsylvania... Sayer and Bennett, London. 1775.

Seale, Richard William. *British cartographer and engraver. Active 1732-1775.*

Seller, Jeremiah. *British instrument-maker. Active 1698-1707.*

Seller, John and Price, Charles. *British map, astronomical chart and globe makers. Active 1658-1701.*
English pilot. J. Darby, London. 1671; J. Larkin, 1692.
Atlas maritimus... James Atkinson, London. 1670? J. Darby, 1675.
Atlas terrestris... London. c. 1685.
Atlas contracta... London. 1695.

Senex, John. *British cartographer, publisher and engraver. Died 1740.*
Map of Africa... By C. Price, London. 1711.
Africa. London. 1725.
Ireland. London. 1712.
A new map of Great Britain. London. 1714.
Poland... London. 1725.
A new general atlas... London. 1721.
Sweden and Norway... London. c. 1725.
The VII United provinces. With J. Maxwell, London. 1709.
An actual survey of all the principal roads of England and Wales... first perform'd and published by John Ogilby Esq. London. 1719, 1742.
Modern geography. T. Bowles and Son, London. 1708-1725.
A map of Virginia. London. 1785.

Servetus, Michael (Villanovanus). *French. 16th century.*
Servetus' woodcut map of the Americas appears in the Trechsel, 1535, edition of Ptolemy. Servatus was burned at the stake by the Lutherans.

Seutter, Albrecht Carl. *French cartographer. 18th century.*
Partie orientale de la Nouvelle France, ou du Canada. T. Conrad Lotter, Augsburg. c.1740.

Seutter, Georg Mattheus the Older; Georg Mattheus the Younger. *German globe makers, cartographers, and publishers. 1678-1757; 1729-1760, respectively.*
Africa juxta navigationes et observationes recentissimae. Augsburg. c.1740.
Recens elaborata mappa geographica regni Brasiliae. Augsburg. c.1720.
Nova designatio insulae Jamaicae. Augsburg. c.1740.
Atlas minor praecipua orbis terrarum imperia. Augsburg. 1744?
Grosser atlas. Augsburg. 1734?
Atlas geographicus... Augsburg. 1725.
Atlas novus sive tabulae geographicae totius orbis... Augsburg, 1741. 2 volumes, Augsburg. 1745

Sharpe, John. *British. 1777-1860.*
Sharpe's corresponding atlas... Chapman and Hall, London. 1849.

Smith, Anthony. *British pilot and cartographer. 18th century.*
A new and accurate chart of the bay of Chesapeake... Robert Sayer and John Bennett, London. 1776.

Smith, Captain John. *British. 1580-1631.*
Virginia. London. 1606 and other editions.
New England... London. 1614.

Smith, Roswell C. *American. 19th century.*
Smith's atlas for schools, academies and families. W. Marshall and Co., Philadelphia. 1835; D. Burgess and Co., Hartford. 1835; Spalding and Storrs, Hartford. 1839; Cady and Burgess, New York. 1839, 1850, and other editions.

Society for the Propagation of the Gospel in Foreign Parts.
The colonial church atlas... Rivington, London. 1842.

Society for the Diffusion of Useful Knowledge (S.D.U.K.).
19th-century learned society that, among other things, was responsible for publishing a major 19th-century atlas of the world. The American branch of the S.D.U.K. was the forerunner of the Smithsonian Institution.

Sotzmann, Daniel Friedrich. *German cartographer and globe maker. 1754-1840.*
Small maps of Prussia and Berlin. Berlin. c. 1814.
Karte von Polen... Berlin. 1793; 1796.
Sotzmann made maps of the American states in the 1790s. These are among the rarest of any American state maps.

Speed, John. *British historian and cartographer. 1552-1629.*
A new description of Carolina. Bassett and Chiswell, London. 1676. (Engraved by Francis Lamb.)
A map of Virginia and Maryland. Bassett and Chiswell, London. 1676. (Engraved by Francis Lamb.)
A prospect of the most famous parts of the world... London. 1646; 1650.
The Theatre of the empire of Great Britaine: presenting an exact geography of the kingdomes of England, Scotland, Ireland, and the isles adioyning: with the shires, hundreds, cities and shire-townes, within ye kingdome of England, divided and described by John Speed. Imprinted at London. Anno, cum privilegio, 1611 [-1612]. And are to be sold by John Sudbury & George Humble, in Popes-head alley at ye signe of ye white horse.
Speed's maps are collected assiduously and can be quite expensive.

Spurner von Merz, Karl. *German. 1803-1892.*
Atlas antiquus... J. Perthes, Gotha. 1850.
Atlas zur geschichte von Bayern. J. Perthes, Gotha. 1838.
Dr. Karl von Spurner's historisch-geographischer hand-atlas. 3 volumes. J. Perthes, Gotha. 1846-1851.
Historisch-geographischer hand-atlas zur geschichte der staaten Europa's... J. Perthes, Gotha. 1846.

Stanford, E. *British publisher. 19th century.*
Stanford's library map of Australasia. London. 1859. Stanford published very high-quality folding linen-backed maps.

Stephenson, John. *British cartographer. Active 1786.*
Laurie and Whittle's channel pilot. London. 1794-1803.

Stieler, Adolf and August. *German cartographers. 1775-1836.*
Schul-atlas. J. Perthes, Gotha. 1841.

Stockdale, John. *British publisher and bookseller. 1739-1814.*
New British atlas... London. 1805.

Strabo. *Roman geographer. c. 63 B.C. to A.D. 25.*
Strabonis nobilissimi et doctissimi philosophi ac geographi rerum geographicarum commentary libris xvii... Henricpetrina, Basle. 1571.

Striedbeck, Johann, Jr. *German printer. 1640-1716.*

Sylvanus, Bernardo. *Portuguese. Edited 1511 edition of Ptolemy which contained the first map printed in two colors.*

Tallis, John B. *British publisher. Active 1830s-1850s.*
London Street Views. London. 1838-1839.
Illustrated Atlas of the World. London. 1851.

Tanner, B. *American engraver working in New York. 1775-1848.*

Tanner, Henry Schenk (Tanner and Marshall; Tanner, Vallance, Kearny and Co.). *American cartographers, engravers, and publishers. 1786-1858.*
Atlas classica... Philadelphia. 1840.
Atlas of the United States... Philadelphia. 1835.
A new American atlas... Philadelphia. 1818-23, 1819-21, 1823, 1825, 1825-33, 1839.
A new pocket atlas of the United States... Philadelphia. 1828.
A new universal atlas... Philadelphia. 1833-34, 1836; Carey and Hart, 1842-43, 1844.
A new general atlas... Philadelphia. 1828.

Tardieu, Ambroise. *French map and globe maker. 1788-1841.*
Atlas universel... Furne et Cie, Paris. 1842.

Tavernier, Melchoir; Gabriel, Jean Baptiste; Daniel, Melchoir Jr. *French cartographers and publishers. 1544-1641, 1605-89, 1594-1665 respectively.*
Carte d'Allemagne... Paris. 1635.
Théâtre géographique du royaume de France... Paris. 1634.

Teesdale, Henry and Co. *British publishers. Active 1828-1845.*
New British atlas... London. 1829, editions to c.1848.
A new general atlas of the world. London. 1835.

Thompson, John. *British cartographer. 19th century.*
A new general atlas. John Stark, Edinburgh. 1830.

Thomson, John and Co. *British publishers. Active 1814-1869.*
A new general atlas... Edinburgh. 1817.

Tirion, Isaac. *Dutch cartographer and publisher. Died 1769.*
Atlas van Zeeland... Amsterdam. 1760.
Nieuwe en beknopte hand-atlas... Amsterdam. 1730-1769, 1744-1769.
Nieuwe en keurige reis-atlas door de XVII Nederlanden... J. de Groot and G. Warnars, Amsterdam. 1793.

Tofiño de San Miguel, Vicente. *Spanish cartographer. 1732?-1795.*
Cartas maritimes de las costas de España. Madrid. 1786-1788.
Plan of the harbour of Cadiz. W. Faden, London. 1805.
Atlas maritimo de España. Madrid. 1787, 1789.

Tyson, J. Washington. *American cartographer. 19th century.*
An atlas of ancient and modern history... S. A. Mitchell, Philadelphia. 1845.

Valck, Gerhard (Valk). *Dutch celestial and terrestrial globe maker, cartographer and publisher. 1651-1726. Father of Leonhard Valck, partner of Schenk. Republished several of Blaeu's plates; republished the Cellarius celestial plates, Harmonia Macrocosmica (1720).*

Valgrisi, Vincenzo. *Italian cartographer and publisher. Active 1561-1562. Published Ruscelli's edition of Ptolemy.*

van den Keere, Pieter (Keer; Kaerius; Kerius). *Dutch engraver, bookseller and globe maker. 1571-1646. Worked in London 1584-93; was Hondius' brother-in-law.*

La Germanie inferieure... Amsterdam. 1621, 1622. (With maps by Emmius, Martini, Surhon and C. J. Vischer.)
Petri Kaerii Germania inferior... Amsterdam. 1622.

van der Berg, Petrus. *Dutch geographer and publisher. 17th century. J. Hondius' brother-in-law.*

van der Aa, Peter (Vander Aa). *Dutch publisher. 1659-1733.*
Atlas nouveau et curieux... 2 volumes, Leyden. 1714, 1728.
La galerie agréable du monde... 66 volumes, Leyden. 1729.
Représentation où l'on voit un grand nombre des isles, cotes, rivières et ports de mer... Leyden. 1730?
Cartes des itineraires et voïages modernes... Leyden. 1707.

van Keulen, Gerard. *Dutch cartographer. Died 1726.*
The new sea map of the Spanish Zee... Amsterdam. c.1690.
De groote niuwe vermeerderde zee-atlas, ofter water-waereld... Amsterdam. 1720.

van Keulen, Joannes I. *Dutch cartographer. 1654-1711.*
The great and newly enlarged sea atlas or water-world... 3 volumes, Amsterdam. 1682-1686.

van Keulen, Joannes II. *Dutch cartographer, map and chart maker. Noted for his superb sailing charts. Died 1755.*
Le grand nouvel atlas de la mer... Amsterdam. 1696.
De groote nieuwe vermeerderde zee-atlas... 5 volumes, Amsterdam. 1682-1684, 1695, 1734.

van Keulen, Joannes III. *Dutch. 1676-1763.*
De nieuwe groote lichtende zee fakkel... Amsterdam. 1753.

van Keulen, (Johannes Gerard, Johannes, Gerard Hulst, Cornelius). *Dutch family of mapmakers, publishers. The firm was in business for over 200 years, beginning in the late 1600s through the late 1800s.*
Well known for publishing sea charts.

van Keulen, Robijn. *Dutch. Active 1682-1698.*
Several atlases, appearing as late as 1736.
The new sea map of the channel betwixt England and France. Amsterdam. c.1700.
Nouvelle carte des costes de Guinée, d'Or... Amsterdam. c.1695.

van Langren, Arnold Florentius. *Dutch cartographer, globe maker. 1580-1644. Made maps for Linschoten. There were several other members of this family active in mapmaking, including Floris, Hendrick, Jacob and Michael.*

van Loon, Joannes (Jan). *Dutch. Active 1661-1668.*
Noort-ster ofte zee atlas. Amsterdam. 1661 (with Gillis van Loon), 1666, 1668.
Die nieuwe groote lichtende zee-fackel... (with C.J. Voogt) J. van Keulen, Amsterdam. 1699-1702.

van Luchtenburg, André. *Dutch mathematician. c. 1700.*

Vancouver, George. *British explorer. 1758-1798.*
Voyage of discovery to the North Pacific Ocean... Atlas. G. G. and J. Robinson and J. Edwards, London. 1798; Paris. 1799.

Vandermaelen, Philippe Marie Guillaume. *Belgian. 1795-1869.*
Atlas universel... Brussels. 1827.

Vaugondy, Gilles Robert (de). *French cartographer, celestial and terrestrial globe maker. 1686-1766. Father of Didier Robert, also known as le Sieur Robert or Monsieur Robert.*
Atlas portatif... Pissot, Paris; Durand, Paris. 1748-1749.
Nouvel atlas portatif... Paris. 1762; Fortin, Paris. 1778; Delamarche, Paris. 1784, 1794-1806.
Tablettes parisiennes... Paris. 1760.
Atlas universel... Boudet, Paris. 1757-1758, 1757-1786, 1783-1799; Delamarche, Paris. 1793.

Vaugondy, Didier Robert (de). *French mapmaker. Didier was Geographer Royal. 1723-1786.*
Atlas Portatif. Paris. 1748 and later.
Atlas Universel. Paris. 1757 and later.

Virtue, George. *Publisher, London. 19th century.*

Vischer, Georg Mathias. *Austrian. 1628-1696.*

Visscher, Nicolaus II (Piscator). *Dutch cartographer and publisher. 1618-1679.*
Atlas contractus orbis terrarum... Amsterdam. 1666? Son of N.J. Visscher.

Visscher, Nicolaus Joannis (Claes Jansz; Piscator). *Dutch cartographer and publisher. 1587-1637.*
Novi Belgii... Amsterdam. c.1651.

Visscher, Nicolaus III (Nikolaas; Piscator). *Dutch publisher and cartographer. 1639-1709.*
Indiae orientalis... Amsterdam. c.1680.
A new mapp of the kingdome of England and Wales... Lea and Overton, London and Amsterdam. 1686.
Novi Belgii... Amsterdam. c.1660-1691, editions to 1717.

von Humboldt, Friedrich Wilhelm Heinrich Alexander. *German explorer. 1769-1859.*
Atlas géographique...de la Nouvelle-Espagne... F. Schoell, Paris. 1811; G. Dufour et Cie, Paris. 1812.
Voyage de Humboldt et Bonpland... Atlas. F. Schoell, Paris. 1805-1839, 1814-1820.

von Littrow, J. *German astronomer. Active 1839.*

von Mädler, Johann Heinrich. *German astronomer. Active 1834.*

von Reilly, Franz Johann Joseph. *Austrian publisher. 18th century.*
Atlas universale... Vienna. 1799.

Voogt, Claes Jansz. *Dutch cartographer. Died 1696.*
Die nieuwe groote lichtende zee-fakkel. G.H. van Keulen, Amsterdam. 1682.

Walch, Jean. *French. Active 1790.*
Charte de l' Afrique. Martin Will, Augsburg. c.1790.

Waldseemüller, Martin. *German cartographer and globe maker. 1470-1518.* Maker of the first map to bear the name "America."

Walker, John; Alexander and Charles. *British cartographers. John: 1759-1830.*
A general map of India. J. Horsburgh, London. 1825.
Map of the United States... (John and Alex.) London and Liverpool. 1827.
Hobson's fox-hunting atlas... (John and Charles with W. C. Hobson) London. 1850; other editions from c.1866-c.1880.
An atlas to Walker's geography... T.M. Bates, Dublin. 1797; Vernor and Hood; Darton and Harvey, London. 1802.
Walker's universal atlas... F.C. Rivington, G. Wilkie. London.

Wallis, James. *British publisher and engraver. Active early 19th century.*
Wallis's new pocket edition of the English counties... London. 1810, c.1814.
Wallis's new British atlas... S. A. Oddy, London. 1812.

Waren, G.K. *American topographical engineer. 19th century.* Worked on Pacific railroad routes.

Weigel und Schneiderschen Handlung (Schneider, A. G. und Weigel). *German publishers. 18th century.*

Weigel, Christoph the Younger. *German cartographer and publisher. Died 1746.*

Weiland, Carl Ferdinand. *German cartographer. Died 1847.*
Atlas von Amerika... Weimar. 1824-1828.

Wells, Edward. *British cartographer. 1667-1727.*

A new sett of maps... London. c.1700; "at the Theater," Oxford. 1700; London. c.1705. Although generally conisdered to be of no great cartographic significance, his maps are both attractive and curious and consequently sought after.

Wheeler, George Montague. *American topographical engineer. 19th century.*

White, John. *British. Active 1585-1593.*
Americae pars... in T. Hariot's *Merveilleux at estrange rapport... des commoditez qui se trouvent en Virginia...* Theodore de Bry, Frankfurt. 1590.

Wild, James. *British cartographer and publisher. 19th century.*
Tasmania or Van Diemen's Land. London. c. 1846.

Wilkinson, Robert. *British publisher. Active 1785.*
Atlas classica... London. 1797, 1797-1805.
A general atlas... London. 1794, editions to 1808?

Willard, Emma Hart. *American. A noted American feminist and educator, founded a female preparatory school. 1787-1870.*
Willard's atlas... O.D. Cooke and Co., Hartford. 1826.

Willetts, Jacob. *American. 18th and 19th centuries.*
Atlas of the world. P. Potter, Poughkeepsie. 1814, 1820.
Atlas designed to illustrate Willett's Geography... P. Potter, Poughkeepsie. 1818.

Wood, William. *British. c. 1580-1639.*
New England prospect... Thomas Cotes for Iohn Bellamie, London. 1634.

Woodbridge, William Channing. *American. 1794-1854.*
Woodbridge's larger atlas... S. G. Goodrich, Hartford. 1822.
Modern atlas on a new plan... O.D. Cooke and Co., Hartford. 1831; Belknap and Hamersley, Hartford. 1843.
School atlas... O.D. Cooke and Sons, Ltd., Hartford. 1821; Beach and Beckwith, Hartford. 1835. Others.

Wyld, James the Elder; James the Younger. *British publishers. 1790-1836, 1812-1887, respectively.* An important publisher of linen-backed folding maps.
A general atlas... J. Thomson, Edinburgh. 1819.
The island of New Zealand. London. c.1851.
Cape district. Cape of Good Hope... London. 1838.
An emigrant's atlas... London. 1848.
A new general atlas... London. 1840?, 1852.

Wytfliet, Cornelis. *Latter half 16th century.* He produced the first atlas dealing exclusively with America.
Descriptionis Ptolemaicae augumentum. Johannis Bogardi, Lovanii. 1597; Gerard Riuj, 1598; Draci Franciscum Fabri, Douai. 1603.
Historie universelle des Indes Orientales et Ocidentales... (with Giovanni Magini). F. Fabri, Douai. 1605, 1611.

Zaltieri, Bolognino (Zaltierius). *Venetian publisher. Active 1550-1580.* Worked with Forlani.

Zatta, Antonio. *Italian cartographer and publisher. 1757-1797.*
Atlante novissimo (4 volumes). Venice. 1779-1785.
Nuovo atlante (with Giacomo Zatta). Venice. 1799, 1800, and other editions.

Zeno, Nicolo. *Italian cartographer. 16th century.* Purported author of the famous "Zeno map" that showed Frisland and other mythical islands in the North Atlantic.
Carta de navegar... woodcut in: *De I commentarii del viaggio in Persia di M. Calterino Zeno...* Francesco Marcolini, Venice. 1558.

266

Appendix B

The Map Collector's Reference Library

In this chapter we list some of the reference works about which serious collectors need to know. As always, I emphasize that this list is not complete; nor can it be. It will, however, introduce you to "what's out there" and give you a start in exploring the literature about maps.

Acquiring a map or two, or even many, is only the first step in collecting. Sooner or later a nascent collector is going to want to learn more about these maps. That wonderful Blaeu map just delivered by UPS — is it 1645 or 1657? What is the meaning of the iconography in the Homann cartouche? What other maps were made of Utah in the 1880s?

One of the nice things about map collecting as opposed to, say, molecular genetics, is that we can all understand the scholarly literature. We may not understand the *context* of the arguments, but it doesn't take years to learn the necessary background in order to understand the articles themselves.

Many books about old maps are out of print, available only on the used book market. While the art collector, for example, can find many periodicals devoted to collecting, the map collector can find but few.

A strong, well-stocked reference library is one of the best investments a collector can make, and I urge collectors to build a large and extensive reference library. Books available today may not be tomorrow. The more you know about even the periphery of your special area of interest, the more sophisticated a collector you will become, and the more you will enjoy your hobby.

JOURNALS AND MAGAZINES

Caert-Thresoor
Published in the Netherlands, in Dutch, articles have a summary in English.

Contact:
> Caert-Thresoor Administration
> P.O. Box 68
> 2400 AB Alphen aan den Rij, The Netherlands

Cartographica Helvetica
Published by the *History of Cartography of the Swiss Society of Cartography*. German language, summaries in English and French.

Contact:
> Hans-Uli Feldmann
> Untere Längmatt 9
> CH-3280 Murten. Switzerland

Imago Mundi
Imago Mundi is a widely respected refereed journal.

Contact:
> Secretary/Treasurer
> Imago Mundi Ltd.
> c/o British Library Map Library
> Great Russell Street
> London WC1B 3DG. England

Mapline
Mapline is a newsletter published by the Hermon Dunlap Smith Center for the History of Cartography at the Newberry Library.

Contact:
> The Editor, Mapline
> The Newberry Library
> 60 West Walton Street,
> Chicago, IL 60610-3380 USA

Mercator's World
Color magazine published bimonthly.

Contact:
> Aster Publishing Corp.
> 845 Willamette Street
> Eugene, OR 97401 USA

267

RELEVANT TRADE PUBLICATIONS

AB Bookman's Weekly. Weekly specialty publication for the second-hand book trade. Contains lists of books wanted and books for sale. Occasionally maps are offered, but generally only inexpensive ones. There is an annual June cartography issue. Subscription only:

> AB Bookman's Weekly
> P.O. Box AB
> Clifton, NJ 07015 USA

Antiques Trade Gazette. This is a British weekly newspaper-format publication for the general antiques trade. Of interest to map collectors since it has good coverage of major auction house book and map sales. Also, maps are just one aspect of the larger antiques market and one can learn a lot by following the market activity in other areas. Subscription only:

> Antiques Trade Gazette
> 17 Whitcomb Street
> London WC2H 7PL England

Bookdealer. This is a British weekly publication for the second-hand book trade; occasionally true antiquarian material appears. Similar to *AB Bookman's Weekly*, this publication has lists of used books for sale and lists of books wanted. It occasionally lists maps, but most interesting are the editiorial parts in the front. Subscription only:

> Bookdealer
> Werner-Shaw Ltd.
> 26 Charing Cross Road (Suite 34)
> London WC2H0DH England

Sheppard's International Directory of Map and Print Sellers This work is updated every few years. It is quite comprehensive but not very selective. Farnham, Surrey: Richard Joseph.

JOURNALS AND MAGAZINES NO LONGER PUBLISHED

The Map Collector's Circle. Published between 1963 and 1975. Complete runs are scarce but do come up at auction and in map dealer catalogues. Worth having.

The Map Collector. A quarterly magazine that ceased publication in 1996. Occasional back issues are available from some dealers. Worth having. Buy a complete run if you find one. In the meantime, buy any loose issues you find; you can always sell them if you locate a complete set.

BOOKS

In recent years the number of map-related books published has increased. The market is unfortunately small and publishers respond by printing only small numbers of each title. This makes the books relatively expensive. Then they go out of print and can only be obtained on the used book market. This makes them more expensive. However, in my opinion, reference books are one of the best investments a collector can make. A hundred dollars or so spent on a reference book is trivial compared to the thousands that many maps cost. Remember also that your public library has funds to buy reference books. Give the library a list of map books you would like it to have. Your tax dollars at work!

In the edited list that follows the symbol: (Δ) means the book is of general interest, a good beginner's book.

GENERAL BOOKS FOR COLLECTORS:

Hill, Gillian. *Cartographical Curiosities.* London: The British Library, 1978.

King, Geoffrey. *Miniature Antique Maps.* Hertfordshire, England: Map Collector Publications, Inc., 1996. A less-than-attractive book, but a good introduction to miniature maps.

ΔMoreland, Carl and David Bannister. *Antique Maps: A Collector's Guide.* Phaidon, Oxford, 1993.

ΔPotter, Jonathan. *Country Life Book of Antique Maps: An Introduction to the History of Maps and How to Appreciate Them.* Secaucus, New Jersey: Chartwell Books, 1989. An excellent standard work.

ΔTooley, Ronald V. *Maps and Map-makers.* London: Batsford, 1949. Fifth edition, 1978.

ΔTooley, R. V., C. Bricker and G.R. Crone. *Landmarks of Mapmaking.* New York: Thomas Y. Crowell Company, 1976.

General readings of a less technical nature

Baynton-Williams, Roger. *Investing in Maps.* New York: Clarkson N. Potter, 1969.

Bettex, Albert. *The Discovery of the World.* London: Thames and Hudson, 1960.

ΔBricker, Charles. *Landmarks of Mapmaking.* Amsterdam: Elsevier, 1968. (In England: *A History of Cartography.* London: Thames and Hudson, 1969.)

ΔBrown, Lloyd. *The Story of Maps.* Boston: Little Brown & Co., 1949. Reprinted by Dover, (New York, 1980).

ΔCampbell, Tony. *Early Maps.* New York: Abbeville, 1981.

Crone, G. R. *Maps and Their Makers: An Introduction to the History of Cartography.* Folkestone, Kent, England; Hamden, Connecticut: Dawson-Archon Books, 1978.

Cumming, W. P., R.A. Skelton, and D.B. Quinn. *The Discovery of North America.* New York: American Heritage Press, 1972.

Gohm, Douglas C. *Maps and Prints for Pleasure and Investment.* London: Gifford, 1969. New York: Arco Press, 1969.

Lowenthal, Mary Alice, ed. *Who's Who In The History of Cartography.* Map Collector Publications, Ltd., 1995.

Skelton, Raleigh. A. *Maps: A Historical Survey of Their Study and Collecting* (Kenneth Nebenzahl, Jr. Lectures in the History of Cartography at The Newberry Library, No. 1). Chicago: University of Chicago Press, 1972.

Skelton, R. A. *Explorer's Maps: Chapters in the Cartographic Record of Geographical Discovery.* London: Spring Books, 1958.

ΔSkelton, R. A. *Decorative Printed Maps of the 15th to 18th Centuries.* London, New York: Staples Press, 1952.

ΔWilford, John Noble. *The Mapmakers. The story of the great pioneers in cartography--from antiquity to the space age.* New York: Alfred A. Knopf, 1981.

Technical references

These books are for the more serious collector. They are recommended for in-depth information about specific topics.

Bagrow, Leo, and Skelton, R. A. *History of Cartography.* London: C. A. Watts; Cambridge, Mass.: Harvard University Press, 1964. Second printing: Chicago: Precedent Publishing, 1985.

Baynes-Cope, A. D. *The Study and Conservation of Globes.* Vienna: Internationale Coronelli-Gesellschaft, 1985.

Campbell, Tony. *The Earliest Printed Maps, 1472-1500.* London: The British Library, 1987.

Harley, J. B. and David Woodward, eds. *The History of Cartography: Cartography in Prehistoric, Ancient, and Medieval Europe and the Mediterranean.* Chicago: University of Chicago Press, 1987.

Note: Volume 1 of a series. A superb, authoritative work that will be the definitive compendium of historical mapping.

Harvey, P. D. A. *The History of Topographical Maps: Symbols, Pictures, Surveys.* London: Thames and Hudson, 1980.

Karrow, Robert K. Jr. *Mapmakers of the Sixteenth Century and Their Maps.* Chicago: Newberry Library, 1993.

Koeman, Dr. Ir C. *Atlantes Neerlandici: Bibliography of terrestrial, maritime, and celestial atlases and pilot books, published in the Netherlands up to 1880; supplement 1880-1940.* Amsterdam: Theatrum Orbis Terrarum Ltd., 1971. The definitive work, now being revised by Peter van der Krogt. An expensive work not for the faint of heart.

Koeman, Dr. Ir C. and H.J.A. Homan. *Atlantes Neerlandici: Supplement 1880-1940.* Alphen Aan Den Rijn: Canaletto, 1985.

Mickwitz, Ann-Mari, et al., comps. *The A.E. Nordenskiold Collection in the Helsinki Library, Annotated Catalogue of Maps made up to 1800.* Helsinki; Atlantic Highlands, N.J.: Helsinki University Library, Humanities Press, 1984.

Nordenskiold, A. E. *Facsimile-Atlas: To the Early History of Cartography, with Reproductions of the Most*

Important Maps Printed in the 15th and 16th Centuries. New York: Dover Publications, 1973.

Pedley, Mary Sponberg. *Bel et Utile: The Work of the Robert de Vaugondy Family of Mapmakers.* Hertfordshire, England: Map Collector Publications, 1992

Phillips, P. Lee. *A List of Geographical Atlases in the Library of Congress.* Amsterdam: Theatrum Orbis Terrarum Ltd., 1971 (reprint).

Robinson, Arthur H. *Early Thematic Mapping in the History of Cartography.* Chicago: The University of Chicago Press, 1982.

Thrower, Norman J. W. *Maps and Civilization: Cartography in Culture and Society.* Chicago: University of Chicago Press, 1996.

Tooley, R. V. *Tooley's Dictionary of Mapmakers.* New York; Amsterdam: Alan R. Liss, Inc., Meridian Publishing, 1979.

Tooley, R. V. *Tooley's Dictionary of Mapmakers-Supplement.* New York; Amsterdam: Alan R. Liss, Inc., Meridian Publishing, 1985.

van den Broecke, Marcel P.R. *Ortelius Atlas Maps — An illustrated guide.* HES Publishers: Westrenen, 1996.

Woodward, David, ed. *Five Centuries of Map Printing.* Chicago: The University of Chicago Press, 1975.

Woodward, David. *Catalogue of Watermarks in Italian Maps c. 1540-1600.* Chicago: The Newberry Library, 1996.

BOOKS ABOUT THE MAPPING OF SPECIFIC AREAS

In this section I list some books that will help the collector learn more about specific maps of specific regions. Although some are illustrated profusely, many assume the reader is able to identify maps by written description and not only by picture. These are important books: the ability for a collector to put maps into historical perspective increases enjoyment enormously. In many instances there is but a single reference book on a subject. Many are out of print and must be obtained on the used book market.

Burden, Philip D. *The Mapping of North America.* Rickmansworth, England: Raleigh Publications, 1996. The new standard work on North America maps.

Clancy, Robert. *The Mapping of Terra Australis.* Macquarie Park: Universal Press Pty. Ltd., 1995.

Cobb, David A. *New Hampshire Maps to 1900: An Annotated Checklist.* Hanover, N.H.: New Hampshire Historical Society; dist. University Press of New England, 1981.

Cortazzi, Hugh. *Isles of Gold Antique Maps of Japan.* New York: Weatherhill, 1983.

Cumming, William P. *British Maps of Colonial America.* Chicago: The University of Chicago Press, 1974.

Cumming, William P. *The Southeast in Early Maps.* Chapel Hill, N.C.: The University of North Carolina Press, 1962.

Deak, Gloria Gilda. *Picturing America 1497-1899* (2 volumes). Princeton, N.J.: Princeton University Press, 1988.

Fite, Emerson D. and Archibald Freeman, eds. *A Book of Old Maps: Delineating American History from the Earliest Days Down to the Close of the Revolutionary War.* New York: Arno Press, reprint 1969 (1926 original).

Gole, Susan. *Indian Maps and Plans, From Earliest Times to the Advent of European Surveys.* New Delhi: Manohar, 1989.

Goss, John. *Braun & Hogenberg's The City Maps of Europe: A Selection of 16th Century Town Plans & Views.* London: Studio Editions, 1991.

Goss, John. *The Mapping of North America: Three Centuries of Map-making, 1500-1860.* Secaucus, N.J.: The Wellfleet Press, 1990.

Guthorn, Peter J. *United States Coastal Charts, 1783-1861.* Exton, Pennsylvania: Schiffer Publishing, Ltd, 1984.

Guthorn, Peter J. *American Maps and Map Makers of the Revolution.* Monmouth Beach, N.J.: Philip Freneau Press, 1966.

Howgego, James. *Printed Maps of London circa 1553-1850.* Kent, England: William Dawson and Sons, 1978.

Hyde, Ralph. *Printed Maps of Victorian London, 1851-1900.* Kent, England: William Dawson and Sons, 1975.

Jackson, Jack, Robert S. Weddle, and Winston De Ville. *Mapping Texas and the Gulf Coast: The Contributions of Saint-Denis, Olivan, and Le Maire.* College Station, Texas: Texas A&M University Press, 1990.

Johnson, Adrian. *America Explored: A Cartographical History of the Exploration of North America.* New York: Viking, 1974.

Jolly, David C. *Maps in British Periodicals. Part I Major Monthlies Before 1800.* Brookline: Jolly, 1990. *Part II Annuals Scientific Periodicals & Miscellaneous Magazines Mostly Before 1800.* Brookline: Jolly, 1991.

Jolly, David C. *Maps of America in Periodicals Before 1800.* Brookline: Jolly, 1989.

Karpinski, Louis C. *Maps of Famous Cartographers Depicting North America.* Amsterdam: Meridian Publishing Co., 1977.

Kershaw, Kenneth A. *Early Printed Maps of Canada, 1540-1703.* Ancaster, Ontario: Kershaw Publishing, 1993.

Laor, Eran. *Maps of the Holy Land: Cartobibliography of Printed Maps, 1475-1900.* New York; Amsterdam: Alan R. Liss Inc., Meridian Publishing, 1986.

Lunny, Robert M. *Early Maps of North America.* Newark, N.J.: New Jersey Historical Society, 1961.

Martin, James C. and Robert Sidney Martin. *Maps Of Texas And The Southwest. 1513-1900.* Albuquerque: University of New Mexico Press, 1984

Matsutaro, Nanba Muroga Nobuo, and Unno Kazutaka, eds.; trans. Patricia Murray. *Old Maps in Japan.* Osaka: Sogensha, Inc. 1973.

McLaughlin, Glen with Nancy H. Mayo. *The Mapping of California as an Island.* San Francisco: California Map Society, 1995.

Modelski, Andrew M. *Railroad Maps of North America: The First Hundred Years.* New York: Bonanza Books, 1987.

Moir, D. G. *The Early Maps of Scotland. To 1850.* Edinburgh: The Royal Scottish Geographical Society, 1973.

Nebenzahl, Kenneth. *Atlas of Columbus and the Great Discoveries.* Chicago: Rand McNally, 1990.

Nebenzahl, Kenneth. *Maps of the Holy Land.* New York: Abbeville Press, 1986.

Nebenzahl, Kenneth. *Atlas of the American Revolution.* Chicago: Rand McNally, 1974.

Norwich, Oscar I. *Maps of Southern Africa.* Johannesburg: Ad. Donker, 1993.

(Norwich, Oscar I.) Stone, Jeffrey C., ed. *Norwich's Maps of Africa: An Illustrated and annotated carto-bibliography.* Norwich, Vermont: Terra Nova Press, 1997.

Palmer, Margaret. *Printed Maps of Bermuda.* London: The Map Collector's Circle; 1964 1st ed.

Phillips, P. Lee. *Maps and Views of Washington and District of Columbia.* Norwich, Vermont: Terra Nova Press, 1996.

Phillips, P. Lee. *Maps and Atlases of the WWI Period.* Norwich, Vermont: Terra Nova Press, 1995.

Phillips, P. Lee. *Virginia Cartography: A Bibliographical Description.* Ann Arbor, Michigan: Arbor Libri Press, 1995.

Phillips, P. Lee. *A List of Maps of America in the Library of Congress.* Amsterdam: Theatrum Orbis Terrarum Ltd., 1967 (reprint).

Phillips, P. Lee. *A Descriptive List of Maps and Views of Philadelphia in the Library of Congress, 1683-1865.* Philadelphia: Geographical Society of Philadelphia, 1926.

Portinaro, Pierluigi and Franco Knirsch. *The Cartography of North America 1500-1800.* New York: Facts on File, Inc., 1987.

Reinhartz, Dennis and Charles C. Colley, eds. *The Mapping of the American Southwest.* College Station, Texas: Texas A&M University Press, 1987.

Reps, John W. *Views and Viewmakers of Urban America.* Columbia, Mo.: University of Missouri Press, 1984.

Ristow, Walter W. *American Maps and Mapmakers: Commercial Cartography in the Nineteenth Century.* Detroit: Wayne State University Press, 1985.

Schilder, Günter. *Australia Unveiled.* Amsterdam: Theatrum Orbis Terrarum Ltd., 1976.

Schwartz, Seymour I. and Ralph E. Ehrenberg. *The Mapping of America.* New York: Harry N. Abrams Inc., 1980.

Shirley, Rodney W. *Early Printed Maps of the British Isles, 1477-1650.* Somerset, England: Antique Atlas Publications, 1991.

Shirley, Rodney W. *Printed Maps of the British Isles, 1650-1750.* Hertfordshire, England: Map Collector Publications, 1988.

Shirley, Rodney W. *The Mapping of the World.* London: The Holland Press, 1987.

Skelton, R.A. *County Atlases of the British Isles 1579-1850.* London, England: Carta Press, 1970.

Suarez, Thomas. *Shedding the Veil: Mapping the European Discovery of America and the World.* Singapore: World Scientific Publishing, 1992.

Thompson, Edmund. *Thompson's Maps of Connecticut.* Two volumes in one (reprint). Norwich, Vermont: Terra Nova Press, 1995.

Tibbetts, G. R. *Arabia in Early Maps.* New York: The Oleander Press, 1978.

Tooley, R. V. *The Mapping of Africa.* London: The Holland Presss Ltd., 1980.

Tooley, R. V. *The Mapping of Australia and Antarctica.* London: The Holland Press Ltd., 1979.

Tooley, R. V. *Collector's Guide to Maps of the African Continent and Southern Africa.* London: Carta Press, 1969.

Verner, Coolie and Basil Stuart-Stubbs. *The Northpart of America.* Toronto: The Hunter Rose Company, 1979.

Wagner, Henry R. *The Cartography of the Northwest Coast of America to the Year 1800.* Berkeley: University of California Press, 1937.

Walter, Lutz, ed. *Japan, A Cartographic Vision: European Printed Maps from the Early 16th to the 19th Century.* Munich, New York: Prestel, 1994

Wheat, Carl I. *The Maps of the California Gold Region, 1848-1857.* San Francisco; Storrs-Mansfield, Connecticut: The Grabhorn Press; (reprint Maurizio Martino)

Wheat, Carl I. *Mapping the Transmississippi West.* 1957. Reprint: Storrs-Mansfield, Connecticut: Maurizio Martino. An important work in the mapping of North America. The original is very expensive; the reprint is well done.

Wheat, James C. and Christian F. Brun. *Maps & Charts Published in America Before 1800.* London: The Holland Press Inc., 1985.

Zacharakis, Christos G. *A Catalogue of Printed Maps of Greece, 1477-1800.* Athens: Samourkas Foundation, 1992.

ASTRONOMY

Bibliography of the Moon. Washington, D.C.: Chief of Engineers, Department of the Army, 1960.

Both, Ernst E. *A History of Lunar Studies.* Buffalo; Buffalo Museum of Sciences, ND (c. 1961)

Brown, Basil. *Astronomical Atlases Maps and Charts.* London; Search Publishing, 1932.

Kopal, Z. and R.W. Carder. *Mapping of the Moon Past and Present.* Boston: D. Reidel Publishing Company, 1974.

Warner, Deborah J. *The Sky Explored: Celestial Cartography 1500-1800.* New York: Alan R. Liss Inc., 1979.

PRICE GUIDES:

American Book Prices Current. (auction records; including maps) Washington (CT); Bancroft-Parkman, Inc. (annual).

Book Auction Records. (auction records; including maps) Folkstone; Dawson. (annual).

Nagel, Gert K. *Alte Landkarten Globen und Städteansichten* Augsburg; Battenberg Verlag Augsburg, 1994

Rosenthal, J. *Antique Map Price Record and Handbook* Amherst, Mass. Kimmel Publications. (Biannual).

Sheets, K.A. *American Maps 1795-1895 A Guide To Values...* Ann Arbor; Sheets, 1994.

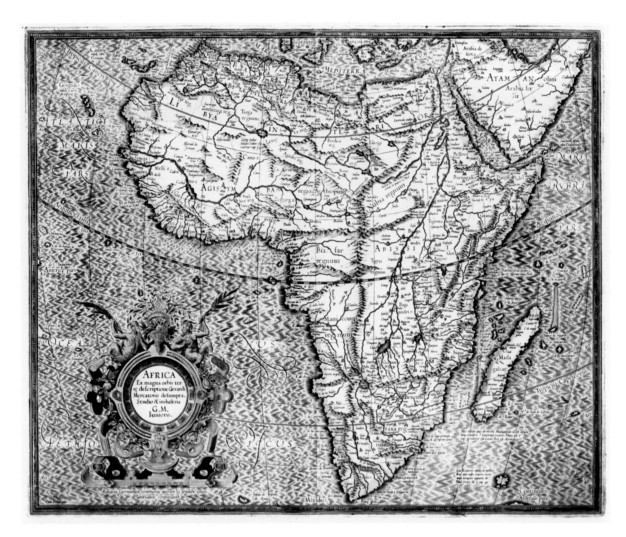

Figure C.1 [Above] The color is early in this copperplate engraved map, *Africa Ex magna orbis terre, descriptione...*, by G. Mercator (Amsterdam, c. 1600), measures 14.75 x 18.25 inches (37.5 x 46 cm). Gerard Mercator extracted this map from his grandfather's eighteen-sheet world map first published around 1569. It was updated and issued as a single-sheet map. See Norwich, *Africa* 21.

Figure C2 [Top, next page] Ortelius' copperplate engraved map of the old world, *Aevi Veteris Typus Geographicus.* Original color. 12 x 17.25 inches (31 x 44 cm). Italian text on verso. Published in Antwerp in 1590, the world is surrounded by a highly decorative strapwork, the oval world is divided into climatic zones, based on those of Mella. Circular medallions in each corner have little maps of the four known continents; North/South America in the lower left. The map shows the world as known to the ancients. Nonetheless, it is contained within a larger oval, such as used by Ortelius for his "modern" world map. This suggests strongly that he means this image to be a small picture of the larger world, a mere fragment of the total knowledge that was to come. Shirley 176.

It has been suggested that this map was drawn by Ortelius himself. Whatever the case, it is a most decorative old map, and priced at a fraction of the Blaeu world map shown below.

Figure C3 [Bottom, next page] *Nova Totius Terrarum Orbis Geographica Ac Hydrographica Tabula.* Blaeu's 1635 copperplate engraved worldmap is shown with original full hand color. 16 x 21.5 inches (40.8 x 54.5 cm). Dutch text on verso. According to Shirley, this is State 4 (of four). Published in Amsterdam. With three cartouches, two polar hemispheric projections, ships, compass roses, sea creatures and brilliant calligraphy, this is one of the most beautiful world maps made. It is considered a major example of 17th-century Dutch mapmaking.

Some maps, predominantly from the Low Countries, became known for their decorative side panels. Such maps are called *cartes à figures* and, because of their relative scarcity and their perceived beauty are expensive. The panels along the bottom of this map depict the seven wonders; the elements are along the left; seasons at the right, and sun, moon and known planets in panels along the top.

Figure C.2 (above) and Figure C.3 (below) For descriptions, see previous page.

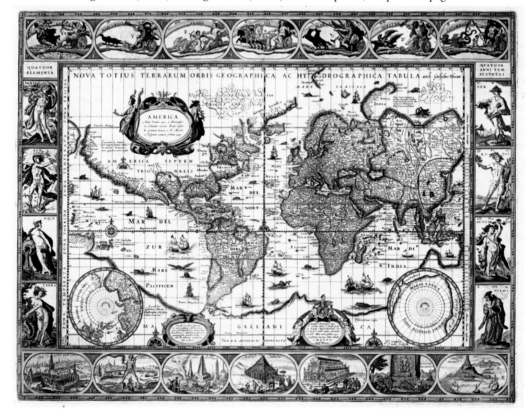

C2

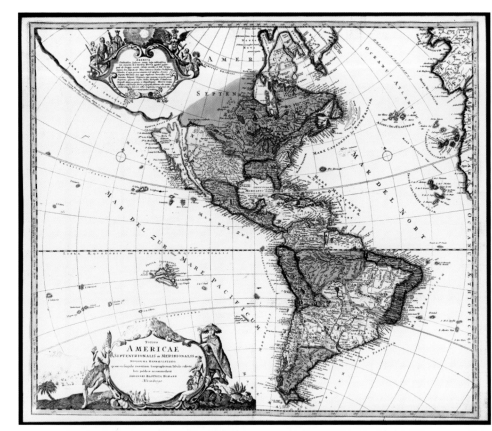

Figure C.4 Homann published maps of America that showed California as an island and, later, as a peninsula. This copperplate engraved map, *Totius Americae Septentrionalis et Meridionalis novissima repraesentio...* shows California as an island on the Sanson model, with some towns added. Published in 1710, the map is much scarcer than the subsequent Homann Americas maps that show California as a peninsula. The colors are usually vivid and the cartouche, copied from De Fer's map of 1699, is generally uncolored.

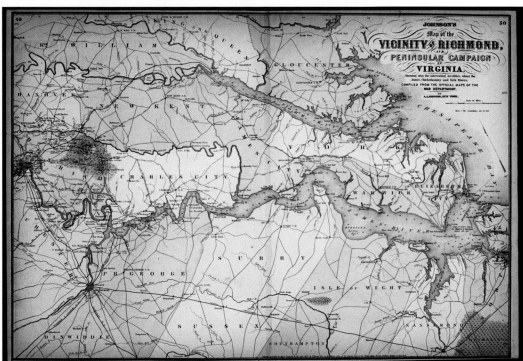

Figure C.5 Included in an atlas, this lithographed map, dated 1862, of one of the American Civil War arenas was published in New York by Johnson, whose maps often are on very brittle paper. In this case, the edges of the paper have age-toned. The image is too large for the sheet size and this map characteristically has very small margins. Size is 17.5 x 26.52 inches (44.5 x 67.5 cm.)

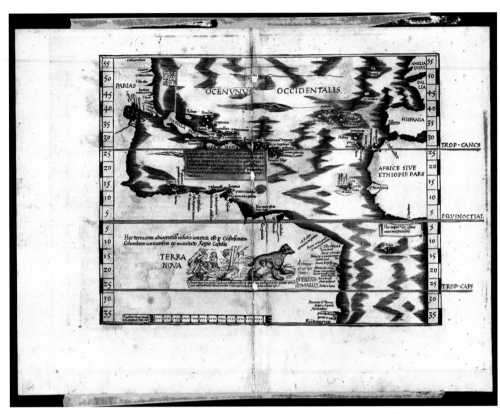

Figure C.6 The recto (above) and verso (below) of the Fries/Waldsemüller map discussed on pages 95-96 show extensive damage, age-toning and discoloration in addition to old masking and black electrician's tape on the verso.

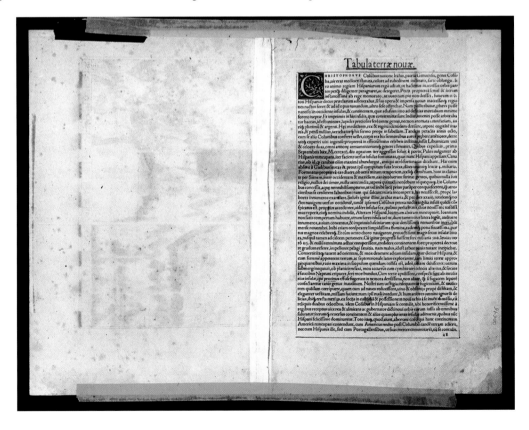

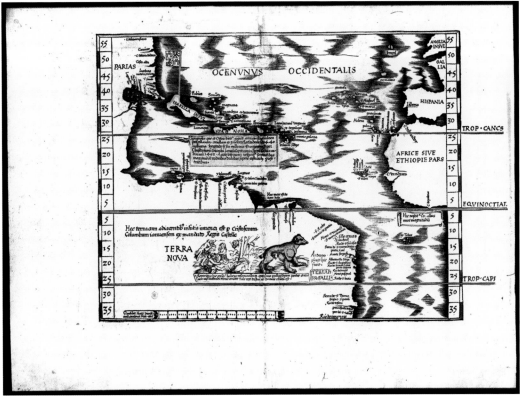

Figure C.7 Careful conservation has removed much of the staining; the holes in the centerfold have been filled and the tape removed. Note that the conservator chose not to attempt to remove all the discoloration; to do so would have jeopardized the original color on the map and given it an unnatural "bleached" look.

Figure C.8 I elected to include this map because it shows particularly severe *offsetting* and *show-through*. The map, a 1630 copperplate engraving by Hondius, has original hand color. The green pigment used (verdegris, or copper acetate) is acidic and over the years damages the paper. The map was folded along the centerfold and the opposing paper surfaces were in intimate contact for a long period of time. This permitted the verdegris to cause browning not only of the region where it was applied initially, but also of the apposing sheet of paper which it contacted. This illustration is a good example of *offsetting*. If you examine it carefully, you will be able to see each half of the map's mirror-image "burned into" the other side.

As noted in the text, some pigments, particularly green, show through on the verso after a long period of time. The illustration to the left shows the verso of the left half of the map above. Note the text, which is letterpress. There is also a series of dark outlines which correspond to the verdegris outlines on the recto. The large amoeba-shaped form is the outline of Peru. You will be able to locate the other images by comparing the recto and verso.

The characteristic show-through of green is often interpreted to indicate that the color is early or original. It is certainly suggestive of early color, but is not a conclusive test. Some early color does not exhibit show-through and some modern colorists have applied false show-through.

Map size: 18.25 x 21.5 inches (46.5 x 54.5 cm).

C6

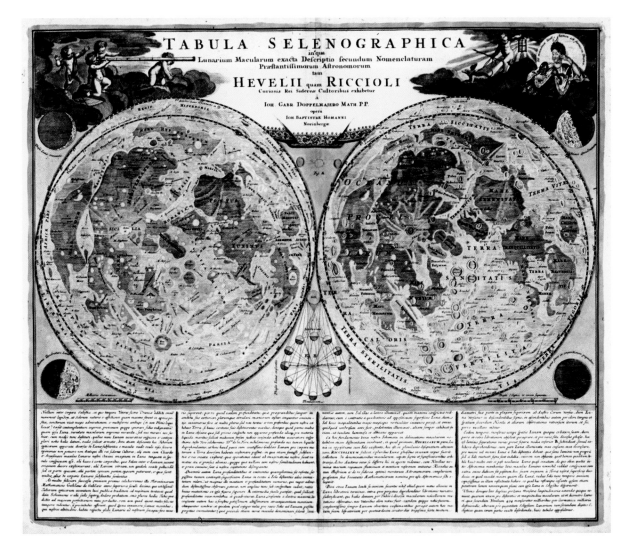

Figure C.9 Moon maps can be quite attractive, as evident in *Tabula Selenographica...,* Doppelmaier's famous double moon, published by Homann in Nuremberg: c. 1730. Copperplate engraved map and Latin text. 19 x 22.5 inches (48 x 57 cm). Each moon map is approximately 11 inches (28 cm) diameter. Original full hand color.

This is a famous double moon map showing the surface according to Hevelius and Riccioli, two of the great early moon mappers (selenographers). The left map is Hevelius' while the right is Riccioli's. Riccioli's nomenclature was adopted and is still in use today, despite Hevelius' greater prominence as the maker of the first true lunar atlas. This is a beautiful plate and an important lunar item.

Telescopes of this time were simple refractors of very long focal length (an attempt to reduce aberrations). They were very unwieldy and produced a very inferior image, so it is remarkable that lunar mapping (selenography) could have achieved these heights. Later improvements permitted astronomers to map the surface in greater detail and the later history of descriptive selenography is a several-centuries effort to draw and measure increasingly fine detail.

Earth-based visual observations were not supplanted by photography and it was only the maps produced by rocket-propelled cameras that replaced the human Earthling.

Figure C.10 This 1859 linen-backed roller map, by Lewis Robinson, shows Vermont and New Hampshire. We note the typical moulding at the top and the dowel with finials at the bottom. Notice the incursions made into the image by moisture, mostly at the top and upper right side. This is unfortunately very common and although restoration is possible, it is probably not warranted in this map since the discoloration is not severe. 29 x 23 inches (73.7 x 58.5 cm).

C8

Appendix C

Roman Numerals

Many maps have dates and other numeric data in the form of Roman notation. I have added this appendix in order to help convert to our conventional system.

Symbols:

i or **I** = one

v or **V** = five

x or **X** = ten

L = fifty

C = one hundred

D = five hundred

M = one thousand

Rules:

Rules are simple:

1. A numeral placed *after* a numeral of higher or equal value is added. Thus **XV** = 10+5; **viii** = 5+1+1+1; **CCV** = 100+100+5; **II** = 1+1.

2. A numeral of lower value placed *before* one of higher value is subtracted. Thus **iv** = 5-1; **IX** = 10-1; **CD** = 500-100.

3. Note that **I** and **C** can repeat, but not generally more than three times. Thus three hundred would be **CCC**. Four hundred would be **CD** or 500-100, although we occasionally see **CCCC**. We sometimes see **I** (or **i**) repeated four times to create the number four (**iiii**) but generally **iv** (5-1) is preferred.

Older texts may use variants. A common one is the replacement of "**M**" (one thousand) with the complex **CIƆ**. Five hundred, (generally **D**) is half of this, or **IƆ**. Fifteen hundred can be written as **MD** or as **CIƆIƆ**.

A century of Roman numbers (plus one).

1700 = MDCC	1751 = MDCCLI
1701 = MDCCI	1752 = MDCCLII
1702 = MDCCII	1753 = MDCCLIII
1703 = MDCCIII	1754 = MDCCLIV
1704 = MDCCIV	1755 = MDCCLV
1705 = MDCCV	1756 = MDCCLVI
1706 = MDCCVI	1757 = MDCCLVII
1707 = MDCCVII	1758 = MDCCLVIII
1708 = MDCCVIII	1759 = MDCCLIX
1709 = MDCCIX	1760 = MDCCLX
1710 = MDCCX	1761 = MDCCLXI
1711 = MDCCXI	1762 = MDCCLXII
1712 = MDCCXII	1763 = MDCCLXIII
1713 = MDCCXIII	1764 = MDCCLXIV
1714 = MDCCXIV	1765 = MDCCLXV
1715 = MDCCXV	1766 = MDCCLXVI
1716 = MDCCXVI	1767 = MDCCLXVII
1717 = MDCCXVII	1768 = MDCCLXVIII
1718 = MDCCXVIII	1769 = MDCCLXIX
1719 = MDCCXIX	1770 = MDCCLXX
1720 = MDCCXX	1771 = MDCCLXXI
1721 = MDCCXXI	1772 = MDCCLXXII
1722 = MDCCXXII	1773 = MDCCLXXIII
1723 = MDCCXXIII	1774 = MDCCLXXIV
1724 = MDCCXXIV	1775 = MDCCLXXV
1725 = MDCCXXV	1776 = MDCCLXXVI
1726 = MDCCXXVI	1777 = MDCCLXXVII
1727 = MDCCXXVII	1778 = MDCCLXXVIII
1728 = MDCCXXVIII	1779 = MDCCLXXIX
1729 = MDCCXXIX	1780 = MDCCLXXX
1730 = MDCCXXX	1781 = MDCCLXXXI
1731 = MDCCXXXI	1782 = MDCCLXXXII
1732 = MDCCXXXII	1783 = MDCCLXXXIII
1733 = MDCCXXXIII	1784 = MDCCLXXXIV
1734 = MDCCXXXIV	1785 = MDCCLXXXV
1735 = MDCCXXXV	1786 = MDCCLXXXVI
1736 = MDCCXXXVI	1787 = MDCCLXXXVII
1737 = MDCCXXXVII	1788 = MDCCLXXXVIII
1738 = MDCCXXXVIII	1789 = MDCCLXXXIX
1739 = MDCCXXXIX	1790 = MDCCXC
1740 = MDCCXL	1791 = MDCCXCI
1741 = MDCCXLI	1792 = MDCCXCII
1742 = MDCCXLII	1793 = MDCCXCIII
1743 = MDCCXLIII	1794 = MDCCXCIV
1744 = MDCCXLIV	1795 = MDCCXCV
1745 = MDCCXLV	1796 = MDCCXCVI
1746 = MDCCXLVI	1797 = MDCCXCVII
1747 = MDCCXLVII	1798 = MDCCXCVIII
1748 = MDCCXLVIII	1799 = MDCCXCIX
1749 = MDCCXLIX	or MDCCIC
1750 = MDCCL	1800 = MDCCC

Woodblock portrait of Sebastian Münster at age sixty. This appeared in a 16th-century edition of his Cosmograpy.

Appendix D

Mapwords: a Foreign Dictionary

The maps most commonly acquired by beginning collectors will generally have caption text in Dutch, English, French, German, Italian, or Latin. It can be very frustrating, for those not familiar with a European language or two, to try to decipher the text in cartouches or other parts of the map. For this reason I include the following foreign language dictionary. It contains most of the words needed to at least get a general notion of the meaning of texts found on most old maps.

In addition, I include, as the first part of this Appendix, the English meanings of some of the more common terms that can appear outside the neatline. The practice of identifying the artist and the engraver, or publisher, on the plate just outside the image was more commonly employed in "artistic" printmaking than in mapmaking. Nonetheless, one can encounter this terminology on maps where it may be found also within the map itself, particularly the cartouche.

cum privelegio This usually indicates copyright.

del., delin., delineavit Literally, *drew*. It is used to refer to the cartographer who drew the map from which the plate was made.

eng., engd., engraved Used to identify the engraver.

exc., exct., excudit Used to identify the publisher. It literally means *made* or *struck out*.

f., fec., fecit Used to identify the maker. It literally means *made* or *did*.

gez., gezeichnet Used to identify the person who drew the map. It literally means *drawn*.

gravée Engraved.

imp. Literal meaning: *printed*. Used to identify the printer, and usually used by intaglio printers. A lithographer may indicate **imp. lith.**

inc., incidit Literal meaning: *incised*. Usually indicates an engraved image.

lith., litho., lithog. This term refers to the lithographic medium.

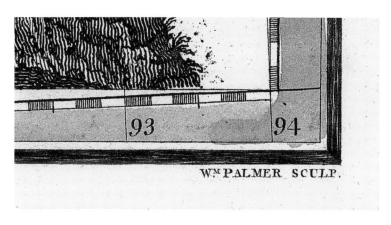

sc., sculp., sculpt. This term means *carved*. It refers to the engraver or etcher when used on intaglio images. The illustration (above) is twice actual size and is from a 1800 map of India published by William Faden in London. The map is engraved; the cartouche is largely etched.

275

Foreign Language Dictionary

This is a very basic dictionary. I have made no attempt to include pronunciation, parts of speech or any other grammatical information. The dictionary, drawn from multiple sources, is designed to provide simple English translations of words most often encountered on old maps.

This is quite adequate for the purpose of identifying the content of most texts on old maps. The Latin dictionary is particularly rich in place names: these contribute much to beginners' problems. Unfortunately I can but include a few.

Dutch–English

aan at
aangrenzend adjacent
aardbodem surface of the earth
aarde earth
acht eight
achter behind
achtste eighth
afbeelding representation
afteekening depiction
al, alle all
algemeen general
als mede as well as
anders otherwise
Azie Asia
baai, bay bay
baan orbit
begrepen included
behalve except, besides
bekend known
berg mountain
beschrijving description
betekenis meaning
beter better
bevaarbaar navigable
bevaren navigate
bevuilen dirty
bezitting possession
bij by, near, as in location
bocht bay, bight, bend
boeckverkooper bookseller
breedte (geografische) latitude
caart map
caert map
canaal channel
cirkel circle
clippen, klippen rocks, reefs
cust coast
d', de the
deel part
derde third
dieplood sounding lead
diepte depth
dorp village
drie three
Duitsland Germany
duizend thousand
een an, one
eerst first

eeuw century
eiland island
elf eleven
elfde eleventh
en, ende and
eyland island
gebeergte chain of mountains
gebetert improved
gebiedt region
gecorrigeert corrected
gedeelte part of
gedruckt printed
geheel whole, entire
gelegen situated
gelijk gradige zee kaart "equal-degree" chart
geografische lengte longitude
gesternte constellation
getrokken extracted
gezigt, gezicht view
graad degree
graadboogh quadrant
graade compass
graveerkunst engraving
groot great
half eilandt peninsula
halfrond hemisphere
handelaar dealer
haven harbor
heilig holy
heilige saint
Heilige Land Holy Land
hemels celestial
het the
Hispangien Spain
hoek angle
hoekmeter quadrant
honderd hundred
hoofdstad capital
ingang entrance
inhoudende containing
inkomen, inkoomen entrance
jaar year
kaap cape
kaart (caert, kaert) map
kaartje small map
kaartverkoper mapseller
kaert map
kanaal canal
keizerrijk empire

klein small
kolonie colony
kompasroos compass card (compass rose)
koningrijk kingdom
krijg war
kust coast
laatste latest
landengte isthmus
landschap landscape
landstreek region
langs along
meer lake
met with
Middellandsche Meer Mediterranean Sea
mijl mile
mitsgaders together with
moeras marsh
mond, mont mouth
na after
naar to, toward
naaukeurige accurate
nabijgelegen, nabygelegen adjacent
neder lower
negen nine
negende ninth
nieuwe new
noord (also nort) north
noordelijk northern
noordelijkste northernmost
noordpool north pole
Nordzee North Sea
oceaan ocean
of, ofte or
om around
omleggende surrounding
onder under
ondiepte shoal
ontdekking discovery
ook also
oost east
oostelijcke eastern
oostelijkste easternmost
op on, upon, at
opdoeningh discovery
opper upper
oud old
over over, across, beyond
papier paper
pascaart sea chart

passaat wind tradewind
perkament parchment, vellum
plaatsnijder engraver
plaeten, platen banks
platte zee kaart chart
pool pole
ree road
reize journey
rievier river
rijk empire
rivier river
rots rock
sande plaeten sand bar
scheepvart naval
schets sketch, outline
schipvaert navigation
spoorweg railroad
stad city, town
Stille Zuidzee Pacific
straat strait
straet strait
strekkende stretching
strom stream
stuck, stuk part
't the
t', te at, in
tegenvoeters antipodes
tegenwoordig present times
tekening sketch
tien ten
tiende tenth
tocht, togt journey
toneel scene
tot as far as
tot aan up to
tropisch tropic
tusschen between
twaalf twelve
twee two
tweede second
uit, uyt out
uitgegeven published
vadem fathom
van from, of
van de polen polar
vandaag today
vande of the
verbeetert improved
verdeelt divided
verdeling division
vereenigd united

Verenigde Staten United States
vermaard famous
verthoonende showing
vertooninghe appearance
vier four
vierde fourth
vijf five
vijfde fifth
vliet brook
vlijt diligence
voerbij before
volgens according to
volksplanting colony
voor for
voor naamste major
waar where, true
waar neming observation
waerachtig true, accurate
wereld world
werelddeel continent
westelijcke western
westelijkste westernmost
wijk district
wild savage
woestyne, woestijn desert
woud forest
zee sea, ocean
zee kaart sea chart
zeevaarder navigator
zeilschip sailing vessel
zes six
zeste, seste sixth
zeven seven
zevende seventh
zijne its
zuid (also zuijd, zuyd zuyt , zur, suyt) south
zuidelijk southern
zuidelijkste southernmost
zuur acid

FRENCH–ENGLISH

a he, she or it has
à at, in, to
abrégé shortened
actuel present time
adjacent adjoining, contiguous
alizé tradewind
alors then
Américain, -aine American
Amérique America
ancien ancient, previous
Anglais (aise) English
Angleterre England
année year
anse cove
appellé (ée) called
après after

aquatique aquatic
archipel, archipelague archipelago
armée army
arpent land measure (approximately an acre)
arpentage survey
astronomie astronomy
au at the, in the, to the
augmenté (ée) enlarged, augmented
aujourd'hui today
auspices patronage
aussi also, likewise
auteur author
autrefois formerly
autres other
avec with
axe axis
azimut azimuth
baie, baye bay
banc bank
bas, basse low
blanc, blanche white, empty
bois forest
bouche mouth
boussole compass
canal channel, canal
cap cape, headland
Cap de Bonne Espérance Cape of Good Hope
carte map
célèbre celebrated
celles these, those
celui, celle this, that
cent hundred
cercle circle
ces these
cet, cette this, that
chemin path, track, route
chemin de fer railroad
chez at the place of
chrétien (enne) Christian
cinq five
cinquième fifth
circonvoisin surrounding
cité city
colonie colony
comme like, as
commissaire-priseur auctioneer
composé composed, constituted
comprenant comprising, covering
comprend includes, covers
comté, comtez county
confins borders
connaissance knowledge
connu known

contenant containing
copié (ée) copied
corrigé (ée) corrected
cosmographe cosmographer
cosmographie cosmography
côte, coste coast
cours course
curieux curious, strange
dans in
de of
déchiffrage legible
déchiré torn
déclinaisons magnetic variation
découverte discovery
décrit (te) described
dédié (ée) dedicated
degré degree
dépôt depository
depuis from
des of the
description description
dessiné (ée) drawn, laid out, designed
dessus upon
détaillé (ée) detailed, itemized
détroit, destroit strait
deux two
deuxième second
devant before
dictioinaire géographique gazetteer
divers diverse, different, varying
divisé (ée) divided
dix ten
dixième tenth
douze twelve
douzième twelfth
dressé (ée) drawn, laid down
du of the
duché, duchez duchy
echelle scale
elle it, she
elles they
embouchure river mouth
en it, at, to
enchères auction
encore again, more
entier (ère) entire
entre between
environs environs
équateur equator
est (he, she, it) is
est east
et and
établissement settlement
état, estat state

été been
être contigu à to border on
étude study, early draft
exactement exactly
extrait extract
extrémite extremity, end of the earth
faisant making
fait, faite made, done
fameux famous
faubourg suburbs, outskirts
feu fire
feuille leaf, sheet
fils son
fleuve river
fond depth, bottom
fort, forte strong, large
général (ale) universal
géographe geographer
globe globe
golfe, golphe gulf, bay
gouvernement government
grand, grande large, great
gravé engraved
guerre war
haut, (e) high
havre harbor
héritier heir
histoire history
historique historical
huit eight
huitième eighth
hydrographie hydrography
il he, it
illustré (ée) illustrated
ils they
imprimé (ée) printed
indiquant indicating
inférieur (ieure) lower
ingénieur engineer
intérieur (ieure) interior, inland
isle, i. le island
isthme isthmus
itinéraire route
jusque up to, as far as
l', la, le the (singular)
La Manche English Channel
lac lake
laquelle who, which
lequel who, which
les Indes the Indies
lesquels, lesquelles who, which
levé (ée) raised
libraire bookseller
lieu place

lieue league, pl. lieues
limite boundary
limitrophe adjoining,
    bordering
lunaire lunar
lune moon
luy he
maison house
manière manner, behavior
manuscrit manuscript
mappe map
mappemonde world map
marchand dealer
marée tide
marge margin
marine navy, marine
marqué (ée) marked
mémoire memoir, report
mer sea
Mer du Nord Atlantic
    Ocean
Mer du Sud Pacific Ocean
Mer Glaciale Arctic
    Ocean
méridional (ale) southern
mettre à la voile sail boat
mille thousand
ministre minister
monde earth, universe,
    world
mont mountain
montagne mountain
montrant showing
mouillage anchorage
navigateur navigator,
    sailor
neuf nine
neuf, neuve new
neuvième ninth
nord north
nouveau new
nouvellement recently
observations observations
occident west
occidental (ale) western
onze eleven
onzieme eleventh
ordinaire ordinary, usual
ordre order
oriental (ale) eastern
ou or
où where
ouest west
ouvrage work
païs, pays land, country,
    nation
par by, at
partie part
pas strait
pas not
passage passage
péninsule peninsula

petit, petite small
peuples peoples
pli fold
plusieurs several
point point
pouce inch
pour for
premier first, foremost
près de near
preséntemente currently
presqu'île peninsula
    (almost island)
principal (ale) principal
principalement major,
    leading, principal
principauté principality
privilège permission (as in
    copyright)
province province, country
publié published
quatre four
quatrième fourth
quay quay, wharf
que that
quelque some
qui who, which, that
recent (ente) recent, new
récif reef
rectifié corrected
recueil collection,
    compendium
reduite reduced
relation narrative, account
retour return
revu revised
riviere river
rocher rock
roi, roy king
route road
royaume kingdom
rue street
sa his, her, its
saint holy
sauvage savage
selon according to
sept seven
septentrional, -ale
    northern
septième seventh
ses his, her, its
situé located, situated
six six
sixième sixth
son his, her, its
sonde sounding
sous under
sud south
suite sequel, continuation
suivant according to
supérieur upper, larger
sur on, upon, above
tableau picture, view

terre earth, land, world
terrestre terrestrial
tiré derived
toise fathom
tout, toute all, whole; pl.
    tous, toutes
traduit translated
treize thirteen
très very
tributaire tributary
trois three
troisième third
trouvé (ée) found
un, une one
uni (ie) united
vaisseau vessel
vallon valley
vent wind
ville city, town
voisin neighboring, nearby
vue view

GERMAN-ENGLISH
Abbildung illustration
Abriß summary, plan,
    outline
Abteilung part
acht eight
achte eighth
allgemeine general
am at the
Amt office
ander other
angrenzend bordering
ans to the
Armee army
auch also
auf on, at, by
Auflage edition
aus from, out of, by
Ausgabe edition, issue
ausgebessert repaired
außere exterior, outside
Aussicht view, prospect
Bach brook, stream
Bad bath
Bahn road, path
Bai, Bay bay
bearbitet repaired
begreifend comprising
bei, bey at, by, with
Belagerung siege
benachbart adjacent,
    neighboring
Berg mountain, hill
berühmt famous
beschnitten trimmed
Beschriftung lettering
Beschreibung description
bis as far as, till
Blatt sheet

blattgroß full page
Breite latitude
Bruch swamp
Bucht bay, inlet, creek
Charte chart
das the
datiert dated
den the
der the
des the
deutsch German
die the
Dorf village
drei three
dritte third
durch through, by means
    of
eigentlich true, proper
ein one
ein, eine a, an
eingerissen torn
Eis ice
Eisenbahn railroad
Eismeer Polar Sea
elf eleven
elfte eleventh
Entdeckung discovery
enthaltend containing
Entwurf sketch, draft,
    plan
Erbe heir
Erde world
Erdteil continent
erforschen explore
Eroberung conquest,
    acquisition
erst first
Erstausgabe first edition
Etlich some or several
Eyland, Eiland island
Falte fold
farbig in color
faltig folded
Fluß river
Flut tide
Fortsetzung continuation
Frankreich France
franzosisch French
frey, frei free
fünf five
fünfte fifth
Fürstenthum principality
gantz, ganz entire
Gasse street, alley
Gebiet district, region
Gebirge mountain range
gegen towards, opposite to
Gegend region, district
gehörig belonging to
Gelobte Land Promised
    Land
gennant called, named

Gestalt shape, form, figure
gezeichnet drawn, designed
Gravieren engraving
groß large
Grundriß ground plan, sketch
Guten Hoffnung, das Kap der Cape of Good Hope
Hafen harbor
Haff lagoon
Halb- half-
Halbinsel peninsula
Halbkugel hemisphere
Haupt principal
Hauptstatt, Hauptstadt capital
heilig holy
heutig nowadays
hundert hundred
ihr its, their
im in the
in der Platte in the plate
ins in the
Insel island
Jahr year
jahrundert century
jenseits beyond
Kap cape
Karte map, chart
Kartusche cartouche
Kaspische Meer Caspian Sea
Keyser emperor
klein small
Kloster monastery, nunnery
Köln Cologne
König king
Königreich kingdom
Kreis circle, district; also: orbit
Kriegsshauplatz theater of war
Kupferstich; Radierung etching, engraving
Kurfurstentum electorate
Kuste coast
lädiert damaged
Lager camp
Land land, country
Landkreis district
Landschaft landscape, estate, district
Länge longitude
Lauf course, route
Markt market, market-town
marmoriert marbled
Maßstab scale
Meer sea, ocean

Meerbusen bay, gulf
Meerenge strait, channel
Meile mile, league
mit with
Mittag noon
Mitternacht midnight
nach after, according to
Name name
nebst besides
neu new
neuest newest
neun nine
neunte ninth
nieder low
Nörd north
nördlich northern
Nördlicher Polarkreis Arctic Circle
ober upper, higher
oder or
ohne Jahr no date
Ort place, region
Ost east
Österreich Austria
Östindien East Indies
östlich eastern
Ostsee Baltic Sea
Pergament vellum
Plan plan, map, chart
Platz place
Preußen Prussia
Rand margin
Reich empire, kingdom
Reichsstadt imperial city
Reise voyage, trip
richtig correct
sampt together with
Scheitelpunkt zenith
schiffbar navigable
Schloß castle
sechs six
sechste sixth
See lake, sea, ocean
Segelschiff sailing ship
sehr very
sein its
sieben seven
siebente seventh
signiert signed
so so, thus
solch such
sowohl as well as
Staat state, country
Stadt, Statt city
Städteansichten city views
Steindruck lithograph
Sternbild constellation
Stille Meer Pacific Ocean
stockfleckig foxed
Straße street, strait
Stuck part

Süd south
südlich southern
Südlicher Polarkreis Antarctic Circle
Südpol south pole
Südsee South Sea
Sund sound, strait
Tafel table, chart
Tag day
tausend thousand
Theil, Teil part
Tote Meer Dead Sea
über over, above
um about, around
und and
Venedig Venice
verbessert improved
vereinigt united
Vereinigten Staaten United States
Vestung fortress
vier four
vierte fourth
Vogelperspektive bird's-eye view
vom from the, of the
von from, of
vor before, in front of
Vorgebirge promontory, cape, headland, foothills
wahr true, correct
wahrhaftig true, genuine
Wald forest
Wappen coat of arms; escutcheon
Wasser water
wasserfleckig waterstained
Wassertiefe sounding
welch which, what, who
Welt world
Weltheil, Weltteil part of the world; continent
Wendekreis des Krebses Tropic of Cancer
Wendekreis des Steinboks Tropic of Capricorn
Westindien West Indies
wie how
Wiek bay, cove
Wien Vienna
zeitgenössisch, der Zeit contemporary, of the period
zehn ten
zehnte tenth
zu to
zum, zur to the
zwei two
zweite second
zwischen between, among

zwolf twelve
zwolfte twelfth

ITALIAN–ENGLISH

a to
adiacente adjacent
ai, agli, alle to the (pl.)
al, allo, alla to the (sing.)
alto high
altrementi otherwise
altro other
antico ancient
anticamente anciently
appresso near
aquaforte etching
arcano arcane, mysterious
asta (vendere all'asta) auction (sell at a.)
atlante atlas
Atlantico Atlantic Ocean
australe southern
baia bay
banco di sabbia sandbank
banditore auctioneer
basso low, lower
Belgio Belgium
bocca mouth
boreale northern
bussola compass
caduta waterfall
canale channel
capitale capital
capo cape
celeste celestial
cento hundred
che that
chi who, whom, whose
chiamato named
Cina China
cinque five
circumpolare circumpolar
citta, civita city
coi, cogli, colle with the, by the (plural)
col, collo, con lo, colla with the, by the (sing.)
colorato colored
cominciare to begin
con with
conosciuto known
contea county, earldom
contenere to contain
corretto correct
corso course
cosmografo cosmographer
costa shoreline, coast
costa coast
costellazione constellation

279

da from
dai, dagli, dalle from the, by the (plural)
dal, dallo, dalla from the, by the (singular)
Danimarca Denmark
decimo tenth
dei, degli, delle of the (pl.)
del, dello, della of the (singular)
delineato drawn
della of
descritta described
descrittione, descrizione description
detto said, called
di of
dieci ten
disegno plan, drawing
dodicesimo twelfth
dodici twelve
ducato duchy
due two
e, ed and
Egitto Egypt
emisfero hemisphere
entrata entrance
equatore equator
esatto exact
est east
fatto done
finire to finish, to end
fino as far as, until
Firenze Florence
fiume river
foce mouth
foglio folio
foglio sheet
fortezza fortress
fra in, between, among
Francia France
Galles Wales
geografia geography
geografico geographical
geografo geographer
gia formerly
Giappone Japan
gli the
golfo gulf
grado degree
grande, gran large
Grecia Greece
i, il the
impero empire
in in
incisione engraving
Inghilterra England
Inglesi Englishmen

intagliata engraved
intagliatore engraver
intorno around
Irlanda Ireland
isola island
isoletta islet
istmo isthmus
la the
lago lake
latitudine latitude
le the
levante east
litografia lithograph
lo the
longitudine longitude
Luigiana Louisiana
maestro main; master
maggiore greater, major
mappa map
mare sea
margine margin
meridiano meridian
meridionale southern
Messico Mexico
meta half
miglio mile
mille thousand
minore lesser, minor
molto much
mondo world
montagna mountain
monte mount
navigare to navigate, to sail
nei, negli, nelle in the (pl.)
nel, nello, nella in the (singular)
nono ninth
nord north
nove nine
nuovamente again, newly
nuovo new
o or
occidentale western
Olanda Holland
Olandesi the Dutch
orientale eastern
originale original
orizzonte horizon
osservazione observation, comment
ostro south
ottavo eighth
otto eight
ovest west
ovvero or
paese country, village
parte part

particolare particular
penisola peninsula
pianta map
piega fold
piu most
polo pole
ponente west
porto harbor, port
Portogallo Portugal
presso at, by, near, in
primo first
principale principal
privilegio copyright
projezione projection
provincia province (pl. -cie)
quarto fourth
quattro four
questo, (a) this
qui here
quinto fifth
rada roadstead
rappresentante representing
redatto written, drawn up
regno kingdom
ridotto reduced
ritratto image
riviera coast
scoperta discovery (pl. ..erte)
scoperto discovered
Scozia Scotland
secca shoal
secondo second; in accordance with
sei six
selvaggio savage, native
sesto sixth
sette seven
settentrionale northern
settimo seventh
sino as far as, until
sobborgo suburb
sonda sound
Soria Syria
sotto under
Spagna Spain
Spagnuola Hispaniola
specchio mirror
spiaggia shore, beach
stabilimento establishment
stampa print
stampa printing press, print
stamperia printing establishment

Stati Uniti United States
strappo tear
strappare torn
stretto straits
sud south
sui, sugli, sulle on the (plural)
sul, sullo, sulla on the (singular)
suo his, her, its
superiore upper
Svezia Sweden
Svizzera Switzerland
tavola map
terra earth
terzo third
tramontana north
tre three
tropico tropic
Turchia Turkey
tutto all
ultimo latest
un, uno, una a
undicesimo eleventh
undici eleven
Ungheria Hungary
universale universal
uno one
vecchio old
veduta view
vento wind
vero true

LATIN–ENGLISH

a, ab from, by
ac and
adjacens adjoining
Aestivarum Insulae Bermuda, (Sommer's Islands)
Africus southwest wind. One of the named winds.
Agrippina colonia Köln (Cologne)
Albania Scotland
Albion Britain
Albis Fluvius Elbe river
aliquot some
Allobrogum colonia Geneva
Alostum Aalst
Alsatia Alsace
alter other
amplissimus most splendid, most glorious; wonderful; esteemed
Amstelodamum Amsterdam

Andegavensis Ducatus Anjou

Andegavum Angers

Andreapolis St. Andrews

Anglia England

annus year

anthropophagi cannibals

Antverpia Antwerp

Aparctias northwind. One of the named winds.

Apelliotes eastwind. One of the named winds.

apud at the establishment of

aqua water

Aquarius (constellation) water-bearer

Aquilo north-by-northeast wind. One of the named winds.

Aquisgranum Aachen

archiducatus archduchy

archiepiscopatus archipiscopate

Archipelagus Septentrionalis Aegean Sea

Argentina Strassbourg

Argestes northwest wind. One of the named winds.

Aries (constellation) the ram

Artesia Artois

atque and

auctor, auctore author or creator of a work

Augusta Augsburg

Augusta Perusia Perugia

Augusta Trebocorum Strassbourg

Augusta Treverorum Treves

Augusta Trinobantum London

Augusta Vangionum Worms

Augusta Vindelicorum Augsburg

Aurea Chersonesus Malay peninsula

Aurelia Orleans

Aurelia Allobrogum Geneva

Auster southwind. One of the named winds.

australis southern

Babenberga Bamberg

Barchino Barcelona

Bardum Barth

Batavia Jarkata

Belgium novum New York

Bellovacum Beauvais

Bercharia Berkshire

Biturigum Berry

Bononia Bologna

Borbetomagus Worms

borealis northern

Boreas north by northeast wind. One of the named winds.

Borussia Prussia

Borysthenis Dnieper river

Brechiniæ Comitatus Brecknockshire

Brema Bremen

Breslanus Breslau, Wroclaw

Britannia Britain, Brittany

Brixia Breschia

Brugae Bruges

Brunopolis Braunschweig

Bruxellae Brussels

Byzantium Constantinople

Cadomum Caen

caelestis celestial

Caesar Augusta Saragossa

Caesarea Insula Jersey

Caesarodunum Turonum Tours

Calatia Ciazzo

Caletensium Calais

Caletum Calais

Cambria Wales

Cancer (constellation) the crab

Candia Crete

Cantabrigiensis Comitatus Cambridge

Cantium Kent

Carinthia Karnten

Capricornus (constellation) the goat or sea-goat

carta map, chart

Casarea Insula Jersey

Cecias northeast wind. One of the named winds.

centum hundred

Ceretica Cardigan

Cestria Chester

chersonesus peninsula

chorographica geography

Chorus northwest wind. One of the named winds.

Cimbrica Chersonesus Jutland

Circius north by northwest wind. One of the named winds.

cis on this (same) side of

citra on this side of

Clivia Cleves

cognitus known

Colonia Cologne

Colonia Munatiana Basle

Colonia Ubiorum Cologne

Comensis Lacus Lake Como

comitatus county

compendiosus abridged

complectens comprising

comprehendens including

Conatia Connacht

conatus effort

Condivincum Nannetum Nantes

confinis neighboring region

confinis (e) adjacent

Connacia Connaught

Constantia Constance

Constantinopolis Constantinople

Constantinopolitanum fretum Bosphorus

continens (-entis) adjacent or neighboring

conventus district, as of a city

Corcagia Cork

Cornubia Cornwall

Corus northwest wind. One of the named winds.

Corvi pons Pontecorvo

Cracovia Cracow; Krakow

cum with

cum privilegio usually indicates copyright

Cumbria Cumberland

Dania Denmark

Darbiensis Comitatus Derbyshire

Daventria Deventer

decem ten

decimus (a; um) tenth

del (see delineavit) drew

delineavit delineated (he)

Delphi Delft

Denbigiensis Comitatus Denbigh Shire

descripsit drew

descriptio description (as a map)

Devonia Devonshire

dicio (ionis) dominion

dioecesis diocese, district

ditio (ionis) dominion

Divio Dijon

divisus (a; um) divided

dominium ownership; rule

Dorcestria Comitatus Dorsetshire

Dordracum Dordrecht

Duacum Douay

ducatus (us) duchy, dukedom

Dunelmensis Episcopatus Durham

duo, duae two

duodecim twelve

Eblana Dublin

Eboracensis Ducatus Yorkshire

Eboracensis nova civ. New York

Eboracum York

editus (a, um) published

Elvetiorum Argentina Strasburg

emendatus (a, um) corrected, improved, amended

Enipontius Innsbruck

episcopatus (us) episcopate, bishopric

Erfordia Erfurt

et and

Euroafricus south by southwest wind. One of the named winds.

Euroauster south by southeast wind. One of the named winds.

Euronotus south by southeast wind. One of the named winds.

Euros southeast wind. One of the named winds.

Eustadium Eichstadt

ex from, out of

ex officina from the workshop of

exactissime most exactly

exactissimus (a, um) most exact

excudebat, excudit excud, exc. made or struck (printer or publisher)

exhibens displaying

Exonia Exeter

facies shape or appearance,

Faventia Faenza

Favonius westwind. One of the named winds.

fecit made, often refers to engraver

fere approximately

finitimus (a, um) neighboring

Fionia Funen Island

Florentia Florence

florentissimus (a, um) most flourishing, eminent or prosperous

flumen (inis) river

fluvius (ii) river

Forum Iulii Friuli

Forum Livii Forli

Frankofortum ad Moenum Frankfurt am Main

fretum strait, channel

Frisia Friesland

Gallovidia Galway

Ganabum Orleans

Gand, Gandavum Ghent

Gebenna Geneva

Gemini (constellation) the twins

Genabum Orleans

geographicus (a, um) geographical

Germania Inferior Netherlands

Glascua Glasgow

Glotta Clyde

Graecia Greece

Gratianopolis Grenoble

Gravionarium Bamberg

Hafnia Copenhagen

Haga Comitis The Hague

Hammona Hamburg

Hanovia Hanau

Hantonia Comitatus Hampshire

Hassia Hesse

Helenopolis Frankfort am Main

Hellas Greece

Hellespontius northeast wind. One of the named winds

Helvetia Switzerland

Herbipolis Wurtzburg

Herculeum fretum Strait of Gibraltar

heres, heredis heir, successor

Hibernia, (ae) Ireland

Hierosolyma Jerusalem

Hispalis Seville

Hispania Spain

hodie present time

hodiernus (a, um) present, modern

Holmia Stockholm

Holsatia Holstein

imago (inis) image, likeness, copy

Imaus Mons Himalayas

impensa (ae) cost, expense

impensis generally indicates publisher

impensus (a, um) expensive

imperium (ii) empire, dominion

in in, on, at

incidit, incidebat (he) cut (engraver)

incola (ae) inhabitant

inferior (ius) lower

insula (ae) island

integro (a, um) whole, entire

Internum mare Mediterranean Sea

inventit devised; usually indicates cartographer or draftsman

inventor usually indicates cartographer

Islandia Iceland

item also

Iuliacensis Ducatus Julich

Iutia Jutland

iuxta, juxta near

Juvavum Saltzburg

lacus lake

Lancastria Palatinatus Lancashire

Larius Lacus Lake Como

Latium Lazio

Legio Leon

Leicestrensis Comitatus Leicestershire

Leida Leyden

Lemovicense Castrum Limoges

Leo (constellation) the lion

Leodium Liege

Leucorea Wittemburg

Libonotus south by southwest wind. One of the named winds

Libra (constellation) the scales, or balance

Libs, Lips southwest wind. One of the named winds.

limes, limitis boundary, route

Lipsia Leipzig

litus, litoris coast, beach

locus (i) place, district

Londinum London

Lotharingia Lorraine

Lovanium Louvain

Ludoviciana Louisiana

Lugdunum Lyons

Lugdunum Batavorum Leiden

Lusatia Lausitz

Lusitania Portugal

Lutetia Paris

Lutzenburgum Luxembourg

Mantua Carpetanorum Madrid

mappa (ae) map

marchionatus, (us) marquisate

mare (is) sea, ocean

Mare Crisium (moon) Sea of Crises

Mare Fecunditatus (moon) Sea of Fertility

Mare Frigoris (moon) Sea of Cold

Mare Humorum (moon) Sea of Moisture

Mare Hyrcanum Caspian Sea

Mare Imbrium (moon) Sea of Rains

Mare Nubium (moon) Sea of Clouds

Mare Rubrum Red Sea

Mare Serenitatis (moon) Sea of Serenity

Mare Tranquillitatis (moon) Sea of Tranquility

Mare Vaporum (moon) Sea of Vapors

Matritum Madrid

Mediolanum Milan

meridonalis southern

Mervinia Comitatus Merionethshire

milia (ium) thousand

mille a thousand

Misnia Meissen

Moguntia, Moguntiacum Mainz

Momonia Münster

Mona Isle of Man

Monachium Munich

Monasterium Münster

Montensis Ducatus Bergh

Montisferratus Monferrato (Italy)

Monumenthis Ducatus Monmouthshire

mundus (i) the world, the universe, the earth

Mutina Modena

Nannetum Nantes

Nassovia Nassau

Natolia Asia Minor

nec, neque and besides, and also

nec non besides, and also

Neocomum Neuchatel

neotericus modern

Nicsia Naxos

nonus ninth

Nordovicum Norwich

Noremberga Nuremberg

Northantonensis Comitatus Northamptonshire

Notus southwind. One of the named winds.

novem nine

Novesium Neuss

Noviodunum Nevers

Noviomagnum Nijmegen

novissimus (a, um) newest, most recent

noviter newly

Novum Eboracum New York

novus (a, um) new

nunc now, nowadays

Nuova Zelanda New Zealand

occidentalis (e) western

oceanus (i) ocean

Oceanus Procellarum (moon) Ocean of Storms

octavus (a, um) eighth

octo eight

officina (ae) workshop, factory

olim formerly

oppidum (i) town

ora border, coast

ora maritima seacoast

orbis globe, orbit

Orcades Orkneys

orientalis (e) eastern

Oxonium Comitatus Oxfordshire

pagus village, province

Palatinatus Rheni Rheinland-Pfalz

Panormum Palermo

Papia Pavia

pars, partis part, region

Parthenope Naples

Parthenopolis Magdeburg

passim ubiquitous

Patavium Padua

Pedemontana Piedmont

per through, by

Perusia Perugia

Pictavium Poitiers

pictor (oris) painter

pinxit (he) drew, painted

Pisces (constellation) the fish

plus, pluris more
Polonia Poland
Pontus Euxinus Black Sea
praecipuus (a, um) excellent, extraordinary, special
praesertim especially
praeter past, beyond
presbiter elder, priest
Presbiter Ioannis Prester John
pretiosus (a, um) valuable, precious
primus (a, um) first
promissionis promise
propius (a, um) special, particular
prout as, just as
Provincia Provence
quartus (a, um) fourth
quattuor four
qui, quae quod who, which, what, that
quinque five
quintus (a, um) fifth
Ratisbona Regensburg
recens (entis) recent
recens, recenter recently, newly
regio (ionis) line, boundary, region
Regiomontium Königsberg
regnum (i) kingdom, dominion
retectus (a, um) opened up, made accessible
Rhedones Rennes
Rhenolandia Rheinland
Rhenus Fluvius Rhine River
Rothomagum Rouen
Rugia Rugen Island
Rupella La Rochelle
Sabaudia Savoy
Sagittarius (constellaton) the archer
Salmantica Salamanca
Salopia Shrewsbury
Salopiae Comitatus Shropshire
sanctus (a, um) holy

Sarnia Insula Guernsey
Saxonia Inferior Lower Saxony
Saxonia Superior Upper Saxony
Scania Zealand Island
Schedamum Scheidam
scilicet certainly, naturally
Scio Chios
Sclavonia Slavonia
Scotia Scotland
Scotia major Ireland
Scorpius (constellation) the scorpion
scripsit (he) wrote or drew; sometimes indicates lettering engraver
sculpsit, sculp, sc (he) carved; usually indicates engraver
secundum according to
secundus (a, um) second
sedes belli seat of war
Senae Siena
septem seven
Septentrio northwind. One of the named winds.
septentrionalis (e) northern
septimus (a, um) seventh
Servia Serbia
seu or
sex six
sextus (a, um) sixth
Sinarum Regio China
Sinus Gangeticus Bay of Bengal
situs (a, um) situated
sive or
Soctia (ae) Scotland
Somersettensis Comitatus Somersetshire
sophus (a, um) wise
Soria Syria
Sorlinges Scillies
stellatus (a, um) starry
subjacens (entis) near
Subsolanus eastwind. One of the named winds.
Suecia Sweden
Suevia Sweden

sumptibus at the cost of; usually indicating the publisher
sumptus cost
superior (is) upper, higher
tabula (ae) map
tam so, so much, to such an extent
tam. . . quam. . . both. . . and. . .
Taprobana Ceylon
Taraco Tarragona
Tarvisium Treviso
Taurica Chersonesus Crimea
Taurinum Turin
Taurus (constellation) the bull
terra (ae) the earth, land
Terra Sancta Holy Land
terrestris (e) terrestrial
tertius (a, um) third
theatrum belli theater of war
Tholosa Toulouse
Thrascias north by northwest wind. One of the named winds.
Ticinum Pavia
Tigurum Zurich
Toletum Toledo
Tornacum Tournai
totus (a, um) all, entire, total; gen. totius
tractus district, region
Trajectum ad Viadrum Frankfurt a. d. Oder
Trajectum, Trajectum ad Mosam Utrecht
trans across, beyond
Transisulana Overijssel
Trebia Trevi
Trecae Troyes
tres, tria three
Treveris Treves (Trier)
tribus (us) tribe
Tricasses Troyes
Tridentum Trent
Turcicum Imperium Ottoman Empire
Turonum Tours
Tuscia Tuscany
typus (i) image, figure

Ultonia Ulster
Ultrajectum Utrecht
Ulyssipo Lisbon
undecim eleven
Ungaria Hungary
universalis (e) universal
unus (a, um) one
urbs, urbis city
Ursina Bern
uterque, utraque utrumque each, both; gen. utriusque
Utinum Udina
Valesium Valois
Vallisoletum Valladolid
Vectis Insula Isle of Wight
vel or
Venetia Venice
Venetum Veneto
ventus (i) wind
verissimus (a, um) truest
vernacule in the vernacular
Veromandua Vermandois
Verona Bonn, Verona
verus (a, um) true, actual
vetus (eris) old
Vicentia Vicenza
vicinus (a, um) nearby
Vienna allobrogum Vienne
Vienna Austriae Vienna, Wien
Vindobona Vienna
Virgo (constellation) the virgin
Vitemberga Wittemburg
Vormatia Worms
vulgo commonly, generally, in the vernacular
Vulturnus southeast wind. One of the named winds.
Wallia Wales
Wetteravia Wetterau
Wigorniensis Comitatus Worcestershire
Wiltonia, Wiltoniensis Comitatus Wiltshire
Zephyrus westwind. One of the named winds.

Note: Virtually all towns, cities, countries and geographical features of the Old World (and many of the New) have Latin or Latinized names. The Latin-English dictionary presented here cannot list all of these and I include only a few of those more commonly found. I recommend *Orbis Latinus* edited by Plechl and Spitzbart (Klinkhardt & Biermann, Braunschweig: 1971) for a comprehensive dictionary of these names.

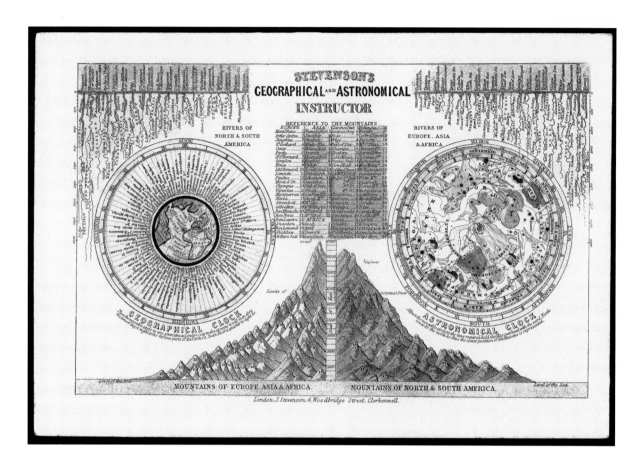

This is an example of a piece of mid-19th-century ephemera that has a map association. Measuring 4.5 x 7.1 inches (11.2 x 18 cm) the lithographed card has comparative lengths of rivers and mountains and two volvelles. The left one can be turned to compare times in different part of the world; the one on the right to show circumpolar constellation positions.

Appendix E

The Substance of Maps: Paper and Vellum

In this Appendix I deal with some of the more common materials used in making maps in the Western world. This information is not only useful in helping us date a map, but necessary if we are to understand the problems a collector may encounter in preserving and caring for a collection, especially in determining how best to prevent further damage or further deterioration. In order to understand the diseases that afflict it, we need to have a basic understanding of the substance of maps.

VELLUM

Vellum is an animal product, made by stretching hide (often sheep or goat) on a frame, scouring it with lime to remove fat, and scraping to remove hair. What remains is essentially the animal's *dermis*. During preparation, the vellum is treated with a variety of substances that gives it its final qualities. Depending upon the intended use of the vellum, different empirically derived concoctions were applied to make the surface more receptive to inks, or to whiten it.

The final product is a robust, flexible and attractive product that was used commonly in Europe prior to the widespread use of paper. Even later, it was used for official documents, and special books such as religious works and musical scores. Larger sheets of vellum were made from ox or cow hides. This latter vellum is sometimes rather coarse and thick and was often used for binding books. There is also a pig vellum, also used frequently in book bindings because of its great strength and wear resistance. The finest vellum, often used for books of hours and fine bibles, was made from the

Figure E.1 The scanning electron microscope, because it uses electrons instead of light, has greater resolution than does the conventional light microscope. This scanning electron micrograph shows the surface of a sheet of 14th-century vellum. The vellum used here is from a music manuscript, but it was made in a very similar manner to that used for early maps. We only show one side, the "light" side, but it illustrates the fibrillar nature of the vellum. The fibrils are bundles of the protein, collagen. This image has been magnified 70x.

skin of fetal animals. This is an exquisite product, white and soft.

Vellum is remarkably robust and its life is measured in centuries. However, since it is an animal product, it is eaten by rodents and other vermin. Vellum and *parchment* are synonyms, despite the confusion in some recent books.

Chemically, vellum is almost pure *collagen*. Collagen is a protein. Unlike the substance of paper, called cellulose, which is made of sugars, collagen is made of amino acids. There are many different types of collagens (for the scientifically-minded reader, each collagen is a specific gene product and we know precisely the sequence of the different amino acids).

Microscopically, vellum shows bundles of collagen fibers arrayed laterally in the plane of the sheet (Figure E.1). This alignment is the result of stretching, since in the untreated dermis (from which vellum is derived) the collagen fibers run in three dimensions. Higher magnification (Figure E.2) reveals the tiny particles of lime and other agents used in preparation. We did not have any vellum available from manuscript maps, but chose

similar vellum for the specimens used for these micrographs. We used vellum from a 14th-century *graduale*, similar in weight and texture to many of the vellums we have seen used for maps.

Vellum expands and contracts by a large amount, depending upon humidity, and has a tendency to curl when very dry.

Vellum was occasionally used for maps, most often manuscript maps. The casual collector will not encounter many vellum maps. Some of them, called *portolans,* were sometimes objects of great beauty, embellished with decorations and gilt. These manuscript maps were much prized by sailors and can command very high prices. In addition, there are lesser maps on vellum, but again, these are unusual. There are some modern papers called "parchment paper" and they should not be confused with authentic animal vellum.

PAPER

Paper is a feltwork of randomly arrayed fibers, each of microscopic thickness. In typical paper the fibers are of plant origin, be it wood (as in many modern

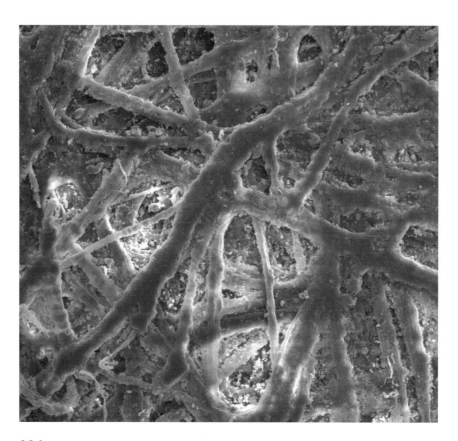

Figure E.2 At 275x magnification, the vellum sample in Figure E.1 reveals more detail. We still note the fibrillar nature of the vellum, but now see more clearly some of the particulate material that represents the lime and other agents used to treat the vellum during its manufacture. These particles are embedded in the spaces between the fibrils.

papers), or linen (as it was in early paper) or any number of other substances including mulberry bark, rice plants, grass, and cotton fibers. Chemically, they are largely *cellulose.*

In principle, papermaking is a simple technique and all paper, from the early handmade papers that were produced in single sheets to the massive rolls of modern paper, is made utilizing identical principles; only the technology is different. Plant material, such as chips of wood or old rags, are beaten to a pulp in water, forming a slurry. The slurry is washed, bleached, dyed, or otherwise treated, and the fibrous matter is separated by a variety of methods to produce a thin layer of wet fibers that is subsequently dried and squeezed to form a uniform, smooth sheet. This process can take place sheet-by-sheet, as in handmade papers or in long, continuous rolls weighing thousands of pounds. The quality of paper is determined by both the starting material and the means by which it is converted into final product.

In making paper by hand, the slurry is picked up on a wire mesh screen, the water allowed to drain, and the resulting feltwork allowed to dry. This process is similar to that used by a cook, who scoops a portion of randomly aligned vermicelli from a pot of water, using a wire strainer. This type of paper is called *laid paper.* When held to the light, laid paper shows a gridwork of lines representing the screen (Figure E.3). The pattern, consisting of *chain lines* and *laid lines,* is characteristic of such handmade paper. The chain lines are heavier, usually spaced about an inch apart and run parallel to the short dimension of the sheet of paper. They appear as the vertical lines in Figure E.3. The laid lines are closer together and run parallel to the long dimension of the sheet. In handmade paper, the wooden frame that holds the

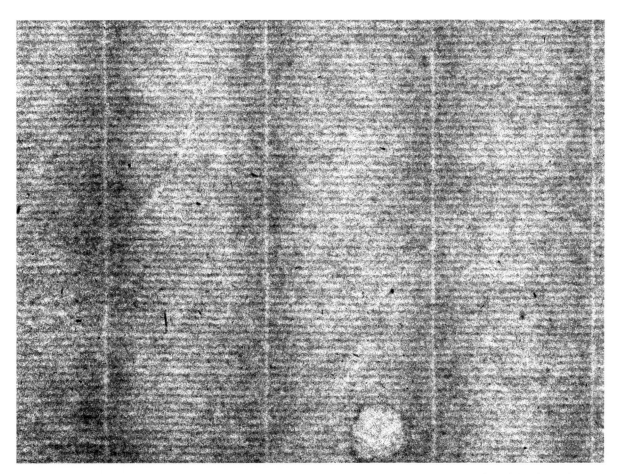

Figure E.3 The chain and laid lines of handmade laid paper are visible in this photograph, taken with transmitted light, of part of a sheet of American-made paper, c. 1800. This characteristic appearance of laid paper can be easily seen if the paper is held to the light. The circular spot near the bottom is a manufacturing artifact. Chain lines are vertical; laid lines horizontal. Enlarged about 1.5x.

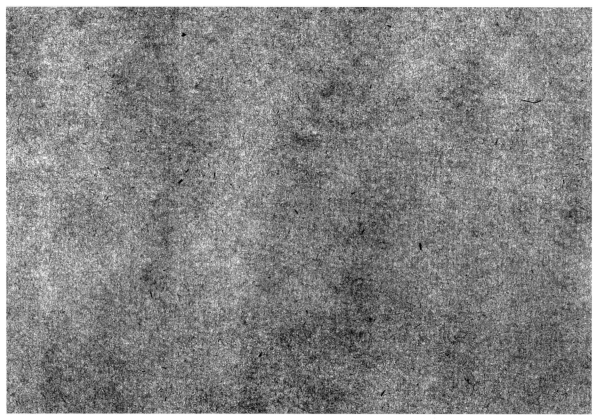

Figure E.4 A photograph of a transilluminated sheet of Whatman *wove* paper made in England in 1802. Note the absence of chain and laid lines. The paper was machine-made using a very fine mesh to produce long lengths of paper. I magnified the image 2x and increased the contrast, enabling us to just barely detect the mesh pattern in the paper. Compare to Figure E.3.

wire meshwork is called a *deckle.* Thus, if the resulting sheet of paper is untrimmed, i.e. has a rough edge, it has a *deckled edge.* Making paper by hand, sheet by sheet, is a laborious, slow (about 4 sheets per minute), and therefore expensive, process.

In the latter 18th century, a papermaking process was invented that did not use a coarse screen, but rather a very fine mesh. This mesh did not impart a readily apparent pattern to the paper. Such paper is a more uniform feltwork, without chain and laid lines, and is called *wove paper* (see Figure E.4). By the early 19th century, machine-made wove paper was commonplace. Both laid and wove paper coexisted throughout the 19th century and both were used in mapmaking during this period, but a map on wove paper is not earlier than about 1800.

Watermarks are designs incorporated into the very body of the paper during manufacture. The paper in the watermark is slightly thinner and more translucent than elsewhere, and the pattern can be discerned readily by holding the sheet to the light. Some papers can be dated readily by their manufacturer's watermarks. For example, the large watermark shown in Figure E.5 is known to have been in use in 1648-49 and the one I photographed appeared in a Dutch atlas of 1650. This is a striking watermark, showing Atlas with the world on his shoulders. It is beyond the scope of this book to go into watermarks in detail, but the reader should be aware that there are several books listing most of the known watermarks. These can be consulted in major art reference libraries.

Paper is a remarkably tough and stable substance. Although it is often eaten by rodents and insects, it can survive centuries without deterioration. The longevity of paper is to a large extent determined by its manufacture and subsequent handling and storage. If made of non-acidic materials and stored in a relatively dry (65% relative humidity has been recommended by conservators) and non-acidic environment, paper should last indefinitely.

288

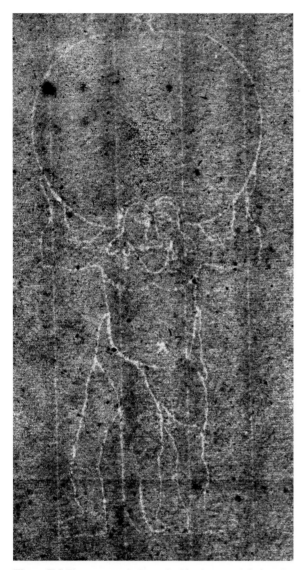

Figure E.5 The watermark shown in this photograph is found on paper made in 1648-1649. It is relatively large for a watermark, measuring about 9 x 4 inches (23 x 10 cm). It was photographed with transmitted light and I enhanced the contrast. Note the prominent vertical chain lines.

THE COMPOSITION OF PAPER

Most papers share some common chemical properties. Plant fibers, be they wood, flax (linen), cotton or mulberry bark are made of arrays of long molecules (*polymers*). Polymers (poly=many; mer=unit) are comparatively large molecules made up of repeating subunits that are, by themselves, molecules. These subunits are linked to each other by means of specific chemical bonds. Rubber, for example, is a polymer, as are plastics, your Formica countertop and the starch in digestible foods that we eat. Starches are polymers whose repeating subunits are sugars. The polymer common to paper, *cellulose*, is also a starch, made of repeating subunits of sugars.

Cellulose is synthesized by plant cells and forms the outside walls of the cells. Obviously, cells cannot synthesize giant polymeric molecules in one piece and then somehow expel them to the outside through their membranes. Instead, cells synthesize the subunits, or *monomers* (mono=one; mer=unit) of these molecules, and then export them outside the cell where they are assembled into the long polymeric fibers. Cotton fibers, flax (linen), and all other long cellulose structures are synthesized and assembled in this way. When paper is made from them, the fibers can still be seen microscopically (Figures E.6, E.7, E.8). Even the machine-made paper of the 19th century (and modern papers of today) still demonstrate this substructure (Figures E.9, E.10).

Although we won't go into this topic here, we can identify many of the fibers in paper by their microscopic appearance.

As an aside, it is interesting to note that whereas we can digest some starches such as potato starch, which is a polymer of the sugar mannose (which in

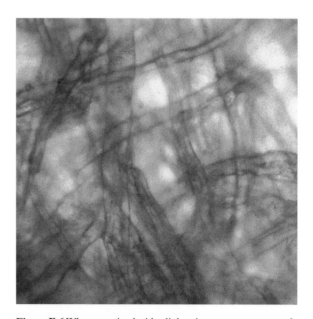

Figure E.6 When examined with a light microscope, paper reveals its fibrillar nature. The cellulose fibers form a random feltwork. This specimen of laid paper used in Kircher's *Mundus Subterraneous,* 1665, was magnified about 435x using phase-contrast optics.

289

Figure E.7 This scanning electron micrograph shows the surface of a sheet of 15th-century hand-made paper used in Schedel's Nuremberg Chronicle, in 1493. A cut edge is in the upper left. Notice the random array of fibers, some of which are sticking out of the plane of the paper. The fibers are cellulose, derived from linen. Magnification is 66x.

turn consists of two molecules of dextrose), because we have the necessary enzymes to cleave the polymeric starch into subunits (i.e. the mannose) which we then can metabolize, we cannot do that for cellulose. We do not have the necessary enzymes. Even termites do not – they have microscopic organisms in their intestines that de-polymerize the cellulose and the termites then digest the resulting sugars. The enzyme that degrades cellulose is called *cellulase* (in biochemistry, the suffix *ase* indicates an enzyme). Cellulose is a polymer of cellubiose, which in turn consists of dextrose. Remember that potato starch also consists ultimately of dextrose molecules. What is different in cellulose is the way they are connected to each other by bonds to form the large polymer. These bonds are very specific to the type of polymer and define some of its properties. As long as these bonds are unbroken, the polymer remains intact. The enzyme present in our saliva, amylase, can, for example, break the bonds that

join the sugars of potato starch, but not the bonds holding cellulose together. Spit on a piece of potato, the starch will immediately begin break down (a process called *hydrolysis*). Spit on your neighbor's 17th-century map and nothing at all will happen to the cellulose in its paper.

However resistant the cellulose of paper is to our amylase, it is not resistant to other chemical agents. It can be broken down by a number of agents, among them acids. It is because of the sensitivity of cellulose to acids that we must take care to keep them apart.

We usually define acids as aqueous (water) solutions containing an excess of hydrogen ions (for a discussion, see Appendix F).

pH

The unit pH (pronounced "pee-aitch") is a measure of acidity. ***The more acidic a solution, the lower its pH.***

290

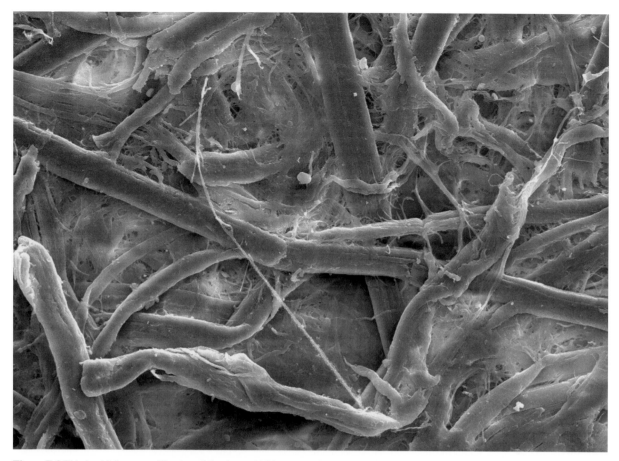

Figure E.8 This is a higher magnification of the sheet of 15th-century hand-made paper seen in Figure E.7. We can see individual fibers partially raised off the surface, creating the texture of the paper. Magnification is 520x.

A pH of 7.0 is neutral. pH values above seven indicate the solution is *basic;* below seven, *acidic.* A more detailed discussion of pH is presented in Appendix F.

Practically, pH is easy to determine in the laboratory. We use a pH meter and a glass electrode that is specially designed for this purpose. If we want to see if a piece of paper is "acidic" we can macerate it in some pure water and insert the electrode and read the pH directly on a meter (these days, generally a digital readout).

Indicator dyes can be used to determine pH. Indicator dyes are complex colored substances that undergo chemical changes at different pH values. These chemical changes produce color changes that we can readily observe. The natural dye, litmus, is perhaps the best known, but it is one of the least sensitive of these indicator dyes (litmus is pink below about pH 7; blue above). *Indicator papers* are available commercially. These are strips of paper containing the dyes and one can dip the strip into solution and determine, approximately, the pH. Although inexpensive, they are not as accurate as using a calibrated pH meter. There are also *indicator pencils* containing these dyes. These can be used to approximately determine the pH by rubbing them onto paper.

Note that I have discussed "acid" in terms of an aqueous solution. In general, water is necessary for acid to play its mischief. This is one reason to keep paper dry.

IS ACID HARMFUL, AND IF SO, WHY?

Deeply embedded in current popular culture is the notion that acidic is bad and basic is good. Much of this is harmless prattle and serves to sell antacid tablets. Nonetheless, in some instances, and paper conservation is one of them, acids are harmful.

291

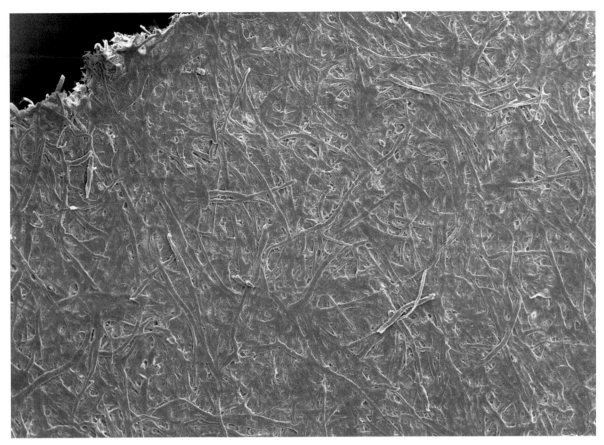

Figure E.9 The scanning electron microscope shows the surface of a sheet of 19th-century machine-made paper used in Johnson's atlas to be remarkably smooth compared to the hand-made paper shown in the previous two micrographs. The paper has been calendered, or pressed between rollers. Magnification is 64x.

Recall that paper is made of cellulose fibers compacted into a feltwork. These fibers, derived from a variety of plant materials, are made up of sugars linked to each other in characteristic ways, to form long molecules, called polymers. The cellulose polymers can be broken, most readily at the bonds linking the substituents. *Acid has the ability to break the chemical bonds holding the subunits together in the cellulose polymer.* Once these bonds are broken the polymer is shorter and some of the strength is lost. Long, flexible fibers are the key to strong paper; short brittle fibers produce a weak paper.

The action of acid on cellulose is water-based. That is, the acid has to be in solution to get into contact with the cellulose subunits. This is why damp is a necessary ingredient for acidic degradation, and if paper is kept dry, it deteriorates only slowly.

WHERE DOES THE ACID COME FROM?

A variety of atmospheric vapors common in industrial areas and cities generally, such as some pollutants, particularly sulfur dioxide, react with water and serve to lower the pH. The cellulose of woody plants such as trees, used widely in papermaking in the past hundred years or so, is embedded in a substance called *lignin*. In order to free the cellulose fibers for papermaking, the lignin has to be removed. Lignin is an acidic substance and traces of it can remain in wood-based papers.

Lignin is often cited as a reason for not storing maps in wooden plan chests. Indeed, paper in *direct* contact with wood for any long period of time will deteriorate. However, if we place the maps first in acid-free folders (which we should do anyway), then the wood will not be in contact with them.

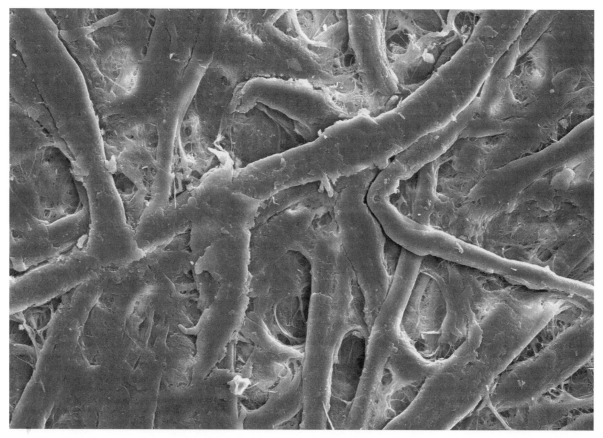

Figure E.10 This scanning electron micrograph shows the surface of a sheet of 19th-century machine-made paper used in Johnson's atlas. We see the flattening, or compression of the fibers resulting from the calendering of the sheet by high-pressure rollers to make the surface smoother. This is still a rag-based paper. We have not seen any microscopically identifiable tree-derived fibers. Magnification is 530x.

BUFFERS

For the purposes of this discussion we can think of *buffers* as sponges that can absorb acid, preventing significant changes in pH. Buffers are molecules that react with acid to form new compounds, "binding" the acid into a form where it cannot be released. Thus, the *buffer removes acid from the environment by making it into a new harmless compound*.

The subject of buffers can become complex, but only a few of their general properties are relevant here. Buffers work in relatively narrow pH ranges, so the buffer has to be chosen with the desired final pH in mind. In addition to simple inorganic buffers such as calcium carbonate, $Ca(CO_3)$, there are very complex organic molecules such as the buffers HEPES and PIPES. In all instances buffers must be selected according to the pH range one wishes to establish and maintain.

Conservators often add a buffer to the final rinse water when washing a map. Such aqueous buffers work well and the treatment is not a difficult or expensive one. However, if one cannot wet a map because of color or other considerations, it is still possible to apply a buffer to neutralize the pH. Such buffers are non-aqueous.

Some manufacturers add buffers to the paper pulp as they manufacture it. This is probably better than not adding it, but nonetheless it implies that the pulp was acidic to begin with. Far better to start with a non-acidic pulp. This consideration achieves practical importance when selecting mat board and backing board for displaying maps. A good neutral rag board is best, since it was never acidic to begin with. Sometimes buffer is added to a neutral pulp as a matter of course. Mat board made this way will be identified as such.

293

Joah. Georg. Berckhmüller inven. et delin. Aug. Vind.   Carolus Dipus Sculp. Lutetiæ Parisiorum.

Many early atlases contain splendid engraved frontispieces. This example is from the Weigel *Atlas Scholasticus et Iterarius* published in Nuremberg. The frontispiece measures about 13.25 x 15.5 inches (33.5 x 39.5 cm) and is a very deep, rich image showing evidence of both etching and engraving. Uncolored, it is a powerful scene filled with many icons of cartography and exploration. Many map collectors collect frontispieces for their intrinsic beauty. Some frontispieces have original color; some are highlighted with gold leaf.

Appendix F

Chemistry for Map Collectors

Each equation...in the book would halve the sales.
Stephen Hawking, *A Brief History of Time* (1988)

Fortunately, at least the curve becomes asymptotic.
F.J. Manasek, *A note to his accountants, 1997*

INTRODUCTION

This appendix presents some very elementary concepts of acids, bases, and buffers.

When I showed the typescript of this book to several colleagues, they almost all thought this material too technical to be in the main body of the book. It is indeed technical, but it is not difficult. I think what "turns people off" is the use of symbols. Once we get beyond that, the material is no more "technical" than our discussion of the names of parts of maps, no more technical than learning the difference between lithography and engraving, and no more difficult than remembering the names of a dozen mapmakers.

It is not necessary to know this material in order to enjoy map collecting, but over the years I have observed that many collectors speak lightly of acid, "acid-free" paper, buffers, and the like, without really knowing what is meant. The information in this appendix is presented for those who may wish to work through it.

ACIDS

Common wisdom dictates that we should avoid an acidic environment for paper. "Acid-free" mats, "acid-free" backing, and "acid-free" drawers in our plan chests are all catchwords. Mat board is advertised as "acid-free," "buffered pH neutral," or "pH 7.6" or some other seemingly magical or arcane number. What is the problem with acid and just what is an acid? What is this pH number that is so often quoted? The word "acid," much like the word "calorie," is used widely by advertisers and others who haven't the foggiest idea what the word means.

Before we begin this discussion about acids, I shall offer a few definitions. The superscript "+" indicates a positive charge; the superscript "–", a negative charge. *Ions* are atoms with either a negative or positive charge. Square brackets, [ ], indicate that we mean the *concentration* of a substance. (The concentration is in units, called *moles*, contained in a liter [L] of solution. Moles are a quantity used by chemists; we cannot go into their meaning and importance here, but the reader should be aware of their existence if he wishes to pursue this topic more.) Other definitions will be added as we work through this exercise.

WHAT IS AN ACID?

There are several definitions of acid. They are each correct, and do not contradict each other; they are simply different useful ways to think about the same thing.

Most common, an *acid is considered to be a substance (usually in aqueous solution) "rich" in hydrogen ions* (or protons), chemically indicated as H^+. These ions are available for chemical reactions. Any solution with excess hydrogen ions is, according to our definition, an acid, since it is a potential proton-donor (see below).

A hydrogen atom consists of a nucleus which has a single proton (with a single positive [or $^+$] charge) and a single electron (with a single negative [or $^-$] charge) going around the nucleus in a path called an *orbital*. The single $^+$ charge and the single $^-$ charge cancel each other and the atom is electrically neutral. If it loses its negative charge (loses its electron in the orbital) then it is called an ion. The hydrogen ion is a single proton, and has a positive charge. Most definitions of acids, since they involve aqueous solutions, consider an acid to be any compound that can donate hydrogen ions, or protons. *Any substance that is a proton-donor is an acid.* As a corollary, *any substance that is a proton-acceptor is its opposite, a base.*

If a solution contains equal amounts of proton-donors and proton-acceptors, the solution is neutral. Pure water, with equal amounts of H^+ and OH^-, is neither acidic or basic, but is neutral. (A note here: as in most things, this statement is not entirely correct, but for practical purposes it works).

Just to confuse the issue a bit, let me introduce another concept about acids. While it is probably true that most of the time we think of acids in terms of hydrogen ions in solution, there is a more fundamental way to think of acids. The hydrogen ion is positively charged. That is, it lacks an electron. Because it lacks an electron, it can accept one.

Here then, is another definition of an acid: *an acid is an electron acceptor.* The useful aspect of this concept in light of this entire discussion is that there are *many substances, in addition to hydrogen ions, that are electron acceptors, and they are all acidic.*

pH, ACIDITY AND NEUTRALITY

The chemical world is dynamic. Molecules "come apart" and "come together" all the time. Water molecules are constantly coming apart in solution, to make ions, only to rejoin and again make a water molecule. This coming-apart is called *dissociation*. The coming apart and coming together reach an *equilibrium*. At equilibrium the number of molecules coming apart equals the number coming together, and the respective concentration of "together" molecules and "apart" molecules (ions) doesn't change.

We all know that water's chemical formula is H_2O. This tells us that a single molecule of water contains two atoms of hydrogen (H) and one of oxygen (O). However, expressing the molecular composition this way doesn't tell us much about the structure of the molecule, nor does it give us insight into some of its possible reactions. We can just as well write the formula for water as H(OH). Here, we still note two hydrogens and one oxygen, but this way of writing it tells us more.

Water comes apart, or dissociates, in the following fashion.

$$HOH \rightarrow H^+ + OH^-$$

This equation tells us that molecular water (HOH) comes apart to form two sub-molecular species, the hydrogen ion (H^+) and the hydroxyl radical (OH^-).

The substituent H^+ is, as we have seen, called a hydrogen ion. The OH^- is called a *hydroxyl radical.* You will note that for every hydrogen ion produced by the dissociation of water, there is one hydroxyl group produced, and the solution remains electrically neutral. The reverse reaction occurs also:

$$H^+ + OH^- \rightarrow HOH$$

Hydrogen ions and hydroxyl groups are constantly rejoining to form molecular water.

The hydroxyl group being a proton acceptor is a base. Look at the above equation. The OH^- group accepts, or combines with a H^+ to form a neutral water molecule.

In pure water, the dissociation of the water molecule into hydrogen and hydroxyl and its reassociation into water is constantly occurring and an *equilibrium* exists. We can measure this equilibrium (or, if you will, the dissociation constant, k) and it has been determined to be the number 1×10^{-14} (we read and speak this number as "one times ten to the minus 14"), that is, a decimal point followed by thirteen zeros followed by a 1. Very small indeed! Keep in mind, however, that for every "acid" proton produced, a "base" hydroxyl is produced, so pure water is virtually neutral. It has neither an excess of acidic nor basic components.

The acidity of a solution depends upon its hydrogen ion concentration. Let us do a bit of elementary algebra on the dissociation of water:

$$[HOH]\ (k) = [H^+] + [OH^-]$$

This equation tells us that the concentration of water times its dissociation constant *(k)* yields a concentration of hydrogen ions and hydroxyl groups.

Since k is so small, HOH can be considered, for practical purposes, unity, or 1.

Then:

$$k = [H^+] + [OH^-]$$

296

Since:

$$[H^+] = [OH^-]$$

Remember: square brackets mean concentration. This equation states that the *concentration* of hydrogen ions equals the *concentration* of hydroxyl radicals; it does not mean that hydrogen ions are the same as hydroxyl radicals.

then:

$$[H^+]^2 = 1 \times 10^{-14}$$

or:

$$[H^+] = 1 \times 10^{-7}$$

Pure water therefore has a hydrogen ion concentration of 1×10^{-7} moles/L.

Remember that this is a neutral solution, because there are an equal amount of OH⁻ radicals present.

pH – WHAT DOES IT MEAN?

Chemists have found it convenient to express the acid/base characteristics of a solution in terms of its hydrogen ion concentration $[H^+]$, ignoring for the most part the basic elements of the system. But rather than deal with negative exponents, we use units, called *pH units,* which are the negative log of the hydrogen ion concentration. *Thus, since in neutral water the $[H^+]$ is 1×10^{-7} moles/L, we call this pH 7.*

The number, 1×10^{-7}, can be expressed as the decimal 0.0000001, a very small number indeed. If we had more hydrogen ions in solution, the number would be larger. For example, 4 zeros followed by 1 (0.00001) would be expressed as 1×10^{-5} or pH 5. It is seen that the pH system of describing the acidity or alkalinity of a solution is based quite firmly on physical chemistry. It is *not*, as often stated in popular literature, an arbitrary system.

Anything that *increases* the $[H^+]$ of a solution *lowers* its pH and makes it, by definition, more acidic.

If we were to add a bit of hydrochloric acid (HCl) to water, it would change the pH substantially. This is because HCl exists almost entirely in dissociated form as H^+ and Cl^-, and essentially all of its hydrogen is available as hydrogen ions. We call such acids *strong acids*. If, on the other hand, we were to add the same amount of acetic acid, for example, the pH change would be less. This is because acetic acid is an example of a *weak acid*. It is not as strongly dissociated and does not have its hydrogen available as readily as a strong acid. Note that the concept of a "strong" acid and a "weak" acid refers only to the availability of its hydrogen ions, it has nothing to do with the concentration or amount of acid.

BUFFERS

We can neutralize (i.e. have equal amounts of H^+ and OH^-) a solution, or a substance, by adding either acids or bases until the pH is neutral (pH 7). However, the result is often very sensitive to change. A slight alteration in the concentration of either results in large changes in pH. We can add substances called *buffers* that will moderate these swings in pH. We will not go into the chemical mechanism of action of buffers here, but instead will qualitatively describe them and try to give some indication of how they work.

We can think of a buffer as a sort of chemical sponge that can absorb H^+ or OH^- without significant changes in pH. This effect is achieved by the buffer reacting with H^+, creating new compounds that do not dissociate readily. Thus, if we add H^+ to a solution containing an appropriate buffer, the concentration of H^+ in the solution will not increase significantly, since the H^+ reacts with the buffer to create a new molecule. In this new molecule the hydrogen is more tightly bound and not free to dissociate and therefore not free to again become part of the H^+ in solution. Therefore, the concentration of H^+ in solution does not go up and the pH does not go down. If we were to continue to add H^+, eventually all the excess buffer would become used up; there would be no more remaining to react with (and remove from solution) the H^+ we are adding and consequently all the additional H^+ would remain in solution. The pH would then go down.

Conservators often add a buffer to the final wash water when washing a map. Such aqueous buffers work well and the treatment is not a difficult or expensive one. However, if one cannot wet a map because of color or other considerations, it is still possible to apply a buffer to neutralize the pH. Such buffers are applied as complex organic molecules in a non-aqueous base. They are available from specialty dealers in conservation materials.

Some manufacturers add buffers to the paper pulp as they manufacture it. This is probably better than not adding it, but nonetheless it implies that the pulp was acidic to begin with. Far better to have a non-acidic pulp. This consideration achieves practical importance when selecting mat board and backing board for displaying maps. A good neutral rag board is best, since it was never acidic to begin with. Buffered board is next-best.

Curious miscellany – 12   An unusual image relating to maps or mapmaking

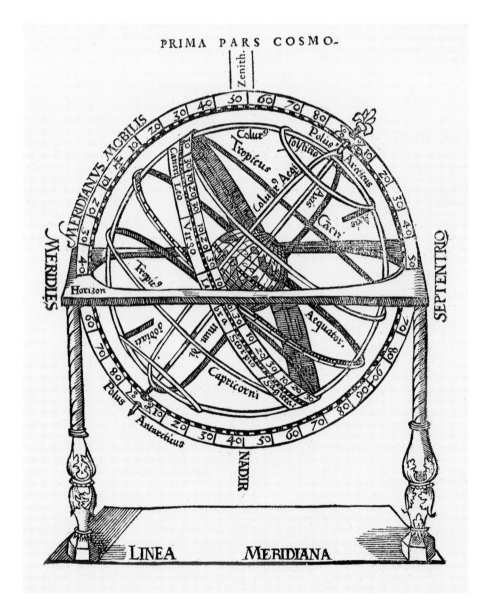

This woodblock print, from Peter Apianus' *Cosmographia* of 1574, shows the earth in relation to the various circles, such as equator, ecliptic, the tropics and the meridian. Original is about 6.5 x 5.25 inches (16.5 x 13.5 cm).

Appendix G

Useful Addresses and Sources

INTRODUCTION

This appendix lists organizations and groups worldwide that map collectors may find of interest. **Only non-commercial organizations are included, with the exception of dealer trade and fair organizations.**

Many academic and not-for-profit organizations do not have the staff to deal with exhaustive enquiry and the beginner is urged to familiarize him/herself with basic literature which will answer most questions. In general, although the staff at academic institutions are unable to do exhaustive research for individuals, my experience has been that they try hard to help. As a kindness to the personnel, I suggest that inquiries be specific and brief.

THE INTERNET

The Internet and the World-Wide Web are valuable resources. Trade organizations, dealers, museums and libraries have Web sites and Internet addresses. Many are too ephemeral to list here, but any search engine can quickly get you to more sites than you'll want. A word of caution – the Web and the Net are a bit like the wild west. Get independent confirmation and check out references before sending anyone any money or buying anything from an Internet address.

My favorite old map site is MapHist. This is a discussion group for individuals who wish to exchange ideas, ask questions and post messages relating to old maps. It is strictly non-commercial. Subscribe by sending command "subscribe maphist" to: **listserv@harvarda.harvard.edu**

Roadmap collectors can enjoy the action on **roadmaps-l.** To join, send an e-mail message to: **majordomo@teleport.com** and type this message: "subscribe roadmaps-l". This list is almost

exclusively devoted to roadmaps of the United States and it accepts "maps wanted" and "maps offered" something MapHist does not do.

Map curators have a list, **maps-l.** To subscribe send e-mail to **listserv@uga.cc.uga.edu** and type in the body: "subscribe maps-l [your name]". This is a very serious list and items such as storage and display are discussed.

There is a Web site that lists current information and e-mail addresses for map societies: **http://www.csuohio.edu/CUT/MapSoc/index.htm**

MAP COLLECTORS' ORGANIZATIONS

Map societies are the place where collectors can link up with kindred spirits and learn about maps, local suppliers, and even go on field trips. Many societies also sponsor meetings and lectures.

The **International Map Collector's Society (IMCoS)** is *the* international association of map people. It publishes a quarterly journal and holds its meetings in different parts of the world. IMCoS also sponsors map fairs (see below).

Contact the IMCoS membership secretary at:

> Jenny Harvey
> 27 Landford Road
> Putney, London SW15 1AQ England

There are a number of other associations, international in scope, that have more specific interests. An asterisk (*) after the name indicates they publish a magazine or newsletter:

> **Association of Map Memorabilia Collectors***
> Siegfried Feller
> 8 Amherst Road
> Pelham, MA 01002 USA

> **Carto-Philatelists***
> Miklos Pinther
> 206 Grayson Place
> Teaneck, NJ 07666 USA

Internationale Coronelli-Gesellschaft für Globen-und Instrumentenkunde*
(International Coronelli Society for the Study of Globes and Instruments)
Rudolf Schmidt and/or Heide Wohlschläger
Coronelli-Gesellschaft
Dominikanerbastei 21/28
1010 Wien, Austria

International Society for the Curators of Early Maps*
Bob Karrow
The Newberry Library
60 West Walton Street
Chicago, IL 60610-3380 USA

Society for the History of Discoveries*
Eric W. Wolf, Secretary-Treasurer
6300 Waterway Drive
Falls Church, VA 22044-1316 USA

There are many local groups that hold meetings, publish newsletters and sponsor the occasional guest lecture. Many members of such groups are not collectors, per se, but do have a keen interest in maps and mapping. I have included only those societies that have a physical mailing address. Many map societies have their own Web sites and e-mail addresses. There is a Web site that lists current information and e-mail addresses for map societies: **http://www.csuohio.edu/CUT/MapSoc/ index.htm**

AMERICAN MAP SOCIETIES

The following list is up to date as of mid-1997. An asterisk (*) following the name of the association indicates that it publishes a bulletin or newsletter.

Canada:

Map Society of British Columbia
P.O. Box 37109,   2930 Lonsdale Avenue
North Vancouver, B.C. V7N 4M4 Canada

United States:

Arizona:

Arizona Map Society*
Jack Mount
Science-Engineering Library
University of Arizona
Tucson, AZ 85721

California:

California Map Society
Bill Warren
1109 Linda Glen Drive
Pasadena, CA 91105

Colorado:

Rocky Mountain Map Society
Wes Brown, President
1736 Hudson Street
Denver, CO 80220

District of Columbia:

Washington Map Society*
Ed Redmond
Geography and Map Division
Library of Congress
P.O. Box 2149
Washington D.C. 20013-2149

Illinois:

Chicago Map Society*
The Newberry Library
60 West Walton Street
Chicago, IL 60610-3380

Kansas:

Road Map Collectors of America*
Dave Schul, President
2214 Princeton Blvd.
Lawrence, KS 66049

Massachusetts:

Boston Map Society
David A. Cobb
Harvard Map Collection
Harvard College Library
Cambridge, MA 02138

Michigan:

Michigan Map Society
P.O. Box 1201
Ann Arbor, MI 48106

New York:

New York Map Society*
Alice Hudson, Chief, Map Division
NYPL, The Research Libraries
5th Avenue & 42nd Street
New York, NY 10018-2788

Ohio:

Northern Ohio Map Society*
Maureen Farrell
Map Department, Cleveland Public Library
801 Superior Avenue
Cleveland, OH 44111

Pennsylvania:

Map Society of the Delaware River
Free Library of Philadelphia
1901 Vine Street
Philadelphia, PA 19103

Virginia:

Northeast Map Society*
Eric Riback
2506 Hillwood Place
Charlottesville, VA 22901

Wisconsin:

> Map Society of Wisconsin
> Sharon Hill
> American Geographical Society Collection
> P.O. Box 399
> Milwaukee, WI 53201

EUROPEAN MAP SOCIETIES

Cyprus:

> Cyprus Association of Map Collectors
> Dr Andreas Hadjipaschalis
> P.O. Box 4506
> Nicosia, Cyprus

Germany/Austria/Switzerland:

> Arbeitsgruppe D-A-CH*
> Thomas Klöti
> Stadt-und Universitätsbibliothek Bern
> Sammlung Ryhiner
> Münstergasse 61-63
> 3000 BERN 7 Switzerland

Germany:

> Arbeitskreis Geschichte der Kartographie
> Prof. Dr. W. Scharfe
> Weimarische Straße 4
> 10175 Berlin, Germany

> Freundeskreis für Cartographica in der
> Stiftung Preußischer Kulturbesitz* (Circles of
> Friends for Cartographica in the Foundation
> Cultural Heritage of Prussia)
> Staatsbibliothek zu Berlin, Kartenabteilung
> Unter den Linden 8
> 10102 Berlin, Germany

Great Britain:

> Charles Close Society*
> Society for the Study of Ordnance Survey Maps
> Roger Hellyer, Membership Sec.
> 60 Albany Road
> Stratford-on-Avon, Warwickshire, CV37 6PQ. UK

Greece:

> Society for Hellenic Cartography
> Themis Strongilos
> 6 Patriarchou Ioachim Street
> GR-106 74 Athens, Greece

Netherlands:

> Werkgroep voor de Geschiedenis van de
> Kartografie (Working Group for the History of
> Cartography)
> Marc Hameleers
> Topografisch Historische Atlas
> Gemeentelijke Archiefdienst Amsterdam
> Postbus 51 140
> 1007 EC Amsterdam, The Netherlands

Sweden:

> Kartografiska sällskapet, historiska sektionen
> (Historical Section of the [Swedish]
> Cartographical Society)
> Göran Bäärnhielm
> Kungl. Biblioteket (Royal Library)
> P.O Box 5039
> 102 41 Stockholm, Sweden

Switzerland:

> Arbeitsgruppe für Kartengeschichte*
> (Study Group for Map History)
> Prof. Arthur Dürst
> Promenadengasse 12
> 8001 Zürich, Switzerland

OTHER REGIONAL MAP SOCIETIES

Australia:

> Australian Map Circle*
> John Cain, Secretary
> Department of Geography, University of
> Melbourne
> Parkville, VIC, 3052 Australia

Israel:

> Israel Map Collectors' Society*
> Eva Wajntraub
> 4 Brenner Street
> IL 92103 Jerusalem, Israel

New Zealand:

> New Zealand Map Society*
> Phil Barton
> P.O. Box 10-179
> Wellington, New Zealand

TRADE ORGANIZATIONS AND FAIRS

There are several trade organizations to which most of the prominent map dealers belong. These are really rare book dealer's associations, but map dealers have traditionally associated themselves with these organizations.

International League of Antiquarian Bookdealers (ILAB). This is the international umbrella organization of national antiquarian bookdealer associations comprised of fifteen autonomous national groups. Many of the national asociations maintain a Web site and by contacting the national group you can get the Web address. *Each national association sponsors one or more fairs; there is a biennial ILAB fair.* Bookfairs sponsored by the ILAB and its associated national organizations are generally among the more upmarket fairs.

You can contact the ILAB through your national organization of rare book dealers, or by contacting:

> **Alain Nicolas**
> 41, Quai de Grands Augustins
> F-75006 Paris, France

The ILAB issues a large, printed directory (there is a charge for this) that lists, by country, the world membership. In this book, virtually all of the world's major book and map dealers (well over 1500) can be found in one place. It is especially useful if you are planning a foreign trip and would like to visit dealers abroad.

The largest national organization is the **Antiquarian Booksellers' Association of America (ABAA),** headquartered in New York. (20 West 44th Street, New York, NY 10036-6604. Tel: 212-944-8291; e-mail: abaa@panix.com; http://www.abaa-booknet.com). The ABAA sponsors about 6 international book fairs a year in different parts of the country. They are a splendid place to buy old maps. The other member groups of the ILAB are:

Australia and New Zealand:

> The Australian and New Zealand Association of Antiquarian Booksellers
> 24 Glenmore Road
> Paddington, NSW 2021

Austria:

> Verband Der Antiquare Österreichs
> Grünangergasse 4
> A-1010 Wien, Austria

Belgium:

> Chambre Professionnelle Belge de la Librairie Ancienne et Moderne
> 53 Boulevard Saint-Michel
> B-1040 Bruxelles, Belgium

Brazil:

> Associação Brasileira de Livreiros Antiquarios
> Rua Visconde de Caravelas 17
> 22.271 Rio de Janiero, Brasil

Canada:

> Antiquarian Booksellers' Association of Canada
> P.O. Box 323, Victoria Station
> Montreal, Quebec
> H3Z 2V8 Canada

Denmark:

> Den Danske Antikvarboghandlerforening
> P.O. Box 2184
> DK-1017 København K   Denmark

Finland:

> Suomen Antikvariaattiyhdistys Finska Antikvariatföreningen
> Runeberginkatu 37
> SF-00100 Helsinki, Finland

France:

> Syndicat National de la Librairie Ancienne et Moderne
> Rue Git-le-Cœur
> F-75006 Paris, France

Germany:

> Verband Deutscher Antiquare E.V.
> Braubachstraße 34
> D-6000 Frankfurt 1 Germany

Great Britain:

> Antiquarian Booksellers' Association
> Suite 2
> 26 Charing Cross Road
> London WC2H 0DG England

Italy:

> Associazione Librai Antiquari d'Italia
> Via Jacopo Nardi 6
> I-50132 Firenze, Italy

Japan:

> The Antiquarian Booksellers' Association of Japan
> 29 San-ei-cho
> Shinjuku-ku
> 160 Tokyo, Japan

Netherlands:

> Nederlandsche Vereeniging Van Antiquaren
> Jansweg 39
> 2011 KM Haarlem, The Netherlands

Norway:

> Norsk Antikvarbokhandlerforening
> Ullevålsveirn 1
> N-1065 Oslo 1 Norway

Sweden:

> Svenska Antikvariatföreningen
> Box 22549
> S-104 Stockholm, Sweden

Switzerland:

> Vereinigung Der Buchantiquare Und Kupferstichhändler In Der Schweiz
> Kartausgasse 1
> CH-4005 Basel, Switzerland

United States of America:

> Antiquarian Booksellers Assn. of America, Inc.
> 20 West 44th Street, 4th Floor
> New York, NY 10036-6604 USA

In every country there are additional bookseller organizations not affiliated with the ILAB. In the

United States, for example, there are state and regional used bookdealer organizations. The list is too large to include here, but the easiest way to contact them is enquire of a local secondhand bookseller. Most belong to the local organization and most local book dealer organizations publish a list; many hold annual or semiannual bookfairs. Many used bookdealers also have old maps; many local map dealers belong to the local book dealers' associations and exhibit at the local fairs. These are good places to acquire more common old maps.

MAP FAIRS

While there are numerous international, national and local groups sponsoring book fairs, there are only three regularly held fairs devoted to old maps.

They are:

IMCoS. This fair is held annually in London, generally in mid-June in association with the numerous book fairs that are held at this time in London. Write to IMCoS for venues. Additionally, IMCoS holds an annual meeting hosted in different countries. Usually there is a map fair organized in conjunction with these meetings.

Bonnington Map Fair. This fair is held each month (second Monday) at the Bonnington Hotel in London. Contact: Antique Map Fair, 26 Kings Road, Cheltenham GL52 6BG, UK for dates.

A newcomer to the map fair circuit is the **Miami Map Fair** held each January or February in Miami, Florida in the USA. This event, starting modestly a few years ago, is growing into a major map fair with international participation. Contact: Map Fair Coördinator, Historical Museum of Southern Florida, 101 West Flagler Street, Miami, FL 33130 USA

SUPPLIES

One of the more vexing problems for the collector is finding the right sort of material in which to store and protect the collection. All large cities have art supply stores that carry acid-free papers and folders, but the more esoteric materials may be hard to obtain, especially in smaller towns and in rural areas.

There are several large suppliers of archival storage materials and conservation supplies that put out detailed catalogues of their products and ship internationally. Because they are commercial ventures, they are not listed here, but a collector can find them easily by contacting local libraries (they also sell book conservation materials), museums, bookbinders or historical societies. Most of these places will have catalogues from the major suppliers. If you join, or contact, a local map society, its members will certainly have discovered reputable sources of supplies.

INSTITUTIONS WITH MAP COLLECTIONS

It was my original intention to include a list of libraries and museums that have map collections open to the public. The list is so large that it is not feasible to include in this book. Worldwide, there are many hundreds of libraries and museums that have collections of old maps. Not all of these are open to the general public: some are, rather, research collections. If I edited the list, I would have to delete too many institutions with smaller, yet fascinating map collections.

Instead, collectors are urged to contact libraries directly. Many large city libraries have map collections. Many colleges and universities have significant collections of old maps. Local historical societies and town halls often have extensive maps of local areas.

It is always best to write or telephone or e-mail ahead. Many collections held by private institutions are not open to the public; others have restricted times.

This is a colored drawing from a manuscript atlas, *Atlante Novissimo Delineato Dal Nobil Giovane & Giovanni Almoro Tiepolo Patrizio Veneto...* The atlas was done by young Tiepolo in his fourteenth year, in 1783. The manuscript title page indicates he was a Venetian of noble birth. The image size is 7.5 x 10.5 inches (19 x 26.5 cm). Note the laid paper: the prominent chain and laid lines are quite visible. The sphere is a Copernican model.

Appendix H

Glossary

Acid. Any substance that lowers the pH of an aqueous solution. Note: This definition is adequate for the purposes of this book. The reader is referred to Appendix F for a more extensive discussion. See *base*.

Acid, strong. An acid that is almost entirely dissociated; therefore most of its hydrogen ions are available.

Acid, weak. An acid that is not highly dissociated; since all the hydrogen ions are not available a weak acid does not lower the pH of a solution as much as the same concentration of a strong acid.

Age-toning. See *browning*.

Anon. Anonymous.

Atlas. A (generally) bound assemblage of maps.

Backing (backed, back). A substance affixed to the verso of a map, using a variety of adhesives, with the purpose of strengthening or flattening the map. See *laid down*. It is noted that some maps, particularly sea charts, were issued on a double paper; this is their normal condition.

Barrier sheet. A sheet of acid-free paper used to separate, with the intention of isolating, one sheet of paper from another, especially if one is not acid-free. Also called *barrier paper*.

Base. A substance that raises the pH of an aqueous solution. Bases react with acids to form salts. A base is any substance that accepts hydrogen ions.

Binder's guard (guard, binder's stub). A strip of paper glued to the map along one side of the centerfold. The map is sewn into the atlas using the guard, obviating the need to make stitcholes in the map itself.

Bird's-eye view. See *three-quarter view*.

Bleaching. The process of whitening or lightening age-darkened paper, or removing stains.

Blindstamp. An impression made in paper using a die that leaves a raised mark.

Border. A (usually) decorative device used to embellish the outer limits of a printed map image.

Breaker. A book or atlas that is destined to be taken apart for its maps or plates. The act is called *breaking*.

Broadside (broadsheet). A single complete sheet of paper printed on one side only.

Browning. The darkening of paper over time. Age-toning. May be indicative of poor paper or improper storage conditions.

Buffer. A chemical substance that prevents large changes in pH. Different buffers stabilize the pH at different values.

Burin. Tool used by engravers to incise the lines. A graveur.

Buyer's premium. A surcharge, imposed by auction houses, on the hammer price.

Cartes à figures. Maps with figures in small panels around the outer sides. Characteristic of some 17th century maps.

Cartouche. Emblem-like device containing title or other information. May be very elaborate or very simple.

Centerfold. A crease running through the center of a map where it was folded to be inserted into an atlas.

Cerography. "Wax printing." A technique whereby the image is engraved in wax and a stereotype plate made by electroplating the incised wax.

Chain lines (chain marks). Marks left in laid paper by the coarse components of the wire mesh of the deckle. The chain marks run parallel to the short dimension of the original sheet and are widely spaced, usually about 2-3 cm. See *laid lines*.

Charts. Maps used for navigation.

Chemise. A separate enclosure that wraps around a folded map and permits it to slide easily into and out of a slipcase.

City plan (town plan). A map of a city or town as viewed from directly above (normal to) the streets.

Coated paper. Paper that has a smooth surface resulting from a very thin coating of a substance (often a clay) that fills the pores of paper and gives it a better surface for modern printing processes.

Color, contemporary. Color that was applied at or about the time the map was printed. Note that the word contemporary refers to the time the map was made, not to the current time period.

Color, full. Color applied to all the parts of a map (except possibly the sea).

Color, later. Color applied to a map after it was printed, but may suggest that the color is not recent or modern.

Color, original. Color applied to the map at or about the time it was printed.

Color, outline. Applied or printed color delineating or accentuating geopolitical boundaries (such as countries) or geological interfaces (such as land/sea) only.

Color, recent (modern). Color applied to an uncolored map recently, as in last week or last year.

Color, wash. Broad areas that are painted are said to have wash color, often to distinguish the color from outline.

Coloring. Any chromatic hue, other than black, added to the map after printing. To be distinguished from *printed color*.

Compass rose. The elements of a compass card shown on a map. Rhumb lines radiate from the compass rose. See also *wind rose*.

Composite atlas. An atlas, other than a regular edition, that contains an idiosyncratic collection of bound maps not necessarily by the same cartographer.

Condition. The overall physical quality of a map.

Conservation. Any deliberate action taken to preserve or prevent further damage.

Contemporary. The period at or about the time the map was published; not contemporary in the sense of being current or modern.

Copper plate (copperplate). The plate, made of copper, that is incised in making copperplate engravings.

Curiosa. An unusual subject, possibly erotic, scatalogical, or simply bizzare or decidedly uncommon. Facetiæ.

Dampstaining. Discoloration of a map resulting from water.

Deckle. The wood frame holding the wire mesh onto which the slurry is scooped in making laid paper. The wood frame gives an uneven edge to the sheet of paper: the so-called deckled edge.

Dissected. Cut into parts. Large maps were sometimes dissected into rectangles and backed with linen; also used to indicate jigsaw puzzle maps.

Edition. The concept of "edition" is difficult to apply to maps. It can be defined by imprint, if one exists. Maps are sometimes described as "...being from the x edition of the atlas..."

Engraving. Type of printing process wherein the ink is retained in grooves cut with a tool (burin) into a plate. The paper is pressed onto the plate and picks up the ink. An intaglio process.

Ephemera. Printed material intended for transient use. It is not made with the intention of having long or lasting value. Examples of ephemera are bus tickets, timetables or route maps.

Etching. An intaglio process, but the grooves in the plate are formed by the action of acid rather than by the engraver's burin.

Facsimile. The manuscript restoration of missing text or image; a close or exact replica of an original. Facsimile maps, when printed on an aged, or appropriate paper can cause fits.

Folio. A book consisting of leaves folded once. In maps, it refers to the size of a map, generally about 20 x 25 inches (51 x 64 cm). It is incorrect, but often done, to use this term to describe size.

Forgery. A facsimile or other copy made for purposes of deception.

Foxing. Brownish (fox-colored) spots on paper indicating damp-related damage.

Frontispiece. The decorative image facing the title page.

Gathering. A group of leaves formed after the printed sheet has been folded to the correct size for inclusion in a book. See signature.

Gold leaf. An exceedingly thin sheet of gold that is applied to maps as decoration.

Gore. Diamond-shaped sections of maps that are designed to be wetted and laid down on a spherical surface to make a globe.

Graveur. Burin.

Gravure. A reproductive printing technique using plates made photochemically.

Guard. See binder's guard.

Gum arabic. A plant derivative that forms mucilage when dissolved in water; in very dilute concentration it is used to highlight colored areas of maps to make them appear "wet" or more intense.

Halftone. A process of simulating a continuous tone image by breaking up the image into small dots, whose size and/or spacing determines image density.

Hammer price. The price at which a bidder acquires a lot at auction, before various other charges such as buyer's premium are added.

Hinge. A tab, often of long-fibered paper, used to affix a map to the mat.

Hypochlorite. Common laundry bleach used to bleach the paper of maps.

Impression. Each individual copy of a map is an impression. Also refers to the quality of the printing, as in "That's a nice impression."

Imprint. Printed data regarding the publisher, date of publication and place.

Intaglio printing. Printing from a plate where the part to print is below the surface of the plate, as in engraving or etching.

Laid down. A map that is backed is said to be laid down.

Laid lines. Lines in laid paper that run lengthwise in the original sheet and at right angles to the chain lines. They are finer than the chain lines and more closely spaced.

Laid paper. Paper made by hand by straining a slurry through a wire mesh held in a wooden deckle. The wire mesh imparts the chain and laid lines to the paper.

Latitude. Angular distance, measured on a meridian, with zero at the equator and 90 degrees at each pole.

Leaf casting. The technique of creating small areas of new paper by using a liquid slurry of macerated paper. This is used most to repair holes or add new margins to valuable maps.

Lignin. A substance that bind together the cellulose fibers of plants.

Lined. Backed. Lined is a term used more often in relation to paintings than to maps.

Linen-backed. Backed with linen.

Lithography. A form of printing invented in 1799. The surface of a plate is treated chemically to accept the inked image which is transferred to paper directly or via an intermediate substrate, as in the case of offset lithography.

Longitude. Angular distance, measured westward from a prime (zero) meridian.

Loxodromes. Lines that intersect meridians at equal angles. Akin to rhumb lines.

Manuscript. Written by hand.

Marbled paper. Paper colored in such a manner to resemble the grain of marble.

306

Margin. That portion of the paper outside the neatline. The margin is not the border.

Mat (mount). A heavy paper surround that is placed between the map and the glazing. *Mount* is the British term.

Meridian. A circle connecting the poles and passing through any given point on the earth's surface. See *Prime meridian.*

Mildew. A mold infecting paper.

Mint condition. A term sometimes used to imply, generally in an excess of exuberance, that a map is in "new" condition. But maps are not minted.

nd (ND; N.D.). No date.

Neatline. The printed line that defines the outer perimeter of a map, not to be confused with *border.*

Octavo. A book made from paper folded three times, creating eight leaves. Maps from such a book are also sometimes described as being octavo (abbreviation: 8vo). Modern hardback novels are generally octavo, which should give you an appreciation of size, but strictly speaking, 8vo is not a dimension.

Offsetting. An image transferred from one surface to another.

On approval. Sending an item to a potential customer for his/her approval. If the item is "approved" prompt payment is expected, otherwise prompt return is expected.

Original. "Original" in the sense used here implies that the map was printed from the original plate at the time the rest of the edition was printed.

Panorama. A view that encompasses a large amount of azimuth.

Paper. A feltwork of randomly arrayed fibers, generally of cellulose.

Parchment. Vellum.

Parchment paper. A modern paper that is brown and mottled and is supposed to look like parchment (vellum) but only does so to those who have never seen vellum. Used to print reproductions and the fake treasure maps sold in tourist shops.

pH. A numerical scale that indicates the concentration of hydrogen ions in solution. The lower the pH the more ions in solution and the more acidic the solution. A pH of 7.0 is neutral; above 7.0 is basic.

Planigraphic. Printing from a flat surface, such as lithography.

Plate. The plate is the prepared surface from which prints are taken. It may be stone, as in lithography, or copper, as in copperplate engraving. The term plate also refers to the actual printed image.

Platemark (plate mark, plate line). The interface between the paper fibers compressed as the plate was printed and the uncompressed fibers that did not get squeezed against the plate as it was printed.

Pocket map. A folding map, generally not linen-backed, usually with self-covers and designed primarily for travelers.

Portolan (portolano). A manuscript sailing chart, generally of the Mediterranean regions, usually from the 13th through 16th centuries.

Prime meridian. The circle connecting the poles that is defined as longitude zero.

Printer's crease. A wrinkle or crease introduced into the paper at the time of printing. Usually minor.

Projection. The means, or transform, by which features on the earth's surface can be represented on a flat surface. Each projection introduces some distortion.

Proof. An impression taken from the plate as the work progresses to check the image.

Quarto. A book made from paper folded twice (i.e. folding a sheet twice creates four leaves). A map extracted from such a book is sometimes called quarto, but in this case it refers to size. This book, on the basis of size, would be considered quarto. Abbreviated 4to.

Recto. The right page of an opened book; the front surface of a map or leaf; the obverse. In Oriental books, which read in reverse, the recto is the left page.

Register. The precise alignment of the paper on the several different plates needed to print in colors.

Relief map. A map that tries to convey elevation differences by adding a third dimension.

Relief printing. Printing from a raised surface, such as woodblock or letterpress.

Remargined. Describes a map that has had its margins replaced or substantially widened.

Reproduction. A copy of an original map. Although some reproductions are very good, they are generally not made to deceive. In that case they would be fakes.

Reserve. A pre-agreed sum, below which an auction house will not sell a consigned lot.

Rhumb lines. Lines emanating from the compass rose. Sometimes called "wind lines."

Roller map. A wall map mounted on a roller so that it can be rolled up when not in use.

Selenography. The mapping of the Earth's moon.

Self-cover. Stiff sections glued to sections of a folding map. When the map is folded properly, the covers protect the map.

Separate issue. A map that was issued by itself; not as part of a book or atlas.

Short title. A contracted version of the full-length title of a work. Only enough is included in a short title to unambiguously identify the work.

Show-through. The image on one side of a sheet of paper visible from the other side, through the paper.

Signature. The letter or number printed in the margin of the first leaf of a gathering of a book. This letter or number is of some use in identifying the issue or edition of some maps. The term also refers to the actual gathering.

Slipcase. An outer case, either box or envelope-like, into which a folded map can be inserted for protection.

State. Each significant alteration to a printing plate creates another state of the image.

Steelplate. Engraved on steel rather than copper. Sometimes a copper plate that has been iron-plated is called steelplate.

Stitch holes. Small holes in the centerfold made where the map was sewn into the book or atlas. To avoid stitch holes, many binders attached binder's guards to the map and sewed through these.

Tab. See *binder's guard.*

Three-quarter view. A perspective view of a town or countryside as seen from an elevation of about 45 degrees. See *bird's-eye-view.*

Tipped in. A map or sheet of paper affixed to a larger one (or inserted into a book) by means of small amounts of adhesive at the corners or along an edge is said to be tipped in.

Tissue guard. Sheet of tissue paper bound (or tipped) into a book to separate the recto of a print or illustration from the adjacent page to prevent or retard offsetting.

Trimmed. A map whose margins have been cut back after printing is said to be trimmed.

Trophy map. Any of a number of famous, important, beautiful and much sought after maps. These are "highlight" maps that often form the nucleus of serious collections.

Vellum. A prepared and treated flexible sheet derived from animal dermis. Used for writing and binding.

Verso. The left page of an opened book; also the "back" side of a map or leaf, or the side on which the image does not appear; the *reverse*.

Volvelle. A circular analogue computer, or nomogram, used to perform calculations, generally of an astronomical nature.

Wall map. A large map designed to be displayed on a wall, often for didactic purposes.

Watermark. An integral design in paper. Best seen with transmitted light.

Wood engraving. The technique of printing from a wooden block that was carved on the end grain. Typical engraving tools, such as the burin, are used and the detail is finer than can be achieved with woodblock technique. Wood engraving is a relief process.

Woodcut (woodblock). An image printed in relief from a carved wooden block. The block is carved, with knives and gouges, on the side grain and blank areas are cut away, leaving raised lines which hold the ink and provide the image. This is relief printing.

Worming. Holes made in paper by the worm-like larvae of insects.

Wove paper. A machine-made paper that is cast on a very fine mesh. On casual inspection one does not see the mesh lines as one does with laid paper. Wove paper came into use in the late 18th century.

Zenith. The point directly overhead.

Index

Numbers in bold indicate the entry is an illustration or map. "C" refers to color section.

311

312

Colophon

This book was manufactured in The United States of
America. It is printed on acid-free paper.

The text is set in the Adobe family of Caslon type.
Figure legends and Index are in Adobe Times Roman.

Jacket photograph by Robert Gere.
Jacket layout by Carrie Fradkin.

Alan Berolzheimer was the proofreader
and copy editor.

MDCCCCXCVIII